NIETZSCHE AND SCIENCE

Nietzsche and Science

Edited by
GREGORY MOORE
THOMAS H. BROBJER

Routledge
Taylor & Francis Group
LONDON AND NEW YORK

First published 2004 by Ashgate Publishing

2 Park Square, Milton Park, Abingdon, Oxon OX14 4RN
711 Third Avenue, New York, NY 10017, USA

Routledge is an imprint of the Taylor & Francis Group, an informa business

First issued in paperback 2017

British Library Cataloguing in Publication Data
Friedrich Nietzsche Society. Conference (11th : 2001 :
Emmanuel College
Nietzsche and science
1. Nietzsche, Friedrich Wilhelm, 1844-1900 – Congresses
2. Nietzsche, Friedrich Wilhelm, 1844-1900 – Knowledge of science – Congresses
I. Title II. Moore, Gregory, 1972- III. Brobjer, Thomas H.
193

Library of Congress Cataloging-in-Publication Data
Nietzsche and science / edited by Gregory Moore and Thomas H. Brobjer.
p. cm.
Includes bibliographical references.
ISBN 0-7546-3402-7 (alk. paper)
1. Nietzsche, Friedrich Wilhelm, 1844-1900–Contributions in theory of knowledge. 2. Knowledge, Theory of. 3. Nietzsche, Friedrich Wilhelm, 1844-1900–Contributions in philosophy of science. 4. Science–Philosophy. I. Moore, Gregory, 1972- II. Brobjer, Thomas H.

B3318.K7N53 2003
193–dc22

2003063711

ISBN 978-0-7546-3402-7 (hbk)
ISBN 978-1-138-27758-8 (pbk)

Typeset by Owain Hammonds, Ceredigion.

Contents

Acknowledgements

Most of the essays collected in this volume started life as papers delivered at the Eleventh Annual Conference of the Friedrich Nietzsche Society, which took place at Emmanuel College, Cambridge between 7–9 September 2001. The conference was sponsored by the British Academy, whose financial support we here gratefully acknowledge. Thanks also to Sarah Lloyd at Ashgate, who first invited us to put together this book.

Abbreviations

Nietzsche's works are abbreviated as follows:

A	*The Antichrist*
AOM	*Assorted Opinions and Maxims*
BGE	*Beyond Good and Evil*
BT	*The Birth of Tragedy*
CW	*The Case of Wagner*
D	*Daybreak*
DS	*David Strauss, the Writer and Confessor*
EH	*Ecce Homo*
FEI	*On the Future of our Educational Institutions*
GM	*On the Genealogy of Morals*
GS	*The Gay Science*
HC	*Homer's Contest*
HH	*Human, All Too Human*
HKW	*Werke und Briefe: Historisch-kritische Gesamtausgabe*
HL	*On the Use and Disadvantage of History for Life*
KGB	*Briefwechsel: Kritische Gesamtausgabe*
KGW	*Werke: Kritische Gesamtausgabe*
KSA	*Werke: Kritische Studienausgabe*
NCW	*Nietzsche contra Wagner*
PTA	*Philosophy in the Tragic Age of the Greeks*
RWB	*Richard Wagner in Bayreuth*
SE	*Schopenhauer as Educator*
TI	*Twilight of the Idols*
TL	*On Truth and Lie in an Extra-Moral Sense*
WS	*The Wanderer and his Shadow*
Z	*Thus Spoke Zarathustra*

List of Contributors

Christa Davis Acampora is Assistant Professor of Philosophy at Hunter College of the City University of New York. She is the author of numerous articles on Nietzsche's philosophy, published in journals such as *Nietzsche-Studien*, *International Studies in Philosophy*, and *Nietzscheforschung*, and she is co-editor of *A Nietzschean Bestiary: Animality Beyond Docile and Brutal* (Rowman & Littlefield, 2003). She is also nearing completion of a book manuscript on Nietzsche's conception of competition.

Babette E. Babich is Professor of Philosophy at Fordham University and Adjunct Research Professor of Philosophy at Georgetown University. Her books include *Nietzsche's Philosophy of Science* (Albany, NY: SUNY Press 1994) and (as editor): *Nietzsche, Theories of Knowledge and Critical Theory*, *Nietzsche, Epistemology and Philosophy of Science* (Dordrecht: Kluwer, 1999) and *Hermeneutic Philosophy of Science, Van Gogh's Eyes, and God* (Dordrecht: Kluwer, 2001). She is also editor of the journal *New Nietzsche Studies.*

Thomas H. Brobjer, Lecturer and Researcher in the Departments of the History of Ideas at Uppsala University and Stockholm University, is the author of *Nietzsche's Ethics of Character* (Uppsala: Uppsala University Press, 1995). His more recent work has been based on exhaustive research into Nietzsche's reading and extant library, the results of which have been published in *Nietzsche-Studien*, *Journal of the History of Philosophy*, *Journal of the History of Ideas*, *Journal of Nietzsche Studies*, *New Nietzsche Studies*, *International Studies in Philosophy* and as contributions to a number of books.

Richard S.G. Brown, Associate Professor of Philosophy at Brock University, has written widely on Nietzsche and on Indian philosophy. His essays on these topics have appeared in numerous collections.

Christian J. Emden is Assistant Professor of German at Rice University. He has published extensively on Walter Benjamin, Aby Warburg and Max Frisch in journals such as *Deutsche Vierteljahrsschrift für Literaturwissenschaft und Geistesgeschichte*, *Zeitschrift für deutsche Philologie* and *Oxford German Studies*. He is the author of *Nietzsche on Language, Rhetoric, and the Mind* (Urbana, IL: University of Illinois Press, forthcoming in 2004) and is currently completing another book on Walter Benjamin's historical anthropology of modernity.

Nadeem J.Z. Hussain is Assistant Professor of Philosophy at Stanford University. His research interests include, in addition to Nietzsche, metaethics, philosophy of action, moral psychology and Islamic philosophy.

Duncan Large is Senior Lecturer in German at the University of Wales, Swansea. He is the author of *Nietzsche and Proust* (Oxford University Press, 2001), has translated *Twilight of the Idols* for Oxford University Press's World Classics and is completing a translation of *Ecce Homo* for the same series. He is co-editor of the forthcoming *The Nietzsche Reader* for Blackwell.

Gregory Moore is Lecturer in German at the University of Wales, Aberystwyth and Secretary of the Friedrich Nietzsche Society. He has published articles in *Nietzsche-Studien*, *Journal of Nietzsche Studies* and *German Life and Letters*, and is the author of *Nietzsche, Biology and Metaphor* (Cambridge University Press, 2002).

Robin Small, who teaches at Auckland University, has published articles on such diverse thinkers as Hegel, Marx, Husserl and Kafka. He is the editor of *A Hundred Years of Phenomenology* (Aldershot: Ashgate, 2001) and Paul Rée's *Basic Writings* (Urbana and Chicago: University of Illinois Press, 2003). His most recent book is *Nietzsche in Context* (Aldershot: Ashgate, 2001).

Tracy B. Strong is Professor of Political Science at the University of California, San Diego. His best-known work includes *Friedrich Nietzsche and the Politics of Transfiguration* (Berkeley: University of California Press, 1975, 1988) and *Towards New Seas: Philosophy, Aesthetics and Politics in Nietzsche* (University of Chicago Press, 1988), as well as many other books on philosophy and political theory.

Introduction

Gregory Moore

Surveying the Nietzsche cult that was already in full swing by the late 1890s, the sociologist Ferdinand Tönnies felt moved to denounce the unwholesome influence of Nietzsche's 'opinions and errors' on impressionable young minds.[1] To be a philosopher at all in the nineteenth century, Tönnies believed, one had to embrace the brave new world of science. But this is precisely what Nietzsche was unable to do. Though he may, on occasion, have paid lip-service to the spirit of the age by wearing the mask of the cheerful free-spirit and celebrating the virtues of rationality and exactitude, he did so anaemically, without conviction. Though he may have appealed to biology and sociology to lend his ideas a patina of credibility, ultimately this amounted to little more than flirting with scientific concepts. Rather, the true Nietzsche was above all a poet, a visionary, a prophet who promised redemption from the moribund modern world. It was precisely in the vatic imagery and intoxicating rhythms of his writing that the immensely seductive power of his thought lay. But empty bombast was not true knowledge; overheated emotionality no substitute for philosophical profundity. What was necessary on the cusp of a new century was not adolescent exuberance, but rather self-possession, healthy scepticism and sobriety: in short, the scientific ethos.

The picture of Nietzsche which Tönnies draws is one that is still recognizable today. We have come to see Nietzsche as the arch-critic of science, someone who, in the struggle between what C.P. Snow famously called the 'two cultures', comes down firmly on the side of art. But we require a more nuanced understanding of Nietzsche's position than this image implies. Nietzsche's thought may not have the rigorously scientific qualities that Tönnies desired, but this does not mean, of course, that it was not shaped by contemporary scientific debates. It would have been inconceivable for a philosopher like Nietzsche writing towards the end of the nineteenth century *not* to have taken account of the advances of science, even in the most superficial way, let alone engage with it in the complex and probing manner that characterizes Nietzsche's approach to the topic. Certainly, Nietzsche's wide reading about subjects ranging from embryology to meteorology demonstrably influenced his thinking, providing him not only with novel insights and theories that he explored both in his notebooks and published work, but also striking images and metaphors with which to dress his ideas. We need only think of his much-discussed relationship to Darwinism, for example, and his long-standing interest in biology and physiology. Or his characteristic reliance on the rhetoric of health and sickness. Or his critique of the mechanist paradigm of contemporary physics. Or his doctrines of the will to power and eternal recurrence of the same, which – at least in part – owe their conception to ideas current in nineteenth-century science. This book accordingly explores Nietzsche's knowledge of science and some of the ways in which science impacted on his thought.

I

The nineteenth century has often been called the 'age of science',[2] and with good reason. For the years from 1820 to 1900 witnessed unprecedented scientific discoveries and technological advances. Within the space of a single lifetime – a period that had itself been extended by new developments in medicine and social hygiene – an overwhelmingly rapid succession of innovations transformed the world and the way people understood it: in engineering, the development of mechanized industry, the railways, telegraphy, synthetic dyes, the telephone, gas and electric lighting; in physics, the articulation of thermodynamics and the rise of statistical mechanics; in the life sciences, the discovery of the cell, evolutionary theory, and experimental physiology.

But in the eyes of many the greatest achievement of science was to end religion's monopoly on 'truth' and to arrogate to itself the cultural supremacy previously enjoyed exclusively by Christianity. 'We claim,' declared John Tyndall in his famous 'Belfast Address' of 1874, 'and we shall wrest from theology, the entire domain of cosmological theory.' For Thomas Huxley, Darwin's chief evangelist in England and one of the leaders of what he liked to call the 'Church Scientific', science 'is teaching the world that the ultimate court of appeal is observation and experiment, and not authority; she is … creating a firm and living faith in the existence of immutable moral and physical laws, perfect obedience to which is the highest possible aim of an intelligent being'.[3] To a freethinker like Huxley – or his German colleague Ernst Haeckel – science would not only cleanse the world of religious chimeras; it would also provide another basis of value. Neither in a groundless belief in God nor in the empty speculations of theology, but rather in the certainties of scientific knowledge, whose efficacy could be seen with one's own eyes, modern men and women would find a firm foundation for those moral values that, for Huxley, and virtually all his contemporaries, remained indispensable.

This kind of faith in science was not restricted to its practitioners and apologists. For if the nineteenth century was the age of science, it was also the era of popular science. Its significance was acknowledged by educated people throughout Europe. Families crowded into the new museums of natural history; amateurs collected fossils and butterflies, observed the migration of birds and sketched wildflowers. The latest discoveries were reported in journals such as the *Westminister Review*, *Daheim* and possibly the most important of the weekly illustrated magazines in Germany, *Die Gartenlaube*, which, by 1880, had a circulation of nearly 400 000, giving perhaps 2 million readers per issue. With its articles on steam and magnetism, on the workings of the mechanical plough and the physiology of the screech owl, *Die Gartenlaube* perfectly encapsulates the heady contemporary belief in both science and public education. The same applies to cheap popular imprints like the International Scientific Library (many of whose books Nietzsche himself owned in the German translation of the series) and Joseph Mayer's 'Penny Library' (motto: 'knowledge is freedom, freedom is power'), and to the numerous scientific popularizers such as Ludwig Büchner, Ernst Haeckel, Balfour Stewart and Grant Allen, whose writings met with astonishing success. 'Science,' observed G.H. Lewes in 1873, 'is penetrating everywhere, and slowly changing man's conception of the world and of man's destiny.'[4]

Of all the scientific discoveries made during the nineteenth century, one of the most momentous breakthroughs occurred in the Moravian town of Brünn, where an Augustinian monk named Gregor Mendel laid the basis for modern genetics by experimenting with the peas he cultivated in the monastery's garden. But Mendel, working in isolation and not fully appreciated until after his death, was an exceptional figure in every way. Most scientists were part of a burgeoning institutional network, members of the same learned societies and disciplinary organizations, subscribers to the same periodicals, participants in congresses where ideas and research findings could be exchanged. Equally important, the institutional locus for science became the research laboratory, where a team of investigators employed the latest technology to study the natural world; in the laboratory, apprentice scientists could participate in group projects, learn research techniques firsthand, and acquire the values and habits on which professional science was based. As a powerful instrument of both innovation and socialization, the laboratory combined research and teaching, the accumulation and dissemination of knowledge. The laboratory marked the end of the Humboldtian ideal of *Bildung* – the moral and intellectual self-realization of the individual – through the acquisition of knowledge (*Wissenschaft*) and the rise of what Weber called 'science as vocation', the value-free advancement of science for its own sake. Science was no longer the domain of gentleman amateurs but of professional experts.

Indeed, the very term 'scientist' was an invention of the nineteenth century. The adoption of a name 'by which we can designate the students of the knowledge of the material world collectively' was seen by the Cambridge philosopher William Whewell as a modest contribution towards overcoming the rapid fragmentation and specialization of knowledge attendant upon the professionalization of science. At the meeting of the British Association for the Advancement of Science in the early 1830s, he recalled, members tried out different translations of the German *Naturforscher*, including 'such undignified compounds as *nature-poker* or *nature-peeker*', which, unsurprisingly, were 'indignantly rejected'. Earlier, Whewell remembered, 'some ingenious gentleman' had proposed that, 'by analogy with *artist*, they might form *scientist*', and although the new word was 'not generally palatable', the term was in the end accepted.[5] But for all his concern about the widening fissure in the field of knowledge, Whewell nevertheless recognized that 'science' had gradually acquired a specific meaning differentiating it from the arts, from philosophy and from knowledge in general: 'The *Sciences* to which the name is most commonly and unhesitatingly given, are those which are concerned about the material world ... their doctrines are obtained by a common process of collecting general truths from particular facts, which process is termed *Induction*.' Other categories of the human quest for knowledge were, Whewell believed, not keeping pace with those 'inductive sciences' that deal with material reality. Thus, while 'important truths' had perhaps emerged from 'Political Economy, or Philology, or Morals, or the Philosophy of the Fine Arts', these were at best only 'supposed sciences' in contrast to the 'material sciences'.[6]

But Whewell's claim that older domains of knowledge could not lay claim to his more narrow conception of 'science' does not take into account the extent to which practitioners of these disciplines strove to reinvigorate their fields by appropriating the scientific method. Even in the first half of the nineteenth century, science – in particular, its dispassionate means of inquiry, its effective combination of observation, logical reasoning and experiment – had begun to influence many other branches of human activity. What the scientific method was supposed to provide was certitude. 'I do think that our age has learnt many lessons from the physical sciences,' wrote Hermann von Helmholtz. 'The absolute, unconditional reverence for facts, and the fidelity with which they are collected, a certain distrustfulness of appearances, the effort to detect in all cases relations of cause and effect, and the tendency to assume their existence, which distinguish our century from preceding ones, seem to me to point to such an influence.'[7] Take psychology, for example, which was transformed from a largely intuitive and subjective enterprise by the development of experimental techniques by Alexander Bain, E.H. Weber, and Wilhelm Wundt, or Quetelet's use of statistics to study human behaviour, or Alexander von Humboldt's elevation of physical geography to the rank of an empirical science. Even socialism, Friedrich Engels thought, had become a science, thanks to Marx's discovery of class conflict as the law of development of human history (a discovery comparable to Darwin's discovery of the law of development in the natural world).

Indeed, as an example of the way in which scientific knowledge and methodologies spilled over into wider cultural debates, we need look no further than biology, which, along with energy physics, emerged in the middle decades of the nineteenth century as perhaps the most vigorous and culturally significant of all the sciences. In the wake of the *succès de scandale* of Darwin's *The Origin of Species* in 1859 and the consequent rapid popularization of the idea of evolution, biological models and metaphors were soon deployed in numerous disciplines and discourses far removed, at first glance, from the narrow scope of biology itself: in political science and sociology (Walter Bagehot's *Physics and Politics, or Thoughts on the Application of the Principles of 'Natural Selection' and 'Inheritance' to Political Society*, 1872), aesthetics (Hippolyte Taine's *Philosophie de l'art*, 1865 and Wilhelm Bölsche's *Die naturwissenschaftlichen Grundlagen der Poesie*, 1887), morality (Spencer's *Data of Ethics*), epistemology and metaphysics (for example, Eduard von Hartmann's attempts to underpin Schopenhauerian idealism and pessimism with evolutionary biology and neurophysiology), history (Ottokar Lorenz's equation of 'social selection' (*sociale Zuchtwahl*) with the historical process), and literary criticism (Erich Schmidt, son of the Darwinist Oscar Schmidt, believed that literary history ought to be seen as investigating inheritance and adaptation in the 'evolutionary history of the intellectual life of a people').

The influence of Darwinism and similar biological metaphors came in part from a belief that the methods of natural science could and should be applied to all forms of inquiry. Science had achieved so much in so little time that it seemed unstoppable, a dynamic, creative force that promised continually to enrich and expand human knowledge; it was the very embodiment of Progress, the great god of the age. In 1870 Wilhelm Scherer, Professor of German Language and Literature at the University of Vienna, paid handsome tribute to the revolutions science had wrought in European culture:

> The same power which roused the railways and the telegraph into life; the same power which brought about an unprecedented blossoming of industry, increased the comfort of life, shortened wars; in short, advanced man's supremacy over nature by leaps and bounds – this same power also governs our intellectual life: it is eliminating dogmas, reshaping the sciences, leaving its stamp on poetry. The *natural sciences* ride like a conquering hero in the triumphal chariot to which we are all bound.[8]

It seems inevitable, then, that many of Nietzsche's own writings should bear witness to the extraordinary cultural impact of the natural sciences in the late nineteenth century. Throughout his career, from his time as a student (take, for example, his aborted doctoral dissertation on Kant and teleology) to the last entries in his notebooks in 1888, Nietzsche wrestled with the question of science, reflecting on its significance, value and relationship to culture, often changing his mind as his thought matured, wavering between outright rejection, enthusiasm and ambivalence, between criticism and cautious acceptance. His early incarnation as a Schopenhauer-influenced metaphysician and propagandist for Richard Wagner in *The Birth of Tragedy* saw him rail against the Socratic rationalism which choked the healthy creativity of Greek tragic culture and caused it to degenerate into abstraction. Yet even here, in a work that ostensibly celebrates the life-affirming qualities of art and vilifies the corrosive effects of science, Nietzsche seems unwilling to dismiss completely the latter's claims – after all, towards the end of *The Birth of Tragedy* he imagines a future synthesis between art and science: the ideal of a 'music-making Socrates' (see *BT* 15). Nietzsche's subsequent break with Wagner between 1875 and 1878 saw him take the necessary steps towards regaining philosophical health, seeking the antidote for his youthful romantic pessimism in what is, perhaps somewhat misleadingly, often referred to as a new 'positivist' outlook (after all, Nietzsche writings have precious little in common with those of Comte, say) – or what he himself, acknowledging his intellectual debt to his friend, the writer Paul Rée, called his 'Réealism'. This phase of Nietzsche's thought, during which time he composed *Human, All Too Human*, *Daybreak* and *The Gay Science*, is characterized by a new confidence in the 'steady and laborious process of science' (*HH* 16), such as 'the physiology and history of the evolution of concepts and organisms' (*HH* 10), whose task it now is to peel away the layers of error and illusion which have encrusted our conception of the world and to provide naturalistic explanations of the origin of religion, art and morality without recourse to 'absolute truths' and metaphysical fictions such as the thing-in-itself. Above all, Nietzsche celebrates, during this period, a free-spirited scepticism and pragmatism, an intellectual integrity and rigour which he sees best embodied in the figure of the scientist:

> we must become the best learners and discoverers of everything that is lawful and necessary in the world: we must become *physicists* in order to be able to be *creators* in this sense – while hitherto all valuations and ideals have been based on *ignorance* of physics or were constructed so as to *contradict* it. Therefore: long live physics! And even more so that which *compels* us to turn to physics – our honesty! (*GS* 335)

Nietzsche's thought may be viewed as a symptom of the wider Victorian crisis of faith engendered by the inexorable rise of science (and, in particular, the revolution in human self-understanding necessitated by Darwinism and the second law of thermodynamics). But he was never content, like most of his contemporaries, merely to mourn or gloat over the death of God. Not for him the easy pessimism of Philip Mainländer or the naïve optimism of Herbert Spencer. Only Nietzsche recognized that removing the divine keystone would bring the entire moral architecture of the universe crashing down with it. The naturalism that typifies his cast of mind during the 'middle period' – but to which he would remain committed for the rest of his intellectual life – is, as Laurence Lampert has argued, a naturalism that is not only 'post-Christian' but 'post-humanist', and which aims to free us from all forms of humanism based on myths of special origins that confer on humankind special rights of dominance and mastery over nature.[9] 'My task: the dehumanization of nature and then the naturalization of the human after it has gained the pure concept of "nature"' (*KSA* 9, 11[211]).

But by 1882–83, Nietzsche had once again moved towards a more critical stance vis-à-vis science as such, returning to the tragic view of existence and consequently rejecting what he saw as the desiccated rationalism of modern culture. The scientist, who lives by an ethic of exalted immolation of self in the pursuit of 'truth', is a servant of the ascetic ideal, and is thus unable to furnish meaningful values. The 'scientific interpretation of the world' would be the 'poorest in meaning', the picture of an 'essentially meaningless world' (*GS* 373), and, in its attempt to stamp a particular interpretation on the world, would rob life of its contingent, perspectivally rich character. Nietzsche becomes, during this period, increasingly preoccupied with the question of the value of science, and in two ways: on the one hand, he concludes that it only has value in so far as it can serve life and, on the other, he is concerned to reveal the extent to which modern science, far from representing a break with religious ideals, is in fact suffused with the values of the Judaeo-Christian tradition. Though scientists such as Huxley and Tyndall, for example, may reject God, modern science is nevertheless unable to break with the Western intellectual tradition's dogmatic faith in the absolute and unconditional value of truth, which governs every inquiry and serves as its incontestable goal; which receives its justification solely from the otherworldly domain of Christian morality. For Truth is being (*Sein*) and cannot be therefore sought in the realm of ceaseless change and radical contingency which constitutes the world – it must therefore reside in another, metaphysical world, an eternal world of necessary and absolute fixity.

But while Nietzsche may condemn its residual metaphysics, he is still not wholly dismissive of science: for all its complicity in Christian morality, science, he suggests in a late note, nevertheless embodies a questioning attitude. 'It is not the victory of science that distinguishes our nineteenth century, but the victory of scientific method over science' (*KSA* 13, 15[51]). It is this scepticism – its means, not its ends or results; its interrogation of all truths, even of those which underpin it – that represents the greatest legacy of science. But science thus understood is still unable to fulfil a creative, positive function. Ultimately it is not science that Nietzsche views as a means of fashioning alternative frameworks for understanding and shaping the world, but art (or at least a synthesis of art and science). Famously,

Nietzsche returns in his late thought to his original emphasis on art (understood in the widest possible sense) as the antidote against, the antithesis to, the otherworldly claims of metaphysics and theology. Art counteracts the ascetic demand for extirpation of the animal instincts by indulging the senses and passions. Art is alone capable of creation and transformation; only it can affirm appearance and consequently life, and as such it does more justice to the contingent, provisional nature of the world.

What is at stake, then, for Nietzsche, throughout his career, is the proper relation between science, philosophy and culture. Of course, Nietzsche's concerns about the role and purpose of philosophy in an age of rapid scientific progress were hardly unique; it was an issue that inevitably preoccupied many thinkers throughout the period. Herbert Schnädelbach has spoken of an 'identity-crisis' afflicting nineteenth-century philosophy.[10] In the early 1820s – at least in Germany – philosophy still enjoyed its status as the 'queen of the sciences'. Long after the methods of exact science were being used in Paris and London, Schellingian *Naturphilosophie* and Hegelianism continued to hold sway in German universities, each rejecting the empirical study of the natural world in favour of deductive reasoning and the construction of elaborate metaphysical systems. As Hermann von Helmholtz later remembered, natural scientists, frustrated with both the vapid speculations of *Naturphilosophie* and Hegelianism's illegitimate pretensions to subordinate to itself all other disciplines, eventually felt compelled to reject philosophy as at best useless, at worst 'mischievous dreaming'.[11] With Justus von Liebig's establishment of the world's first research laboratory at Giessen in 1826, the experimental method was finally given an institutional base in Germany, leading to further endowments of laboratories and an unprecedented systematic organization of research. The subsequent rapid expansion of knowledge and increased specialization meant philosophers could no longer keep up with scientific advancement as they had done in Schopenhauer's and Hegel's day.

Cut loose from science, German philosophy appeared increasingly sterile. It was forced to fight a rearguard action against obsolescence by seeking out different ways in which to carve out a meaningful role for itself. Academic philosophy began slowly to view itself as a science – a human science or *Geisteswissenschaft*. This was helped, for example, in the first decades after Hegel's death by the rise of the history of philosophy, as practised by men such as Eduard Zeller and Kuno Fischer. But by the late 1840s a younger generation of philosophers had emerged – chief among them Jakob Moleschott, Ludwig Büchner and Carl Vogt – who viewed the relationship between philosophy and science rather differently. They worshipped at the altar of natural science, uncritically seizing upon the latest advances in physiology and chemistry to underpin a crudely materialist and mechanist world-view, according to which all natural forces were reducible to inertia and motion. Where Büchner reduced consciousness to molecular configurations, Moleschott famously explained mental activity by using the physical analogy that the brain secretes thoughts just as the liver secretes bile and coined the phrase that Feuerbach would later popularize: 'Man ist, was man ißt'.[12] Immensely popular and populist (Büchner's *Kraft und Stoff* (*Force and Matter*, 1855) had by 1898 gone through 19 editions), they were (or at least thought they were) the mirror image of the *Naturphilosophen*, claiming that, in systematizing the results and discoveries of the natural sciences, they had finally put an end to philosophy; philosophy was nothing more than the theoretical aspect of the empirical sciences.

But the gross simplifications and naïve realism of the materialists, as well as the doctrinal disputes between the Hegelians, led more sophisticated thinkers like Liebig and Helmholtz – significantly, both natural scientists rather than philosophers – to attempt to overcome the schism between science and philosophy and to question the foundations of scientific knowledge. This process of reappraisal led them back to Kant. For the sole object of Kant's critical philosophy, according to Helmholtz, 'was to test the sources and the authority of our knowledge, and to fix a definite scope and standard for the researches of philosophy'. It was this, 'the criticism of the sources of cognition, and the definition of the functions of the intellect',[13] that was the proper office of philosophy, and thus it could no longer be ignored by natural scientists, especially as Kant's theory of the *a priori* categories of human thought seemed to have been proved scientifically by the physiological investigations of Johannes von Müller and by experimental psychology. Alois Riehl, a later neo-Kantian (and, incidentally, an interpreter of Nietzsche), hailed Helmholtz as the inaugurator of the 'epoch of scientific philosophy'. Under his influence, and that of subsequent thinkers such as Ludwig Boltzmann, Henri Poincaré and Ernst Mach, '[s]cience and philosophy, the two hostile powers that had long been divided, had not only been reconciled, but, above and beyond this, had entered into a personal union'.[14]

Though Nietzsche obviously arrives at radically different conclusions, his own extensive writings nevertheless owe something to the preoccupation of mainstream German philosophers, following Helmholtz's resurrection of Kant, with the conditions and limits of knowledge. This ought not to be surprising given his familiarity with the works and ideas of early neo-Kantians such as Otto Liebmann and, more famously, Friedrich Lange, who, in his *History of Materialism* (1866), appropriates and develops Helmholtz's physiological Kantianism in the service of a critique of vulgar materialism. Nietzsche's own scathing indictment of the shallow materialism of David Friedrich Strauss's *The Old Faith and the New* (1872), that literary monument to the bourgeois worship of science, was an expression of the same mood that fostered the neo-Kantian critique of positivism and materialism. One might even argue that the various positions vis-à-vis science which Nietzsche adopts throughout his life recapitulate in a single career the different stages in the development of the relationship between science and philosophy in the nineteenth century. If he began by indulging in metaphysical speculation and romantic *Naturphilosphie*, then, like the early neo-Kantians, Nietzsche believed in his middle period that science would account for all that hitherto had been thought to demand a philosophical explanation. (His occasional and controversial lapses into medical materialism sometimes smacks of the vulgar reductionism of Moleschott *et al.* – see Chapter 2 of this book). The reinterpretation of epistemological problems in terms of the physiology of perception provided Nietzsche with the empirical corroboration of the Schopenhauerian interpretation of Kant. His psychological (and biological) analyses of the problem of knowledge and value are an expression of the same anti-philosophical animus borne by the modern scientific spirit. What is more, his ultimate position, that science ought to be subordinated to philosophy, though of course very different in many ways, nevertheless resembles to some extent the stance of later neo-Kantians, such as Heinrich Rickert, who rejected Helmholtz's notion of philosophy as handmaiden to science in favour of the view that science was ancillary to philosophy as the transcendental knowledge of the validity of logical categories and cultural values.[15]

Nietzsche's lifelong meditation on the meaning and validity of science is not the only way in which the cultural pre-eminence of the sciences in the nineteenth century manifests itself in his work. His writing – even (and especially) that of the late 1880s, when he began to question once again the very presuppositions of science – is also characterized by a rich vein of scientific metaphors, by a habitual appeal to the idiom, concepts and models of contemporary science. We see this not just in such well-known slogans as the 'chemistry of concepts and sensations' and the 'physiology of art', but also in his frequent deployment of a whole range of scientific ideas, such as the struggle for existence and the conservation of energy, and some of which, as products of Victorian pseudo-science, are long obsolete: neurasthenia, moral insanity, degeneration and so on. With this in mind, even those modern readers and critics who are apparently only concerned with him as, say, a philosopher of culture (whose cultural criticism is informed by the nineteenth-century sense of decadence and predicated on values of health and sickness) or religion (*The Antichrist*, for example, is littered with references to contemporary psychiatry), cannot avoid taking into account his attitude towards science. This is not to say that contemporary science in some way constitutes the core of Nietzsche's philosophy. His use of scientific ideas certainly differs from contemporaries such as Hartmann, Dühring and Engels, who exploit the latest theories and discoveries in a much more systematic and detailed fashion to underpin their philosophies. But the natural sciences *did* exert a powerful influence on Nietzsche; they *did* provide an important framework within which he articulated his ideas.

The starting-point of this volume of essays is, accordingly, the assumption that any account of the development of Nietzsche's thought remains incomplete, indeed in some respects unintelligible, until the significance of his philosophical preoccupation with science in general and his reception of, and recourse to, particular scientific discourses has been properly investigated. The essays included in this book explore Nietzsche's knowledge and understanding of specific disciplines, such as physiology and chemistry, for example; the influence of particular scientists on his thought; and the function and status of scientific theories and models in his writings. The various contributions also seek to present and discuss both Nietzsche's criticism and deployment of scientific theories, not only within the general historical context of the nineteenth century, but also within the framework of his own intellectual development. To what extent was Nietzsche influenced by the overwhelmingly scientistic culture of his day? What is its significance for his philosophy? These are some of the questions addressed by this book.

II

To date, work on the subject of Nietzsche and science has been patchy, leaving a vacuum which it is hoped this present volume will go some way towards filling. To be sure, some of Nietzsche's earliest interpreters in the late nineteenth and early twentieth centuries explored the links between his thought and science, but in recent years this important topic has not attracted sufficient scholarly interest and certainly lacks any systematic treatment. With the Colli-Montinari critical edition of Nietzsche's writings and the ongoing re-evaluation of his unpublished notes, scholars have, since at least the late 1980s, made efforts to ascertain and quantify the influence of particular authors – including, of course, a large number of scientists or science writers – whose books Nietzsche is known to have read or owned, among them Afrikan Spir, Otto Caspari, Alfred Espinas, Charles Féré.[16] But as yet there has been no concerted attempt to synthesize or evaluate these painstaking researches. Though modern critics – some of them contributors to this volume – have produced a number of important studies exploring facets of Nietzsche's relation to natural science, the only substantial monograph to aim at comprehensiveness is still Alwin Mittasch's 1952 book *Nietzsche als Naturphilosoph* (Stuttgart: Kröner, 1952), which, for all its virtues, is marred by a lack of historical and philological sensitivity.

Nietzsche wrote when philosophers and scientists were exercised by two recent scientific discoveries of profound significance. One was Darwin's theory of evolution by natural selection, which undermined traditional religious ideas about – in Huxley's phrase – 'man's place in nature' and encouraged a plethora of naturalistic accounts of human being and corresponding theories of culture and morality. The other important debate of the time concerned the second law of thermodynamics and its promise of the heat death of the universe.[17] Both of these scientific controversies preoccupied German thinkers during the latter half of the nineteenth century and both figure prominently in Nietzsche's own reading in natural science.

Inevitably, perhaps, Nietzsche's complex and shifting relationship to Darwinism has attracted the most critical attention, though few commentators have added much to the arguments of Helene Druskowitz, one of the first critics to interpret the *Übermensch* narrowly as a product of future biological evolution. Her judgement that the 'idea underpinning *Zarathustra* is the consequence of Darwinism' – perhaps the inspiration for Nietzsche's vitriolic remarks in *Ecce Homo* about those 'learned cattle' who suspect him of Darwinist sympathies (*EH* 'Books' 1) – was shared by a great many subsequent commentators and remains, stubbornly, a popular misconception.[18] This *idée reçu* prompted social Darwinists and eugenicists both in Britain and Germany, such as Alexander Tille and Maximilian Mügge, to claim Nietzsche as one of their own, so that by 1917 the Belgian writer Charles Sarolea could write: 'Nietzsche is the spiritual father and forerunner of the Eugenists [*sic*].'[19] At least some early twentieth-century critics were prepared to embark upon more careful investigations of Nietzsche's relationship to Darwinism and to biological concepts in general. Claire Richter's 1911 book *Nietzsche et les théories biologiques contemporaines* argues that Nietzsche's thought (albeit unconsciously) owes more to a Lamarckian rather than a properly Darwinian understanding of evolution (an argument influenced by the then widespread tendency in France to downplay the achievements of Darwin at the expense of his supposedly more illustrious Gallic forerunner). In his monumental six-volume study, *Nietzsche, sa vie et sa pensée*, Charles Andler also explored the influence of what he called 'neo-Lamarckian' (or what might be more accurately termed pre- and non-Darwinan) biologists on Nietzsche's ideas such as Ludwig Rütimeyer (a colleague of Nietzsche's at Basle), William Rolph and Wilhelm Roux. A more recent and detailed discussion of Nietzsche's extensive borrowings from Roux was undertaken by the late Wolfgang Müller-Lauter in his classic study 'Der Organismus als innerer Kampf: Der Einfluß von Wilhelm Roux auf Friedrich Nietzsche', one of the most important modern contributions to the debate on Nietzsche's relationship to Darwinism. Müller-Lauter's attempt to trace and contextualize Nietzsche's borrowings from the biological sciences has been extended by Gregory Moore in his *Nietzsche, Biology and Metaphor*, in which he not only attempts to provide a definitive account of Nietzsche's understanding of organic change in general, but also considers the importance of his evolutionary perspective for the development of his moral and aesthetic philosophy. A different and philosophically more sophisticated approach has been attempted by Keith Ansell Pearson in his *Viroid Life*, a reflection on the ideas of Nietzsche and Deleuze in relation to Darwinism, neo-Darwinism and complexity.[20]

If the doctrine of the *Übermensch* has often been seen in the light of biology, then the question of whether other central Nietzschean concepts such as the will to power and eternal recurrence are compatible with nineteenth-century physics or whether they even foreshadow more modern discoveries has also regularly been the subject of lively debate. Such discussions have been prompted above all by Nietzsche's well-known notes on eternal recurrence in which he invokes the principle of the conservation of energy and his remark that eternal recurrence is 'the *most scientific* of all possible hypotheses' (*KSA* 12, 5[71], p. 213). Walter Löb was, in 1908, one of the first critics to discuss this idea in terms of thermodynamics, but after thoughtfully weighing up its tenability concluded that eternal recurrence lacked any scientific basis, no matter how much Nietzsche might claim otherwise.[21] And yet, as Robin Small has most recently and compellingly shown in his illuminating book *Nietzsche in Context*, Nietzsche's complex ideas about time did indeed emerge through a dialogue with a number of contemporary scientists and scientist-philosophers such as Johann Gustav Vogt, Johann Friedrich Zöllner and Robert Mayer (see, in particular, Small's chapter 'The Physics of Eternal Recurrence').[22] In fact, Robert Mayer has often been linked with the development of Nietzsche's thought, ever since Alwin Mittasch (a chemist by training) first pointed to jottings inspired by Mayer's paper *Mechanik der Wärme* in Nietzsche's notebooks; rather less convincingly, perhaps, he also argued that Nietzsche's concept of will to power anticipates Wilhelm Ostwald's energetics. Another natural scientist with whom Nietzsche has frequently been associated is the eighteenth-century Jesuit physicist Roger Boscovich, who attacked the mechanist theory of atoms and argued instead for centres of force without extension. In this respect, Andler was again the first critic to discuss in detail Nietzsche's reading of Boscovich, and subsequent scholars have continued to tease out the full significance of Boscovich for Nietzsche, including Anni Anders and Karl Schlechta, George J. Stack, Alistair Moles and Greg Whitlock.[23]

After physics and biology, the other scientific field with which Nietzsche has regularly been associated is psychology (leaving aside sociology, many of whose founding fathers, such as Weber and Simmel, owed much to Nietzschean stimuli).

After all, Nietzsche liked to style himself as above all a psychologist, and was quick to praise the psychological acumen of men such as Pascal and Dostoevsky. William James may have had little time for his thought, but it comes as no surprise that Nietzsche excited a great deal of interest amongst early twentieth-century psychologists. Ludwig Klages paid tribute to Nietzsche's achievements as a psychologist in his *Die psychologischen Errungenschaften Nietzsches* (1926), but it is inevitably Nietzsche's impact on the ideas of Sigmund Freud – despite the latter's coyness in acknowledging his indebtedness to the philosopher – and the wider psychoanalytical movement that has proved to be the most significant instance of his influence. This relationship has long been recognized, and Reinhard Gasser, Ronald Lehrer and Paul-Laurent Assoun have all published several studies on this topic in recent years. Paul Bishop has given a comprehensive account of Carl Jung's reception of Nietzschean ideas, including his famous seminar on *Also sprach Zarathustra*.[24]

Other scholarly works which have broached the theme of Nietzsche and science in some way in recent years have included George J. Stack's important book *Lange and Nietzsche*; Laurence Lampert's *Nietzsche and Modern Times: A Study of Bacon, Descartes and Nietzsche*, which argues that Nietzsche came to recognize the dangers attendant upon modern science and set about to remedy them with a new understanding of science based upon a more adequate understanding of nature; and Christoph Cox's *Nietzsche: Naturalism and Interpretation*, which seeks to resolve the problem of relativism thrown up by his post-metaphysical epistemology and ontology by exploring Nietzsche's radical naturalism. Perhaps the critic who has devoted most attention to the topic under consideration here is Babette E. Babich. In her *Nietzsche's Philosophy of Science* and two collections of essays that she has edited, *Nietzsche, Theories of Knowledge, and Critical Theory* and *Nietzsche, Epistemology, and Philosophy of Science* (which appeared under the umbrella title *Nietzsche and the Sciences*), Babich has exhaustively explored the relationship between art and science in Nietzsche's thought, as well as the repercussions of his critical epistemology for rethinking the philosophy of science.[25] The emphasis of the present volume – to which Babich is a contributor – is rather different. We are less concerned with the preconditions of knowledge and Nietzsche's critique of modern science in general than with the issue of his familiarity with, and relationship to, particular scientific disciplines; with his response to the hegemony of science in nineteenth-century culture; and with his deployment of whatever scientific understanding he had in his writings. It is these questions that the essays in this collection will address. To this end, the book is divided into two sections, moving from the specifics of the influence of particular scientists on his thought in the first section to a consideration of the more general issues involved in Nietzsche's understanding of science in the second.

In the late nineteenth century no philosopher aiming to present a comprehensive interpretation of the world could afford to neglect the sciences. But because his background was in classical philology (a *Wissenschaft* but no *Naturwissenschaft*), Nietzsche lacked any formal training in the natural sciences and was often acutely aware of his ignorance in this area. As a consequence, he sought, on several occasions throughout his life, to compensate for his shortcomings and read widely, if not always deeply, on numerous topics of contemporary scientific interest, ranging from popular treatises on thermodynamics to criminal psychiatry. At times, he saw this programme of reading as absolutely necessary to his philosophical project – hence the aborted plan to study for ten years at the University of Vienna with Rée and Lou Salomé to lay the foundations for the next phase of his philosophical development. It seems obvious, then, that any exploration of Nietzsche's relationship to science must of necessity begin by addressing the question: how much science did he understand? This question is answered by Thomas Brobjer in his essay, which aims to give a broad, chronological overview of Nietzsche's reading of scientific works and discusses specific thinkers and writers from Nietzsche's extant library, many of whom, if now largely forgotten, nevertheless often played a crucial role in determining his understanding of science. Brobjer's work, which is based on several years of careful study of Nietzsche's papers and reading habits, traces the ebb and flow of his interest in natural science and relates it to the development of his thought from boyhood to his breakdown in 1889.

In his essay Richard S.G. Brown addresses the issue of Nietzsche's well-known propensity, particularly in his later thought, to invoke physiology as the court of last appeal. Brown starts by exploring why Nietzsche described the Buddha in *Ecce Homo* as 'that profound physiologist', before moving on to ponder whether Nietzsche is just as deserving of that epithet. He discusses a number of contemporary critics who have, to some degree at least, treated Nietzsche as a 'physiologist' and who serve as a springboard for Brown's own reflections on the role physiology and its cognates play in Nietzsche's writings.

Gregory Moore explores Nietzsche's enthusiastic reading of nineteenth-century medical handbooks in order to describe the scientific and cultural context in which Nietzsche understood his own chronic illness and bodily processes, and the ways in which this impacted upon his writing. He concentrates in particular on Nietzsche's stomach disorders (and his consequent obsession with digestion), understanding them in terms of a wider nineteenth-century preoccupation with 'inner hygiene', and considers Nietzsche's interest in meteorology, particularly in its application to medicine sparked by his avowed sensitivity to weather conditions, against the background of the growing fashion for 'climatic health resorts'. The importance of the right diet and climate for a healthy body and mind, however, is Nietzsche's answer to the question 'why I am so clever' in *Ecce Homo*. Moore concludes by exploring the extent to which Nietzsche's own diatetics are influenced by, and a reflection of, ideas and trends in nineteenth-century hygiene.

Christian J. Emden offers a novel perspective on Nietzsche's belief that knowledge is ultimately dependent on metaphor, a linguistic process of translation. Emden argues that Nietzsche's ideas about the metaphoricity of language and knowledge are inextricably linked to his reception of nineteenth-century neurophysiology and psychology. Drawing on a vast amount of background material, Emden is able to show that not only do arguments and ideas derived from this cutting-edge science allow Nietzsche to enrich his thoughts on rhetoric, leading him to a radical reinterpretation of 'truth', but that Nietzsche also comes to an understanding of mental activity which interweaves rhetorical thought and neurophysiological concepts such as 'electromotoric transmission', 'localization' and 'induction'.

Nadeem J.Z. Hussain attempts in his contribution to reconcile the apparently contradictory strains in Nietzsche's thought: on the one hand, his admiration for both scientific methodology in general and for particular scientific disciplines and, on the other, his apparently sceptical claims about the status of scientific theories. Where some critics have assumed that taking seriously Nietzsche's positive claims about science requires explaining away the negative comments, Hussain suggests that we learn a very different lesson when we look closely at Nietzsche's historical context: in the late nineteenth century there were many views that were at once both science-friendly and, in some ways, quite sceptical. As an example of this tendency, he focuses on the physicist Ernst Mach, arguing that reading Nietzsche through Mach allows us to see how this tension in Nietzsche's thought can be resolved.

Babette E. Babich returns, in her essay, to the question of Nietzsche's repudiation of science in favour of art and life. She begins by locating Nietzsche's understanding of the concept of 'science' or '*Wissenschaft*' in its proper cultural and historical context, carefully teasing out its various significances to ask whether Nietzsche's problematization of the scientific enterprise, his questioning of the very value and rationale of science, can be utilized to rethink the aim of the philosophy of science. In the second part of her chapter, she moves from a discussion of Nietzsche's critique of logic and pure reason to show the manner in which he exposes the limitations of natural science as a discipline with pretensions to 'know' the real world. Such knowledge is, of course, impossible, for Nietzsche views nature as a swirling, generative chaos and believes that any act of cognition or conceptualization involves throwing a net of illusions over the world. Science, Babich therefore concludes, must acknowledge the therapeutic resources of art; science, which purports to be an objective, value-neutral (and hence nihilistic) enterprise, must be re-established on the ground of what makes it possible: art

Robin Small sets out to shed new light on Nietzsche's attitudes towards science by locating them in the context of the scientific rivalry between Germany and Britain during the nineteenth century, which was reflected in disputes on a theoretical and ideological level – for example, over the implications of the law of entropy and Darwin's theory of natural selection. Small argues that Nietzsche's position in these debates often uncovers the deeper issues, even if this was not always evident to Nietzsche himself. His varying attitudes towards 'English' thought appear with special clarity in the course of his relationship with Paul Rée. Ultimately, however, Nietzsche aspired to a new European cultural synthesis which would overcome the opposition between British empiricism and German rationalism.

Christa Davis Acampora focuses on Nietzsche's rejection of teleological and mechanistic theories of organic development as a means of exploring his suggestive concept of 'gay science'. Both teleology and mechanism unnecessarily restrict the aims and ends of humankind – neither allows a role for creativity in the developmental process. Acampora argues that Nietzsche seeks to mediate between them, offering what she calls his 'agonistic solution' to the problem of causality. What results is a conception of the human as an arena in which countless drives and desires find their expression, opposition and coordination. She explores how Nietzsche believes it is possible to arrange creatively the forces of an organism for the production of excellence and health in the same way that the Greeks harnessed the power of Eris. Nietzsche's gay science, Acampora concludes, combines both the conscience of the scientist with the exploitation of creative energies in the dynamic process of struggle, resistance and growth that Nietzsche thinks is characteristic of all life.

Duncan Large's essay takes its title from the programmatic opening paragraph in Nietzsche's *Human, All Too Human*, 'Chemistry of Concepts and Sensations'. He argues that Nietzsche's apparently sudden enthusiasm for chemistry here must be appreciated in the context of a preoccupation with the subject which spans his intellectual career, surfacing at strategic moments as a recurrent theme in his writings. Large concentrates on two aspects in particular: Nietzsche's propounding of a new 'gay science' of existential experimentalism in the laboratory of the world and his exploitation of the metaphor of 'sublimation'. Large also explores how 'Nietzsche's conceptual chemistry' points, at the same time, back to the German Romantics and forward to Freud.

Finally, Tracy B. Strong looks at Nietzsche's reflections in the early 1870s on the relationship between science and philosophy, in particular in the light of Nietzsche's repudiation of Aristotle's famous declaration that philosophy has its origin in wonder. His attempt to reground the practice of philosophy goes hand-in-hand with a period of fascination with natural science. Strong pursues Nietzsche's suggestion that science releases us from wonder and argues that, for Nietzsche, science starts and makes possible the modern philosophical enterprise, the very notion of which Nietzsche hopes, as a consequence, to redefine.

Notes

1 Ferdinand Tönnies, *Der Nietzsche-Kultus: Eine Kritik* (Berlin: Akademie-Verlag, 1990), p. 9.
2 See, for example, David Knight, *The Age of Science* (Oxford: Blackwell, 1986).
3 T.H. Huxley, 'A Lobster; or, the Study of Zoology', in *Collected Essays*, 9 vols (London: Macmillan, 1894–1908), vol. 8 (1894), pp. 196–228 (p. 226).
4 G.H. Lewes, *Problems of Life and Mind*, 5 vols (London: Trübner, 1874–79), vol. 1, *The Foundations of a Creed*, (1874), p. 1.
5 Whewell quoted in Sydney Ross, '*Scientist*: The Story of a Word', *Annals of Science*, 18:2 (1962), 65–88.
6 William Whewell, *The Philosophy of the Inductive Sciences founded upon their History*, 2 vols, 2nd edn (London: Parker, 1847), vol.1, pp. 2, 7.
7 Hermann von Helmholtz, 'On the Relation of Natural Science to Science in General', in *Popular Lectures on Scientific Subjects* (London: Longmans, 1895), pp. 1-28 (p. 21).
8 Wilhelm Scherer, 'Die neue Generation', in *Vorträge und Aufsätze zur Geschichte des geistigen Lebens in Deutschland und Oesterreich* (Berlin: Weidmannsche, 1874), pp. 408–14 (p. 411).
9 Laurence Lampert, *Nietzsche and Modern Times: A Study of Bacon, Descartes and Nietzsche* (New Haven, CT: Yale University Press, 1993).
10 Herbert Schnädelbach, *Philosophy in Germany, 1831–1933* (Cambridge: Cambridge University Press, 1984), p. 92.
11 H. Helmholtz, 'On the Relation of Natural Science to Science in General', p. 7.
12 On Büchner, Vogt and Moleschott, see Frederick Gregory, *Scientific Materialism in Nineteenth-Century Germany* (Dordrecht: Reidel, 1977).

13 H. Helmholtz, 'On the Relation of Natural Science to Science in General', pp. 5, 7.

14 Alois Riehl, 'Der Beruf der Philosophie in der Gegenwart', in *Philosophische Studien* (Leipzig: Quelle & Meyer, 1925), pp. 304–12 (p. 304).

15 See Steven Galt Crowell, 'Nietzsche among the Neo-Kantians; or the Relation between Science and Philosophy', in Babette E. Babich and Robert S. Cohen (eds), *Nietzsche, Theories of Knowledge, and Critical Theory: Nietzsche and the Sciences, I* (Dordrecht: Kluwer, 1999), pp. 77–86.

16 See, for example, the work by Hans Erich Lampl, 'Ex Oblivione: Das Féré-Palimpsest', *Nietzsche-Studien*, 15 (1986), 225–49 and Andrea Orsucci, 'Beiträge zur Quellenforschung', *Nietzsche-Studien*, 22 (1992), 371–88. New source-critical studies are published annually in *Nietzsche-Studien* under the rubric *Beiträge zur Quellenforschung*.

17 This was a source of anxiety for the Victorians. The heat death of the universe is when the universe has reached a state of maximum entropy. This happens when all available energy has moved to places of less energy. Once this has happened, no more work can be extracted from the universe. Since heat ceases to flow, no more work can be acquired from heat transfer. The universe is effectively dead, especially for the purposes of humankind.

18 Dr H. Druskowitz, *Moderne Versuche eines Religionsersatzes. Ein philosophischer Essay* (Heidelberg: Weiß, 1886), quoted in Richard Frank Krummel, *Nietzsche und der deutsche Geist*, vol. 2 (Berlin: de Gruyter, 1974), p. 58. Other early discussions of Nietzsche in relation to Darwinism include: Heinrich Gotzes, 'Nietzsche und der Darwinismus', *Natur und Glaube*, 7 (1904), 337–49; Oskar Ewald, 'Darwin und Nietzsche', *Zeitschrift für Philosophie und philosophische Kritik*, Ergänzungsband 1 (1909), 159–79; Ludwig Haas, 'Der Darwinismus bei Nietzsche', PhD thesis, University of Gießen (1932).

19 Charles Sarolea, *German Problems and Personalities* (London: Chatto and Windus, 1917), p. 92. See Alexander Tille, *Von Darwin bis Nietzsche* (Leipzig: Naumann, 1895); Maximilian A. Mügge, 'Eugenics and the Superman: A Racial Science and a Racial Religion', *Eugenics Review*, 1 (1909–10), 184–93. See also Curt Grottewitz, 'Nietzsches Herrenmoral und die Naturwissenschaft', *Das Magazin für Litteratur*, 66 (1897), 1519–24; Wilhelm Schallmayer, *Verberbung und Auslese im Lebenslauf der Völker* (Jena: Gustav Fischer, 1903); James Lindsay, 'Eugenics and the Doctrine of the Superman', *Eugenics Review*, 7 (1915–16), 247–62.

20 Claire Richter, *Nietzsche et les théories biologiques contemporaines* (Paris: Mercure de France, 1911); Charles Andler, *Nietzsche, sa vie et sa pensée*, 6 vols (Paris: Bossard, 1920–31); Wolfgang Müller-Lauter, 'Der Organismus als innerer Kampf: Der Einfluß von Wilhelm Roux auf Friedrich Nietzsche', *Nietzsche-Studien*, 7 (1978), 189–223; Gregory Moore, *Nietzsche, Biology and Metaphor* (Cambridge: Cambridge University Press, 2002); Keith Ansell Pearson, *Viroid Life: Perspectives on Nietzsche and the Transhuman Condition* (London: Routledge, 1997).

21 Walter Löb, 'Naturwissenschaftliche Elemente in Nietzsches Gedanken', *Deutsche Rundschau*, 137 (1908), 264–69.

22 Robin Small, *Nietzsche in Context* (Aldershot: Ashgate, 2001). See also Alistair Moles, 'Nietzsche's Eternal Recurrence as Riemannian Cosmology', *International Studies in Philosophy*, 21 (1989), 21–35.

23 See Karl Schlechta and Anni Anders, *Friedrich Nietzsche: Von den verborgenen Anfängen seines Philosophierens* (Stuttgart: Fromann, 1962); G.J. Stack, 'Nietzsche and Boscovich's Natural Philosophy', *Pacific Philosophical Quarterly*, 62 (1981), 69–87; Alistair Moles, *Nietzsche's Philosophy of Nature and Cosmology* (New York: Peter Lang, 1990), pp. 172–76; Greg Whitlock, 'Roger Boscovich, Benedict de Spinoza and Friedrich Nietzsche: The Untold Story', *Nietzsche-Studien*, 25 (1996), 200–20.

24 See Ronald Lehrer, *Nietzsche's Presence in Freud's Life and Thought: On the Origins of a Psychology of Dynamic Unconscious Mental Functioning* (Albany, NY: SUNY Press, 1994); Reinhard Gasser, *Nietzsche und Freud* (Berlin: de Gruyter, 1997); Paul-Laurent Assoun, *Freud & Nietzsche* (London: Athlone Press, 2000); Paul Bishop, *The Dionysian Self: C.G. Jung's Reception of Friedrich Nietzsche* (Berlin: de Gruyter, 1995). See also Graham Parkes, *Composing the Soul: Reaches of Nietzsche's Psychology* (Chicago: University of Chicago Press 1994); Jacob Golomb, Ronald Lehrer and Weaver Santaniello (eds), *Nietzsche and Depth Psychology* (Albany, NY: SUNY Press, 1999).

25 George J. Stack, *Lange and Nietzsche* (Berlin: de Gruyter, 1983); Laurence Lampert, *Nietzsche and Modern Times*; Christoph Cox, *Nietzsche: Naturalism and Interpretation* (Berkeley, CA: University of California Press, 1999); Babette E. Babich, *Nietzsche's Philosophy of Science: Reflecting Science on the Ground of Art and Life* (Albany, NY: SUNY Press, 1994); Babich and Cohen, *Nietzsche, Theories of Knowledge, and Critical Theory*; idem (eds), *Nietzsche, Epistemology, and Philosophy of Science: Nietzsche and the Sciences, II* (Dordrecht: Kluwer, 1999). There have been numerous studies published which document the relationship between art and science in Nietzsche's thought; see, for example, Mihailo Djuric and Josef Simon (eds), *Kunst und Wissenschaft bei Nietzsche* (Würzburg: Königshausen und Neumann, 1986); Rüdiger Sünner, *Ästhetische Szientismuskritik: zum Verhältnis von Kunst und Wissenschaft bei Nietzsche und Adorno* (Frankfurt am Main: Peter Lang, 1986).

PART I
NIETZSCHE'S KNOWLEDGE OF SCIENCE

Chapter 1

Nietzsche's Reading and Knowledge of Natural Science: An Overview[1]

Thomas H. Brobjer

If we are properly to appreciate Nietzsche's views on science, we must take into account not only his explicit statements on the subject, whether drawn from his published writings or his notebooks, but also the relevant books he read before formulating his position. For a man occasionally dismissive of the virtues of reading (as opposed to original, independent thinking) and notoriously shy about acknowledging his intellectual debts (apart from a select band of forerunners such as the ancients and Goethe), Nietzsche was a voracious reader and his philosophy was profoundly coloured by the works he perused. As I have argued elsewhere,[2] two important tasks face Nietzsche scholarship today: first, to determine what Nietzsche read and when; second, to examine how he responded to his reading and how this influenced the contents of his published works. Only by uncovering Nietzsche's stimuli and sources will we come to understand fully the themes and questions to which his writings are a response. This chapter will focus on the first of these tasks: I shall discuss Nietzsche's intellectual development in terms of his reading of books relating to natural science.

Nietzsche's interest in natural science was limited and came to him relatively late in life. Nevertheless, an examination of his reading suggests that he was better informed and more engaged in questions relating to natural science than has generally been assumed. About one-tenth of the books in his private library (that is, about 100 titles) can be classified as belonging to the field of natural science, though he also read a large number of relevant books no longer extant in his private library. Most of these he bought in 1875 or later. The three disciplines in which he demonstrated the most interest were physics, physiology and Darwinism. By and large, though, his knowledge of physics was sketchy (and his understanding of mathematics was rudimentary). He was altogether better informed – that is to say, he read more – about physiology and about issues relating to Darwinism. As I will show in more detail below, in 1868 Nietzsche planned a programme of study of natural science, but did not carry it out. His first real encounter with natural science was in 1873 when he read several scientific texts in an attempt to understand pre-Socratic thinking; his second was in 1875–76, when he bought and read numerous scientific books in order to compensate for his lack of education in this area. He was interested in natural science primarily during the so-called 'middle period' of his career (1875–82), though his interest in physiology began in 1880–81 and intensified until his mental collapse. His acquisition and reading of scientific books reached a peak during 1875–81. After 1883 Nietzsche continued to read about physiology, but his reading was not concentrated in any particular year. His interest in natural science corresponds neatly with the common division of his intellectual development into three phases, and a causal connection between them does indeed exist – that is, his view of science and truth is closely related to the shifts in his thinking.

It is not my intention here to discuss in detail the influence which individual works or authors exerted on Nietzsche, or to identify the many excerpts and borrowings that can be traced in his notebooks. That would be impossible in the limited space available to me here. Instead, I aim to provide a chronological survey of Nietzsche's reading of natural science throughout his life, based on a close examination of his published writings, letters and notebooks, as well as on the unpublished papers and contents of his extant library in Weimar. That does not mean that I shall be concentrating solely on works by experimental scientists such as Robert Mayer, Wilhelm Roux and Carl von Nägeli. In fact, Nietzsche's views on, and knowledge of, natural science were probably influenced more by philosophers and thinkers such as Arthur Schopenhauer, F.A. Lange and Otto Liebmann than by purely scientific books, and for this reason they will figure prominently in the discussion below. His tastes were eclectic, ranging from eminent Victorians such as Alexander Bain and Herbert Spencer to often quite obscure figures like William Rolph and Afrikan Spir, who even in the nineteenth century were largely unknown. This, then, is what makes a study of Nietzsche's reading so important: through it we are able to reconstruct the idiosyncrasies of Nietzsche's understanding of science.

Nietzsche and Darwinism

Nietzsche's relation to Darwinism is an important and complex issue, and has, over the years, attracted a great deal of attention. For this reason, I shall deal with it only briefly here and in the greater part of this chapter concentrate on the other branches of the natural sciences relevant to Nietzsche's reading and thinking.

Darwin's theory of evolution by natural selection had an incalculable influence on the scientific, intellectual and cultural climate of the second half of the nineteenth century, including the fields of philosophy and literature – those fields in which, broadly speaking, Nietzsche himself worked. As such, Darwinism – or more precisely, evolutionism – represents an important general background for Nietzsche's thought. Some aspects of Nietzsche's philosophy – such as his view of development, progress and history, as well as his concepts of the will to power and the *Übermensch* – cannot be fully understood without taking into account his attitude towards Darwinism. Yet Darwinism was not just a backdrop to Nietzsche's thinking, it also constituted a theory and a world-view which Nietzsche directly discussed and criticized. But for all the discussion there has been of Nietzsche and Darwinism, it is important not to lose sight of the fact that Darwinism never seems to have fully engaged Nietzsche's interest and that he always remained a somewhat superficial and amateur commentator. This was despite the fact that he read a large number of books on the subject, including specialist biological treatises as well as works of popular science (many of these will be mentioned in due course), and the fact that his best friend during his 'middle period', Paul Rée, was a firm supporter of Darwinism. Furthermore, unlike Rée, Nietzsche seems not to have read anything by Darwin himself (nor by major advocates of Darwin such as T.H. Huxley or Ernst Haeckel), with the one possible exception of Darwin's 'Biographical Sketch of an Infant', an article published in *Mind* in 1877 on the development of infants (but of little or no relevance to Darwinism). Furthermore, Nietzsche's interest in Darwinism, such as it was, was almost exclusively restricted to its significance for human affairs and culture; like most non-specialist writers of the time, he was never particularly concerned with it as a narrowly biological theory that sought to explain the origins and diversity of life (although he read several detailed studies dealing with it in this manner). It is hardly surprising, then, that Nietzsche's evolutionism has a greater affinity with Lamarckism (which allows for cultural and intellectual inheritance) than with Darwinism as such.

There are, broadly speaking, two schools of thought on the subject of Nietzsche and Darwinism. On the one hand, Nietzsche has been frequently regarded as a Darwinist and his writings interpreted in this light.[3] There are apparently good reasons for doing so. He introduces his concept of the *Übermensch* with the words: 'You have made your way from worm to man, and much in you is still worm. Once you were apes .... Man is a rope, fastened between animal and Superman' (*Z* 'Prologue' 3–4). Later he will say: 'Our way is upward, from the species across to the super-species' (*Z* i 'Bestowing Virtue'). He often speaks of the crucial importance of physiology and breeding (*Züchtung*); he affirms naturalism and becoming (*Werden*); his well-known emphasis on struggle and competition has, at first glance, much in common with the Darwinian concept of struggle for existence; and he rejects the idea that nature is teleological, that life develops according to some inherent goal or purpose.

However, Nietzsche has been just as often regarded as an anti-Darwinist, and this view also has strong arguments in its favour. His express criticisms of both Darwin the man and Darwinism the theory are many and harsh. He disputes the claim that the main motor of organic change is the instinct for self-preservation (whether formulated as Schopenhauer's 'will to life' or Darwin's 'struggle for existence') and the passive adaptation of an organism to its external environment; instead, he prefers to stress an internal creative force – what he calls the 'will to power'. Furthermore, he rejects the idea of progress implicit in Darwin's theory – that life is advancing to ever greater degrees of perfection – and complains in *Ecce Homo* that his concept of the *Übermensch* has caused 'learned cattle' to suspect him of Darwinism (*EH* 'Books' 1). And despite his use of the term 'super-species' in *Zarathustra*, he claims in *The Antichrist* that he is *not* interested in creating some possible higher species beyond humanity but in producing more noble forms of humanity: 'The problem I raise is not what ought to succeed mankind in the sequence of species (– the human being is an *end* –): but what type of human being one ought to *breed*, ought to *will*, as more valuable, more worthy of life, more certain of the future' (*A* 3).

So which of these interpretations is correct? In a way, both are. In the most general sense in which Darwinism is (and was) understood – that is, as a byword for evolutionism in general – there can be no doubt that Nietzsche accepted it. He seems never to have wavered in his rejection of teleology or to have doubted the basic proposition that human beings evolved from animals. In a well-known note from the year 1872–73 he wrote, for example, of the 'awful consequence of Darwinism, a theory which, by the way, I hold to be true' (*KSA* 7, 19[132]). Not long afterwards, Nietzsche echoes this sentiment in *On the Use and Disadvantage of History for Life*, where he considers 'true but deadly' the 'doctrines of sovereign becoming, of the fluidity of all concepts, types and species, of the lack of any cardinal distinction between man and animal', and argues that, should these ideas be more widely disseminated, the fabric of society will disintegrate as moral and legal codes lose their binding force (*HL* 9).

Furthermore, Nietzsche's antagonism towards Darwinism proper – that is, Darwin's theory of evolution by natural selection — is perfectly compatible with a commitment to evolutionism in general. In fact, many nineteenth-century biologists celebrated Darwin's achievement of putting evolutionism on a scientific basis whilst also rejecting outright, or at least downplaying, the role of natural selection (the absence of a genetic model of heredity undermined its explanatory power). Adaptation and the selective pressure of the struggle for existence – which Nietzsche himself saw as an essentially second-order phenomenon – were viewed by many of the writers in his library as subordinate to other evolutionary mechanisms: a law of development (*Bildungsgesetz*) (F.A. Lange in his *Geschichte des Materialismus*); an internal struggle for existence among the constituent parts of an organism (Wilhelm Roux in *Der Kampf der Theile im Organismus*); life's insatiable tendency to expand itself (W.H. Rolph's *Biologische Probleme*); and an endogenous 'perfection principle' impelling life to ever more complex structures (Carl von Nägeli's *Mechanisch-physiologische Theorie der Abstammungslehre*).[4]

Nietzsche's reading of a number of books about different aspects of Darwinism, many of them critical, is certainly one of the reasons why he takes a more hostile view of Darwin as he moves into the last phase of his thought around 1883. From this point onwards there are indeed a number of notes with titles such as 'Against Darwinism' and 'Anti-Darwin' (see, for example, *KSA* 12, 7[25]; *KSA* 13, 14[123], [133], [137]; also *TI* 'Expeditions' 14). But this opposition to Darwinism is rooted not only in the biological arguments he appropriated from Roux and Nägeli, for example. Internal changes in Nietzsche's thinking, related to his move away from 'positivism' (involving, among other things, the increasing importance of the idea of eternal recurrence and the will to power, and a greater emphasis on the importance of values) resulted in a fundamental shift in attitude towards Darwinism and an awareness that it is based on *values* – especially values such as passivity and utility – which Nietzsche rejected as decadent. Nevertheless, Nietzsche continues during this period to stress 'becoming' rather than 'being', that man is an animal and the importance of struggle and power, and he continues to deny teleology and metaphysics.

The Young Nietzsche and Natural Science, 1844–69

The young Nietzsche had little interest in, and knowledge of, the natural sciences, and read few books in the field. A major reason for this was the fact that at Schulpforta, where he studied from between the ages of 14 and 20, the teaching curriculum was strongly informed by the classics. With the exception of mathematics, for which Nietzsche showed little talent, and two hours of physics per week during his last two years at the school, natural science did not form part of the curriculum. Nietzsche would later regret this lack of a scientific education (see, for example, *WS* 195).

The first serious indications that Nietzsche had begun to realize the importance of natural science can be found in his early essay 'Fatum und Geschichte' ('Fate and History'), which he composed in April 1862. In it he writes:

> How often has the longing for natural science and history crept over me in the course of my fruitless speculations! History and natural science, the wonderful legacies of our past, the harbingers of our future: they alone are the secure foundation upon which we can build the tower of our speculation ... To find the common centre of all oscillations, the infinitely small circle, is the task of natural science. Because man looks for a centre in and for himself, we now know what a unique meaning history and natural science must have for us ... World history is, then, the history of matter, if one takes the meaning of these words in the broadest sense. (*HKW* 2, pp. 55–57)[5]

He talks further of 'uniting natural science with philosophy', and writes excitedly: 'In the middle of the vast ocean of ideas one yearns for solid ground; how often, when engrossed in fruitless speculations, has the yearning for history and natural science not crept over me!' (*HKW* 2, pp. 54, 55). Nietzsche also expresses evolutionary ideas in the text: 'Is man not perhaps the development of stone through the medium of plant or animal? ... Has this eternal becoming no end?' (*HKW* 2, p. 65). This enthusiastic and thoroughly positive evaluation of natural science is not a feature of any of Nietzsche's other contemporary notes, making it difficult to assess to what extent it is mere rhetoric or genuinely meant. What inspired Nietzsche to write so favourably about natural science in this essay is unknown, but he may have been influenced by the journal *Anregungen für Kunst, Leben und Wissenschaft* (see below).

It was at this time that Nietzsche first became preoccupied with the problem of materialism, which would later find expression in his well-known enthusiasm for Friedrich Albert Lange's 1866 book *Geschichte des Materialismus und Kritik seiner Bedeutung in der Gegenwart* (*History of Materialism and Criticism of its Significance in the Present*), in his work on Democritus (and Epicurus), and in his interest in science and physiology. The earliest evidence for this first encounter with materialism is one of those book lists that Nietzsche habitually drew up, and which reveals that he intended to acquire two works by Feuerbach in the autumn of 1861 (*HKW* 1, p. 251), and a jovial and humorous letter to his schoolfriend Raimund Granier, in which Nietzsche writes: 'Ask me what I have been occupying myself with – with refuting materialism, while you seem to believe in it' (*KGB* I/1, 28 July 1862). That this claim should be taken seriously is borne out by a later note, in which Nietzsche summarizes his most important literary and musical activities, apparently covering the 1861–62 school year: 'During the [Easter] holiday [1862] the essay "Fate and History". Also read Emmerson [*sic*], Büchner in the *Anregungen für Kunst*, etc., at Pforta began Schiller's *Aesthetic Education*, etc. ... Materialism' (*HKW* 2, p. 100).

Nietzsche's curiosity about, and knowledge of, materialism at this time was stimluated by two principal sources: his interest in Ludwig Feuerbach, especially his *The Essence of Christianity* (which, along with another work by Feuerbach,

Thoughts on Death and Immortality, Nietzsche had requested for his birthday in October 1861), and the journal *Anregungen für Kunst, Leben und Wissenschaft*, edited by Franz Brendel and Richard Pohl, which appeared in monthly issues between 1856 and 1861. As the title implies, this periodical was a cultural journal with broad intellectual interests, but which focused in particular on music, especially that of Wagner, and materialist philosophy. The cultural society Germania, which Nietzsche founded together with his friends Gustav Krug and Wilhelm Pinder, subscribed to the journal in 1861. There is also an issue from November 1859 extant in Nietzsche's library.

When Nietzsche returned to the topic of materialism a few years later, he was inspired to do so by his discovery of Lange's *History of Materialism* in the second half of 1866. During the next two years Nietzsche would make several references to materialism and scientism in his notebooks, which all seem to have their origin in his reading of Lange (and to a lesser extent Friedrich Überweg's *Grundriß der Geschichte der Philosophie von Thales bis auf die Gegenwart*), who discusses modern materialism in detail. In 1867–68, for example, Nietzsche refers to, and discusses, Laplace, Gassendi, La Mettrie, Büchner and Moleschott and the status of materialism generally.

It seems that it was Nietzsche's growing fascination with philosophy, in particular his first encounter with Schopenhauer in 1865, as well as his reading of Lange in 1866 and 1867, which led him to take a more wide-ranging interest in the natural sciences. Although Schopenhauer's philosophy is, of course, fundamentally metaphysical and speculative in character, throughout his writings he discusses, takes issue with and appeals to a whole host of contemporary scientific theories to underpin his system. He even published a treatise called *On Vision and Colours* in 1816, which was almost entirely a work of physics. Schopenhauer is therefore likely to have been a significant pervasive influence on Nietzsche's attitude towards science and the philosophy of nature.[6] Moreover, as I mentioned above, Lange's *History of Materialism* contains a quite detailed account of the natural sciences and their relation to philosophy. It seems as if Lange was as crucial in introducing Nietzsche to, and awakening his interest in, various scientific debates and theories, as Schopenhauer was for his engagement with philosophy a year earlier. Not only did Lange's book provide him with an outline of the history of philosophy and a defined philosophical position, it also seems to have motivated his subsequent reading of several important philosophical works – volumes, for example, by Kant, Democritus, Überweg and Radenhausen. However, whereas his interest in philosophy developed and flourished (alongside his interest in antiquity, literature and music), this was not the case with the natural sciences. What we see during the two years following his first reading of Lange is an appreciation of the importance of natural science and a stated *intention* to study it and its philosophical–cultural significance. But this intention would remain largely unfulfilled for several years to come.

Both Schopenhauer and Lange built on the philosophy of Kant, and Kant became the next major influence on Nietzsche. An expression of Nietzsche's philosophical interests in the late 1860s is his claim in a letter to Hermann Mushacke: 'Kant, Schopenhauer and this book by Lange – more I do not need' (*KGB* I/2, November 1866). Nietzsche appears to have read little by Kant firsthand (except the third

Critique, which includes extensive discussions of teleology). But he did read a large amount of secondary material about him, including discussions of Kant's significance and relevance for science. For example, a few years later Nietzsche studied 50 dense pages in Zöllner's *On the Nature of Comets* devoted to precisely this topic.

Of major importance for Nietzsche's relation to science and naturalism is his early interest in the ancient philosopher Democritus, who acted as a stimulus for Nietzsche's thinking about atomism, materialism and the question of teleology. Nietzsche's first mention of Democritus occurs in letters and notes during the second half of 1867 – that is, after his reading of Lange, though he would, of course, have known about him before then. Lange praises Democritus fulsomely, and it seems likely that he provided Nietzsche with the impetus to start working on Democritus, first philologically and then with an increasingly philosophical orientation. (This is suggested by the fact that Nietzsche informed his friend Carl Gersdorff of his desire to become personally acquainted with Lange and his intention to send Lange his monograph on Democritus as a sign of gratitude.) Nietzsche's notebooks from 1867 and 1868 contain a great deal of material on Democritus, including drafts and notes for essays.[7] However, the offer of a professorship in Basle in December 1868 seems to have brought the work on Democritus to an end, and Nietzsche never seriously took it up again, although he deals with him in his lectures on the pre-Platonic philosophers in a positive spirit.

Apart from his work on Democritus, the other sign of Nietzsche's concern with natural science was his short-lived intention to write his doctoral dissertation in philosophy rather than in philology. He considered two closely related topics and titles: 'The Concept of the Organic after Kant' and 'Teleology after Kant'.[8] This work was inspired by his reading of Schopenhauer and Lange, who both dealt with these questions in some detail, and Otto Kohl's dissertation *I. Kant's Ansicht von der Willensfreiheit* (1868).[9]

The notes Nietzsche made for this work contain an examination and rejection of the philosophical arguments for a belief in teleology (purpose). Closely related to this are his references to Empedocles' solution to the problem of how something apparently purposive can arise from countless random occurrences. He frequently refers to this point in his notes. Already at this early stage, then, Nietzsche denies both a teleological explanation of nature and that there is any fundamental difference between organic and inorganic matter (for they obey the same laws). These fairly extensive notes, with a continual emphasis on eternal becoming, 'das ewig Werdende', were often based on excerpts from Lange, Schopenhauer, Kuno Fischer's *Geschichte der neuern Philosophie: Immanuel Kant und seine Lehre* (*History of Modern Philosophy: Immanuel Kant and his Teaching*), and Kant's *Critique of Judgement*, to which he added his own reflections and arguments.

Nietzsche's study of Lange led him to go on to read Friedrich Überweg's *Grundriß der Geschichte der Philosophy von Thales bis auf die Gegenwart* (*Outline of the History of Philosophy from Thales to the Present Day*), from which he copied down two lists of books to read (*HKW* 3, pp. 393–94). Most of these titles treat the relation between philosophy and natural science, but a number of them are 'purely' scientific:

Kant, Allgemeine Naturgeschichte und Theorie des Himmels 1755.
Der einzig mögliche Beweisgrund zu einer Demonstration des Daseins Gottes.
Holbach *Système de la nature*.
...
To read
Schopenhauer, Über den Willen in der Natur.
Treviranus [Über] die Erscheinungen und Gesetze des organischen Lebens 1832
Czolbe neue Darstellung des Sensualismus Leipz. 1855
die Grenzen und der Ursprung der menschl[ichen] Erkenntniß Jena und Leipz. 1865.
Moleschott Kreislauf des Lebens 1862.
die Einheit des Lebens Gies[s]en 1864.
Virchow 4 Reden über Leben und Kranksein Berlin 1862.
gesamm[elte] Abhandl[ung] zur wissen[schaftlichen] Med[izin] Frankf. 1856.
Trendelenburg Logische Untersuchungen Leipz. 1862.
Überweg, System der Logik
Helmhol[t]z, über die Erhaltung der Kraft Berlin 1847.
über die Wechselwirkung der Naturkräfte 1854.
Wundt [Vorlesungen] über die Menschen- und Thierseele
Lotze, Streitschriften Leipz 1857
Medicin[ische] Psychologie 1852
...
Schelling Ideen zu einer Philosophie der Natur
Herder Ideen zur Philos[ophie] der Gesch[ichte] der Menschheit.

to read Bichat, [*Recherches physiologiques*] *sur la vie et la mort*.
Joh. Müller / über das organ[ische] Leben.
über die Physiologie der Sinne.

Kant Kr. d. Ur[theilskraft] 1790.
Fries mathem[atische] Naturphilos[ophie] Heidelberg 1822.
Schleiden über den Materialis[mus] in der [neueren] Naturwissensch[aft] Leipz. 1863.
(in Schleiden organisms understood mechanistically)
C. Rosenkranz, Schelling Vorles. Danzig 1843.
Sal. Maimon 1790 (Berl[iner] Journal f[ür] Aufklärung, von A. Riem Bd. VIII St.1.)
Oken die Zeugung 1805
Lehrb[uch] der Naturphilosophie 1809 II Aufl[age] 1843
Carus Grundzüge der vergl[eichenden] Anatomie und Physiologie 1825.

But with the exception of the work by Schopenhauer (which Nietzsche read before 1868) and possibly one or two of the other titles later (such as those by Moleschott perhaps), there is no concrete evidence that Nietzsche actually read any of the above books at this time.

By 1868, then, Nietzsche was already firmly in favour of a dysteleological view of nature and, it seems, to a large extent accepted Darwinism (or at least evolutionism). His strong attachment to Lange is noteworthy.

Nietzsche and Natural Science, 1869–75: Some General Remarks

During the 'early' phase of Nietzsche's career – which spans the years 1869–75 – the most significant influence on his thought was the philosophy of art espoused by Wagner and Schopenhauer. Nietzsche's views with regard to metaphysics at this stage were complicated. At this time Nietzsche had a highly positive view of Kant, for he felt that Kant had shown the limits of reason – and had thus opened up the space for some sort of non-empirical beyond. But though Nietzsche was convinced that a higher realm of values was conducive to, and necessary for, the development of culture, he seems already to have rejected the fundamental idea that underpins traditional metaphysics – that is, the existence of a 'real world' beyond the world of experience. Nietzsche felt that art could and should take the place previously occupied by this noumenal realm in religion, German philosophy and culture generally. This philosophical orientation meant that Nietzsche's interest in natural science waned during this period, though there are a couple of exceptions to this rule. One view regarding natural science that he held at this time follows from the above: he came to regard science and superficial empiricism as a threat to culture. But a more positive attitude towards science arose from his work on the pre-Socratic philosophers.

Nietzsche's rather peripheral interest in natural science at this time sprang from his preoccupation with the question of culture and was based on his reading of cultural and philosophical works rather than scientific treatises. Three dominant themes characterize his thought on the subject. First, Nietzsche sets up a dichotomy between art and science, and expresses his sympathy with art. As he later put it in the new preface to *The Birth of Tragedy* in 1886, Nietzsche saw his task at this time as looking at '*science through the prism of the artist*' (*BT* 'Attempt at Self-Criticism' 2).

Second, and closely related to the first, is his critique of the effect of science on culture: 'The historical and the natural sciences were necessary to overcome the Middle Ages: knowledge against faith. We now direct *art* against knowledge: Return to life! Harnessing the drive for knowledge! Strengthening of the moral and aesthetic instincts!' (*KSA* 7, 19[38]). The criticism that Nietzsche aims at science is that it is superficial – that is, it rejects the idea of a higher realm that can be reached through aesthetic experience (or perhaps through introspection) and takes as its guiding principle the *certainty* of knowledge rather than its importance and indispensability for human beings (*KSA* 7, 19[37]). In other words, science is nihilistic (although he does not yet use this term) and leads to cultural decline (*KSA* 7, 19[171], [198]).

Finally, he regards all forms of knowledge, including of course natural science, as anthropomorphic. Science does not, cannot say anything about objective reality; it is merely symptomatic of human needs and habits of thought. In several important notes (*KSA* 7, 27[37], [77], 29[8], [12]; see also *KSA* 7, 26[11], [12]) he speaks of 'Empfindungsgesetze' – that is, the laws of sensation – as the basis of all science (a view compatible with neo-Kantianism): 'All natural science is only an attempt to understand the human, the anthropological; or, more precisely, to return always to the human via the longest and most roundabout way. The expansion of the human to the macrocosm, in order ultimately to say: "in the end you are what you are"' (*KSA* 7, 19[91]).

Philosophical Influences on Nietzsche's Understanding of Science, 1869–75

The most important new reading in philosophy that Nietzsche undertook during his first years as professor (apart from ancient philosophy, of course) was Eduard von Hartmann's *Philosophie des Unbewussten: Versuch einer Weltanschauung* (*Philosophy of the Unconscious: Attempt at a World-view*, 1869), which he read intensively and carefully in 1869 and 1870, and again, together with other works by Hartmann, in 1873, 1874 and probably again in late 1887 or 1888. Although from about 1874 onwards Nietzsche became distinctly critical of Hartmann, he was initially much more enthusiastic: 'I read him often, for he has the most beautiful knowledge,' he told his friend Erwin Rohde (*KGB* II/1, 11 November 1869). Indeed, with the exception of Schopenhauer and Lange, Hartmann is likely to have been the philosopher from whom Nietzsche learned most – not just about philosophy in general but about natural science as well. For, like Schopenhauer, Hartmann incorporated a great deal of scientific data and theories into his metaphysical system, from dynamic physics (including a discussion of atoms as points of force, though without reference to Boscovich) to nerve physiology. His book *Wahrheit und Irrthum im Darwinismus: Eine kritische Darstellung der organischen Entwicklungstheorie* (*Truth and Error in Darwinism: A Critical Account of the Theory of Organic Evolution*), which Nietzsche bought in 1875 but which is no longer extant in his library, contains a detailed account and critique of Darwinism, and is likely to have been one of Nietzsche's primary sources for information about the debates surrounding the various competing theories of evolution.

Another major and pervasive influence on Nietzsche's attitude towards natural science was the famous biblical scholar David Friedrich Strauss. In 1872–73 Nietzsche entered into a critical debate with Strauss in his notebooks, dealing with his view of the role of science in modern culture, among other things. At this time he was reading Strauss's bestselling book *The Old Faith and the New* (he would reread it several times thereafter), and possibly several other works by him.[10] In *The Old Faith and the New* Strauss argues that modern Europeans are no longer Christians, that the Christian conception of the universe is no longer tenable and that we should embrace the new scientific world-view that energy physics and Darwinism in particular have done so much to construct. Much of Nietzsche's subsequent essay *David Strauss: The Writer and Confessor*, the first of the *Untimely Meditations*, is devoted to exposing Strauss's stylistic flaws and upbraiding Strauss for his German nationalism and 'philistine optimism'. Nietzsche almost completely ignores Strauss's critique of Christianity and his views on science. He thus avoids an explicit discussion of the role of natural science, although his harsh critique of Strauss and his book seems also to imply a rejection of the scientific world-view that he championed.

A further important influence on Nietzsche's thinking concerning philosophy and natural science was the obscure Russian-born Kantian metaphysician Afrikan Alexandrovich Spir (1837–90).[11] By 1872 at the latest Nietzsche had acquired Spir's *Forschung nach der Gewissheit in der Erkenntnis der Wirklichkeit* (*Research into Certainty in our Knowledge of Reality*, 1869), for he took notes on it that year (see *KSA* 7, 19[249]). He seems to have sold his copy of this work, which was superseded by Spir's 1873 book *Denken und Wirklichkeit: Versuch einer*

Erneuerung der kritischen Philosophie (*Thought and Reality: Attempt at a Renewal of Critical Philosophy*) containing extensive discussions of issues relating to epistemology and natural science. Nietzsche borrowed the first edition of Spir's *Thought and Reality* in the year it was published, 1873. He seems to have been particularly interested in Spir's critique of Kant, which influenced Nietzsche's argument in the unpublished (at least during his lifetime) essays *Philosophy in the Tragic Age of the Greeks* and *On Truth and Lie in an Extra-Moral Sense*. Later, in early 1877, Nietzsche bought the expanded two-volume second edition of *Thought and Reality* (1877), the copy of which in his library contains extensive underlinings and marginal notes that indicate he studied it particularly carefully. Nietzsche probably annotated the book when he first read it in 1877, but he may also have done so when he made another thorough study of it in 1881. He returned to the book once more in 1885.

As with Lange, Nietzsche never mentions Spir's name explicitly in his published writings, although he does twice refer to, or quote, Spir without naming him in *Human, All Too Human* (*HH* 18 and 131). Despite this lack of public acknowledgement of his debt, many of Nietzsche's notes and comments regarding epistemological issues were written in response and opposition to his reading of Spir's *Thought and Reality*. Spir's book is almost certainly one of the most important philosophical sources of, and stimuli to, Nietzsche's thinking about questions of epistemology.[12] Spir argued that it is the task of philosophy to seek absolutely true knowledge (and thus he assumes the dichotomy of a true and an apparent reality which Nietzsche came to reject). This could only be done by understanding consciousness and the supreme law of thought, the principle of identity. When Nietzsche reread *Thought and Reality* in 1881, he argued in particular against Spir's assumption that the laws of thought provide us with knowledge of the world.

Nietzsche's Reading of Natural Scientists, 1869–75

As we have seen, Nietzsche did not generally show much curiosity about natural science as such during the years 1869–75. Elisabeth Förster-Nietzsche, though, has claimed that, in 1869, Nietzsche actively entered into the debate about Darwinism, supporting his Basle colleague, the professor of zoology and comparative anatomy Ludwig Rütimeyer, after he had caused controversy with his criticism of Haeckel's *Natural History of Creation*. According to his sister, Nietzsche by and large subscribed to the pre-Darwinian brand of evolutionism which Rütimeyer, a follower of Karl von Baer, espoused.[13] But there is little evidence of this in his writings. (He did, however, own and plan to excerpt two of Rütimeyer's books: *Die Bevölkerung der Alpen* (*The Population of the Alps*, 1864) and *Vom Meer bis nach den Alpen* (*From the Sea to the Alps*, 1854), though these are only tangentially related to evolutionism (see *KSA* 8, 4[2]).) In 1872 and 1873, he considered writing an *Untimely Meditation* on the subject of science; in all likelihood he planned to discuss its influence on culture (see *KSA* 19[330], 29[163], [164], 30[8]; cf. *KSA* 7, 8[82]). He also read a few relevant books on chemistry, physics and Darwinism.[14] Despite this lack of interest in science for its own sake, it is during this time that

Nietzsche developed ideas about various aspects of science to which he would remain wedded for the rest of his life. This is true for his dynamic conception of physics, his view of science as anthropomorphic in character and for his acceptance of, but also critical attitude towards, Darwinism/evolutionism.

It was in 1873 that Nietzsche began for the first time to read genuine natural scientific texts and to become intrigued by the role and content of science. The most obvious stimulus for Nietzsche's interest in science during this time was his work on the pre-Socratic philosophers, both for the unfinished manuscript 'Philosophy in the Tragic Age of the Greeks' and in preparation for his course 'The Pre-Platonic Philosophers', which he delivered for the first time in 1872 (although it had been originally announced for the winter semester 1869–70) and again in the winter of 1875–76 and summer of 1876. These lecture notes contain discussions and allusions to contemporary scientific and epistemological debates. It is quite possible that, of all the courses he taught, this was the one about which Nietzsche was most enthusiastic, for in it he was able to combine his interest in philology, philosophy and natural science.[15] Nietzsche seems to have revised the lecture notes extensively during the spring of 1873, when he also revised the text of *Philosophy in the Tragic Age of the Greeks*. We know this because he borrowed a number of books about modern science from the University Library in Basle in the spring of 1873, and these were subsequently used for the lecture notes. Indeed, almost all the books about natural science that he read during this period seem to be related to this work.

In these lectures Nietzsche advances the following claims: (1) everything in nature is becoming; (2) there is no opposition between material and immaterial; (3) order in nature can arise from chaos without teleology (Newton, Darwin); (4) time is relative. He insists, furthermore, that materialism is an important hypothesis and that mathematics is vital for understanding the world. These conclusions are drawn from a range of sources – some echo the prevailing attitudes among a number of contemporary thinkers, others can be traced back to Nietzsche's earlier studies of Democritus and teleology, Darwinism, Schopenhauer, Kant and Lange. But his ideas were, of course, also influenced by those scientists whom he read in 1873.

The most important of these figures were the German astrophysicist Johann Carl Friedrich Zöllner and the eighteenth-century Croatian physicist Roger Boscovich. Several times during 1872 and 1873 Nietzsche borrowed and read the second edition of Zöllner's *Über die Natur der Kometen: Beiträge zur Geschichte und Theorie der Erkenntnis* (*On the Nature of Comets: Contributions to the History and Theory of Knowledge*, 1872). He later bought the book and it is still in his library. Zöllner's book is a strange mixture of advanced science, epistemology, Kantian philosophy and moral statements. Though Nietzsche's copy does not contain annotations, it nevertheless provided Nietzsche with a detailed account of modern physics – including extensive discussions of those early ideas which would eventually lead to the birth of non-classical physics – and seems to have found its way into his writings.[16]

The other important scientific work which Nietzsche borrowed repeatedly from the library of Basle University in 1873 and 1874 was Boscovich's *Philosophiae naturalis Theoria* (1759) and a commentary to Moses Mendelssohn's critical review of the book, *Anmerkung über den Auszug, und die Kritik eines Berlinischen Herrn Recensenten das Boscovichische System betreffend* (1773).[17] It appears that

Nietzsche was profoundly impressed by Boscovich in the early 1870s, but it was not until the 1880s that he began to mention and praise him in a number of notes and letters.

Philosophiae naturalis Theoria is a complex scientific and mathematical treatise, written in Latin. It is somewhat difficult to believe that Nietzsche, who, as we have seen, lacked any formal training in science, was able, without any assistance, not only to read this book but also to understand it, and even to go on and base some of his views on matter and science on its conclusions. It is likely, then, that whatever source first led Nietzsche to seek out and borrow the book probably argued for Boscovich's significance.[18] The most important conclusion that Boscovich reaches (with the help of advanced mathematics) is this: atoms have no extension, and should be regarded as mathematical points with mass. He further concludes that the number of atoms is finite and that the universe is not infinite in size. Nietzsche seems to have made these ideas his own, except that he regarded the universe as infinite.

The most remarkable thing about Nietzsche's scientific world-view is that, in a number of ways, it foreshadows the birth of modern non-classical physics. This is most evident in his well-known long note about the atomic theory of time and space from early 1873 (*KSA* 7, 26[12]).[19] In this long note Nietzsche writes, for example:

> It is possible
> 1) to reduce the external world into space points
> 2) these can further be reduced to an atomic theory of time.
> 3) The atomic theory of time finally corresponds to a theory of sensations. *The dynamic timepoints* is identical with the point of sensation. Since there exists no contemporaneity of sensations.

Schlechta and Anders argue that Nietzsche is here inspired by Boscovich's mathematical theory of atoms and translates this into a theory of sensations, which he had encountered principally in Spir's work. They also regard Zöllner as a probable important stimulus.[20] This may well be true, but one cannot exclude the possibility that the whole note is an excerpt from some currently unidentified work Nietzsche had read, as Alwin Mittasch has suggested.[21] For it deals with questions about which Nietzsche had previously said little and to which he will not return in any narrow sense.

Throughout 1874 Nietzsche continued to borrow (and presumably read) Spir, Boscovich and Zöllner, but otherwise read little else of relevance to natural science.

Nietzsche and Natural Science, 1875–82 (the 'middle period')

In 1875–76 Nietzsche underwent an emotional and intellectual crisis, which in important ways fundamentally altered his manner of thinking. He broke with the ideas of Schopenhauer and Wagner – the two men who had hitherto exerted the most sustained influence on his intellectual development – and replaced his youthful idealism, pessimism and 'metaphysical-artistic views', which he now described as 'pleasant but untenable' (*KSA* 8, 23[159]), with a free-spirited scepticism, a preference for science over art and a high regard for the Enlightenment. There is indeed an undeniable difference in style and content between the four *Untimely Meditations* (1873–76) and *Human, All Too Human* (1878).[22] In a note from 1875 which is typical of his new way of thinking, he writes:

> One expects to acquire *skills* from studying the ancients: earlier, e.g. to be able to write and speak well. But what does one now expect! – Thinking and drawing conclusions: but one does not learn this *from* the ancients, but at most *through* the ancients, by means of science [*Wissenschaft*]. However, all historical conclusions are very conditional and uncertain: one ought to prefer the natural sciences. (*KSA* 8, 5[88])

The standard view of the breach between the early and 'middle' Nietzsche is that it occurred in August 1876, shortly after the publication of *Richard Wagner in Bayreuth*, when Nietzsche abruptly left the *Festspiele* and 'fled' to Klingenbrunn.[23] The two causes most often cited for his change of heart are his disappointment when he saw what the Wagnerian cultural revival actually meant in practice, and the influence of the French *moralistes* and aphorists such as La Rochefoucauld. A more plausible explanation for the shift in Nietzsche's thought is the influence of Paul Rée, whose book *Psychological Observations* Nietzsche had read and praised in October 1875, and who became a personal friend during the spring of 1876. Rée had joined Nietzsche in Bayreuth, left with him in August, and was to remain Nietzsche's closest friend and mentor until the breakdown of their relationship over Lou Salomé in 1882.

However, Nietzsche's thought had already begun to drift in another direction *before* he wrote *Richard Wagner in Bayreuth* – that is, during the spring and early summer of 1875. It follows that none of the standard explanations can be correct: neither the practical implications of Wagner's cultural reforms, nor the reading of the French *moralistes* (since he only began to study them in 1876), nor the influence of Paul Rée can have been the original cause of the crisis and upheavals in Nietzsche's thinking. What is certain is that he had become increasingly dissatisfied with Christianity and with his earlier idealism. Furthermore, it is possible to argue that in 1875–76 Nietzsche was able to follow up his original, abortive plans to study natural science in 1866–68, which he had been prevented from doing by the influence of Wagner and Schopenhauer and the huge demands placed on him as a university professor. It was not until 1875 that he again had enough time and energy to be able to cultivate interests outside classical philology and the Wagnerian cause. In 1875 we see an intensive buying and reading of books in the fields of natural science, and it was this reading that was an important contributory factor to his break with Schopenhauer and Wagner.

During the spring and summer of 1875 Nietzsche drew up a long list of scientific books, ordered them and began to work his way through them. In a jotting, he tells himself that the first hour of the day is to be dedicated to learning (*KSA* 8, 4[3]). He started to make extensive notes on the positivist and anti-metaphysical thinker Eugen Dühring's *Der Werth des Lebens* (*The Value of Life*, 1865), partly for the purpose of re-examining his own relation to Schopenhauer (*KSA* 8, 9[1], pp. 131–81). Immediately afterwards he excerpted passages from the first chapter of the German translation of the Scottish physicist Balfour Stewart's *The Conservation of Energy* (1872, published in German as *Die Erhaltung der Energie*, 1875) (*KSA* 8,

9[2]). Around the same time, Nietzsche compiled another list, this time of scientific disciplines (including mathematics, physics, chemistry and physiology), which he presumably planned to study; several of these correspond to books he bought in 1875 (*KSA* 8, 10[6]). Nietzsche also wrote a letter to Carl Gersdorff, in which he reveals: 'I am now in the process of belatedly filling the awful gaps in our education (I am thinking of Pforta and the universities among other things) ... one must work one's way through the study of several difficult sciences, especially the properly *exact* ones.' (*KGB* II/5, 21 July 1875).

Under the influence of Paul Rée, Nietzsche was encouraged to go even further in embracing the importance of natural science. For a while he seems even to adopt a view closely related to positivism. 'Positivism absolutely necessary' (*KSA* 8, 20[19]), he writes during the winter of 1876–77. Other notes from the same period express similar sentiments: 'Not to shy away from a temporary coarsening (through the supremacy of natural science, mechanics)' (*KSA* 8, 22[14]). And: 'Necessary to absorb the whole of positivism, and yet to remain a carrier of idealism' (*KSA* 8, 22[37]).

It is well known that Nietzsche was generally rather contemptuous towards 'English' thought (see, for example, Robin Small's chapter in this volume). But between 1876 and 1880 he actually admired and respected many British thinkers, partly because of their scientific approach to certain fields of study. In 1875 Nietzsche began to buy and read a large number of books by British and American writers on science and anthropology, as well as in other areas (such as American humorous literature), and he even expressed increasingly pro-British sympathies. Britain, he thought, was leading the way in 'philosophy, natural science and history'; its tradition of scholarly independence contrasted favourably with the German emphasis on 'education, methodology and schools', which were choking the originality of its best minds (*KSA* 8, 23[170]). And in a letter to Rée written in early August 1877, Nietzsche referred to British philosophers and thinkers as 'the only good philosophical company which exists at present'. In the same spirit he sent another letter to Rée two years later, at the end of July 1879, with the plea: 'Could you send me an informative book, if possible of English origin, but translated into German and with clear and large print?' (*KGB* II/5).

In *Ecce Home* and the preface to the second edition of *Human, All Too Human* (written in 1886) Nietzsche insists that, during this middle period, he read science books almost exclusively. That claim is exaggerated, but it is true that the engagement with science and natural science fundamentally changed Nietzsche's thinking, and in the first work characteristic of this new phase in his development, *Human, All Too Human* (1878), he regarded it as a mark of a higher culture to value 'the little unpretentious truths' won by strict or scientific method more highly than 'the beautifying and brilliant errors' of metaphysical and artistic ages (*HH* 3). In a later aphorism in the book, entitled 'Annual rings of individual culture', he seems to describe not only his positivistically-inspired view of the evolution of the human mind in general, but also his own intellectual development:

> Men at present begin by entering the realm of culture as children affected religiously, and these sensations are at their liveliest in perhaps their tenth year, then pass over into feebler forms (pantheism) while at the same time drawing closer to science; they put God,

> immortality and the like quite behind them but fall prey to the charms of a metaphysical philosophy. At last they find this, too, unbelievable; art, on the other hand, seems to promise them more and more, so that for a time metaphysics continues just to survive transformed into art or as a mood of artistic transfiguration. But the scientific sense grows more and more imperious and leads the man away to natural science and history and especially to the most rigorous methods of acquiring knowledge, while art is accorded an ever gentler and more modest significance. All this nowadays usually takes place within a man's first thirty years. It is the recapitulation of a curriculum at which mankind has been labouring for perhaps thirty thousand years. (*HH* 272)

Apart from Rée's three books (*Psychological Obervations* (1875), *The Origin of the Moral Sensations* (1877) and *Evolution of Conscience* (1885), the last two of which were already available to Nietzsche in the late 1870s and early 1880s), Nietzsche read a number of other works of philosophy which coloured his view of science during this period. These include Afrikan Spir's *Thought and Reality* and several works by Eugen Dühring (1833–1921), including *Der Werth des Lebens: Eine philosophische Betrachtung* (*The Value of Life: A Philosophical Examination*, 1865), *Cursus der Philosophie als streng wissenschaftlicher Weltanschauung und Lebensgestaltung* (*Course of Philosophy as a Strictly Scientific World-view and Way of Life*, 1875) and probably *Kritische Geschichte der allgemeinen Prinzipien der Mechanik* (*Critical History of the General Principles of Mechanics*, 1873). He also read Philipp Mainländer's *Philosophie der Erlösung* (*Philosophy of Redemption*, 1876/1886), Alfons Bilharz's *Der heliozentrische Standpunkt der Weltbetrachtung: Grundlegungen zu einer wirklichen Naturphilosophie* (*The Heliocentric Standpoint in the Understanding of the World: Fundamentals of a Real Philosophy of Nature*, 1879) and his friend Paul Deussen's *Die Elemente der Metaphysik* (*The Elements of Metaphysics*, 1877), a strongly Kantian and Schopenhauerian work. In 1875 Nietzsche also read the Enlightenment physicist and aphorist Georg Christoph Lichtenberg, the British liberal historian of philosophy George Henry Lewes's *Geschichte der alten Philosophie* (*The History of Philosophy from Thales to Comte*, Part 1, 1871) (Lewes wrote in the same 'modern' spirit as the 'Whig' historian of ancient Greece, George Grote, whom Nietzsche had read earlier), and delivered a course of lectures on Democritus which probably reawakened his earlier interest in materialism and atomism. In addition, he read two works by the American positivist John William Draper, *History of the Intellectual Development of Europe* (1864, published in German as *Geschichte der geistigen Entwickelung Europas*, 1871) and *History of the Conflict between Religion and Science* (1872, published in German as *Geschichte der Konflikte zwischen Religion und Wissenschaft*, 1875). In 1875 and 1880 Nietzsche also bought two works by the British philosopher Herbert Spencer, whose contemporary popularity as a thinker was based on his ability to fuse Comtean positivism, utilitarianism and evolutionism: *Data of Ethics* (1879, published in German as *Die Thatsachen der Ethik*, 1879) and the two-volume *Introduction to the Study of Sociology* (1873, published in German as *Einleitung in das Studium der Sociologie*, 1875). Spencer seems to have been a particularly crucial influence in the early 1880s, if the numerous entries in Nietzsche's notebooks taking issue with his thought are anything to go by.

Nietzsche read a large number of books relating to positivism and the philosophy of nature in 1875–76, but few during the period 1877–80, mainly because of his poor health. In 1880–81 he resumed his more extensive reading. In 1880 he read the fourth edition of the positivist Émile Littré's *La science au point de vue philosophique* (1876), which contains a number of interesting essays on different aspects of natural science. Although the only two essays annotated in Nietzsche's copy of the book are the ninth ('De la physiologie') and the sixteenth essay ('De l'histoire de la civilisation en Angleterre, par Buckle'), it seems likely that he read the whole work. He obtained and read two works by the Scottish philosopher and psychologist Alexander Bain (1818–1903), *Mind and Body: The Theories of Their Relation* (1873, published in German as *Geist und Körper: Die Theorien über ihre gegenseitige Beziehungen*, 1874), one of the founding texts of modern brain science, and *Education as a Science* (1879, published in German as *Erziehung als Wissenschaft*, 1880), which invokes recent advances in psychology and physiology in an attempt to put pedagogy on a more systematic footing. Nietzsche also read Henry Maudsley's *Responsibility in Mental Disease* (1874, published in German as *Die Zurechnungsfähigkeit der Geisteskranken*, 1875), Alfred Espinas's *Die thierischen Gesellschaften* (*Animal Societies*, 1879), Karl Semper's *Die natürlichen Existenzbedingungen der Tiere* (*The Natural Conditions of Existence as they Affect Animal Life*, 1880), E.H. Lecky's *History of the Rise and Influence of the Spirit of Rationalism in Europe* (1865, published in German as *Geschichte des Ursprungs und Einflusses der Aufklärung in Europa*, 1873), Friedrich von Hellwald's *Culturgeschichte* (*History of Culture*, 1875), Lange's *History of Materialism*, and, most importantly, works by Otto Caspari and Otto Liebmann.

Though Otto Caspari is likely to have been one of the principal influences on Nietzsche's conception of nature, there has been no thorough examination of Nietzsche's reception of him. Nietzsche ordered Otto Caspari's *Der Zusammenhang der Dinge: Gesammelte philosophische Aufsätze* (*The Relation of Things: Collected Philosophical Essays*, 1881), which contains discussions of contemporary works of natural science and the philosophy of science, from his publisher Ernst Schmeitzner in June 1881 and read it during the autumn.[24] Many of the books Nietzsche read in 1881–83 relating to natural science he learnt about from his reading of this work, which led him thereafter to acquire them. His copy of the book is heavily annotated.

Another major influence on Nietzsche's view of nature and science, and again someone who has received almost no attention in Nietzsche scholarship, is the neo-Kantian thinker Otto Liebmann (1840–1912), who wrote at length on natural science. Nietzsche never mentions him in his published works but studied him intensively. Nietzsche ordered two of his most important books in August 1881: *Kant und die Epigonen* (*Kant and his Epigones*, 1865) and the second edition of *Zur Analysis der Wirklichkeit: Eine Erörterung der Grundprobleme der Philosophie* (*On the Analysis of Reality: A Discussion of the Basic Problems of Philosophy*, 1880). The latter work, Nietzsche's copy of which is heavily annotated throughout and which contains his usual marginal comments such as 'NB' and 'gut', was probably of great importance for his knowledge and views of natural science.

Nietzsche's Reading of Works by Natural Scientists, 1881–82

The third and last major period in which Nietzsche read scientific books in a sustained manner occurred in 1881 (or rather, during the years 1881–83, with the heaviest programme of reading in 1881). The reason for this is tied up with new departures in his philosophy. In August of that year he was struck by the idea of eternal recurrence – probably partly as a result of his reading in natural science – and he thereafter read a number of works in physics in order to explore this idea further. Unlike the idea of eternal recurrence, the concept of will to power developed slowly over a period of time, but by 1880–81 it had become another fundamental concept in his thought, acquiring paramount importance in 1883. For the purpose of investigating, elaborating and validating these two ideas, eternal recurrence and the will to power, Nietzsche began to read more and more scientific books in 1881 and the following years. His reading now became more focused around questions and themes relating to these ideas. Probably the most important works related to natural science for Nietzsche in this period were Robert Mayer's *Mechanik der Wärme* (*Mechanics of Heat*, 1867); J.G. Vogt's *Die Kraft* (*Force*, 1878), Richard A. Proctor's *Our Place among Infinities* (1876, published in German as *Unser Standpunkt im Weltall*, 1877); Otto Schmitz-Dumont's *Die mathematischen Elemente der Erkenntnistheorie* (*The Mathematical Elements of Epistemology*, 1878) and *Die Einheit der Naturkräfte* (*The Unity of Natural Forces*, 1881) (both of these works contain annotations); and Wilhelm Roux's *Der Kampf der Theile im Organismus* (*The Struggle of the Parts in the Organism*, 1881). He was to revisit and reread (or continue reading) several of these authors and books in 1883.[25]

At around the same time, Nietzsche's 'positivism' and faith in facts, truth, science and the Enlightenment tradition underwent a change and became more sceptical and negative, which signalled the beginning of the third phase of his intellectual development. Not only did the concept of truth now begin to appear more complicated for him, he realized that facts, science and truth are essentially value-neutral or 'nihilistic', and thus cannot contribute to the affirmation of life.

In a letter to his friend Franz Overbeck, Nietzsche emphasized his desire and need to read scientific works in 1881.

> *Daybreak* is a beginning of my beginnings – what lies still ahead of me! on me! ... I am at the 'peak' of life, i.e. of my tasks, which life has finally set me, and I must, wherever possible, without any external disturbance, dedicate these next four years to precisely these tasks, and will not think of anything else. Help me in this, you most faithful of the faithful! ... Let me say in confidence: the little work that I can do with my eyes now belongs almost exclusively to physiological and medical studies (I have been so badly educated! – and must really *know* so much! (*KGB* III/1, 20–21 August 1881)

Nietzsche certainly acquired numerous science books that year. He received Mayer's *Mechanics of Heat* from Köselitz in March. In June he ordered from his publisher various books related to physiology and chemistry, including Michael Foster's *A Text Book of Physiology* (1877, published in German as *Lehrbuch der Physiologie*, 1881), Roux's *The Struggle of the Part in the Organism*, James Johnston's *The Chemistry of Common Life* (1853–55, published in German as *Chemie des täglichen Lebens*, 1881), Caspari's *Relation of Things*, as well as several health handbooks and works of popular medicine – Gregory Moore discusses the latter in his contribution to this volume. In the letter to Overbeck quoted above he further requests Caspari's *Die Thomson'sche Hypothese* (1874), Liebmann's *Analysis of Reality*, Adolf Fick's *Ursache und Wirkung* (*Cause and Effect*, 1882), Vogt's *Force*, as well as the journals *Philosophische Monatshefte* and *Kosmos*. In October he asked his friend to send him Pierre Foissac's *Meteorologie* (*On Meteorology*, 1859). It seems as if he was led to many of the titles he ordered by his reading of Caspari's *The Relation of Things*.

For the year 1882, when Nietzsche published *The Gay Science* (in July–August) and a collection of poems, *Idyllen aus Messina*, we have almost no evidence of any scientific reading, except continued reading of the Darwinist Friedrich von Hellwald's *Culturgeschichte* (which he had begun reading in 1877 or 1881) and Robert Mayer. However, he probably read much more: this was also the year of his ill-fated liaison with Lou Salomé, and she recalls that Nietzsche was interested in studying science during this time.[26] And in September that year, he visited Leipzig for the purpose of finding and reading new books, several of which are likely to have been about natural science.

Nietzsche's Reading of Natural Science, 1882–88: Some General Remarks

In the third and last phase of his intellectual development, Nietzsche focused his attention on the problem of value rather than on educating himself, which led to his project of revaluing all values, but also to a decline in his interest in the natural sciences. As in his early phase, he consequently became more critical vis-à-vis knowledge, truth and science.

In 1883, the year when he wrote the first two books of *Thus Spoke Zarathustra* (and began work on the third), Nietzsche continued to read a large number of scientific books, especially about biology and physiology. Some of his reading found its way into *Thus Spoke Zarathustra*, but, since this is primarily a poetic work, its presence is mostly buried or only alluded to. However, the three main themes of *Thus Spoke Zarathustra* – the *Übermensch*, will to power and eternal recurrence – to some extent all owe their conception to natural science: to Darwinism, physiology and to cosmology and physics in particular. His interest in exploring new theories and discoveries in the natural sciences was still manifest in 1883, but after that year this inclination is rather less evident in his writings, letters and reading habits.

In general Nietzsche bought and read significantly fewer scientific books during this period. He seems to have read very few texts in the exact sciences, but continued to read about physiology, biology and Darwinism (again, relating mainly to his concern with the will to power).

Nietzsche's Reading of Philosophical Texts Relating to Natural Science, 1883–88

Nietzsche read numerous philosophical works discussing natural science between 1883 and 1888, but in most cases he had already read them before. The exception was his reading of a trio of important and influential scientist-philosophers with positivist sympathies: Emil Du Bois-Reymond, Richard Avenarius and Ernst Mach (see discussion below). Of especial importance is his continued reading of Lange, Dühring, Hartmann and Rée. Nietzsche seems to have read Lange's *History of Materialism* almost every year during this period. Since he had given away his own copy long before, most of this reading was probably done in libraries. Based on the evidence of the two extracts in his notebooks, he appears to have read Lange in 1883, 1884 and 1885. (However, these extracts do not relate directly to natural science, and so his reading of Lange at this time may have been of little importance for his views on natural science.) In 1887 he bought the new third edition of Lange's *magnum opus* and read and annotated it again. The chapter with the most underlinings is entitled 'Darwinism and Teleology'. Dühring's *Course of Philosophy* is another work which Nietzsche returned to frequently – for example, in 1883, 1885 and perhaps also in 1888 (and earlier in 1881). In the main, Nietzsche had little sympathy for Dühring, but this book nevertheless seems to have acted as a stimulus on him. Nietzsche's copy of the book contains extensive annotations throughout, but it is impossible to determine when these markings were made, since he reread the book so often. The work is positivistic in spirit, and contains a long chapter on the natural sciences, 'Principien des Naturwissens', and some further discussions of science in a later chapter, 'Wissenschaft und Philosophie in der alten und in der neuen Gesellschaft'. Nietzsche's reading of these chapters in 1881 was almost certainly important for developing his idea of eternal recurrence. Eduard von Hartmann was another philosopher whom Nietzsche held in low regard, but nonetheless often revisited. He intensively read, took notes on and heavily annotated Hartmann's *Phenomenology of Moral Consciousness* (1879) in 1883. Though this massive work is largely a study of ethics written from a Schopenhauerian perspective, at least the chapter 'The Evolutionistic Moral Principle' includes some discussion of natural science. Other related books which Nietzsche read or reread between 1882 and 1885 have some bearing on science and are worth mentioning here: Gustav Teichmüller's *Die wirkliche und die scheinbare Welt: Neue Grundlegung der Metaphysik* (*The Real World and the World of Appearance: A New Grounding for Metaphysics*, 1882); Maximilian Drossbach's *Über die scheinbaren und die wirklichen Ursachen des Geschehens in der Welt* (*On the Apparent and Real Causes of Events in the World*, 1884); Paul H. Widemann's *Erkennen und Sein* (*Knowledge and Being*, 1885); and Afrikan Spir's *Thought and Reality*.

During or before 1887 Nietzsche acquired and then later, during the summer of 1887, began to read systematically another book by the neo-Kantian philosopher Otto Liebmann, *Gedanken und Tatsachen* (*Thoughts and Facts*, 1882). From this work, which discusses scientific issues extensively, Nietzsche noted a few remarks that have a bearing on his continued thinking about the will to power (*KSA* 12, 9[92]). Furthermore, his copy of the book is heavily annotated; among the numerous comments written in the margins of pages is the phrase 'Wille zur Macht' on page 11.

Relevant studies in cultural history which Nietzsche read at this time include (once again) Friedrich von Hellwald's *History of Culture* and E.H. Lecky's *History of the Rise and Influence of the Spirit of Rationalism in Europe*.

Nietzsche's Reading of Works by Natural Scientists, 1883–88

After 1883 Nietzsche's interest in science was mostly limited to works of biology and psychology. During this time he also read three of the most significant contemporary philosophers of science – Emil Du Bois-Reymond, Richard Avenarius and Ernst Mach – a fact that has been curiously overlooked.

Nietzsche had first encountered the work of the Berlin physiologist and philosopher of nature, Emil Du Bois-Reymond (1818–96), in the early 1870s. He possessed a bound copy of two of Du Bois-Reymond's essays, *Über eine Akademie der deutschen Sprache – Über Geschichte der Wissenschaft: 2 Festreden* (1874). It seems it was the first of these which interested Nietzsche at that time: he discussed it with the Wagners and mildly criticized Du Bois-Reymond in two notes and in *Schopenhauer as Educator* (*KSA* 7, 32[83], 35[12], p. 815; *SE* 6).[27] We do not know whether Nietzsche read the second essay, but he probably did so. By 1881 he seems to have become more interested in Du Bois-Reymond, and asked Overbeck if there existed any collection of his essays (*KGB* III/1, 20/21 August 1881). Overbeck's reply is lost, but anyway a volume of his collected essays was not published until 1885–87. Later, probably between 1884 and 1887, Nietzsche bought a bound copy of two of Du Bois-Reymond's most well-known essays, *Über die Grenzen des Naturerkennens, Die sieben Welträtsel: 2 Vorträge* (*On the Limits of Our Knowledge of Nature*, *The Seven Riddles of the World: 2 Lectures*, 1884).[28] In the first essay Du Bois-Reymond argues the case for a complete determinism (based on atoms in motion), but also suggests that there are some questions that lie forever beyond the ambit of science. The famous conclusion to his essay, *ignorabimus* (we shall never know), sparked a long-running debate about the limits of human knowledge about the world. The thrust of the second essay is similar: Du Bois-Reymond discusses seven philosophical puzzles, including the origin of matter, motion, life and consciousness, and decides that, whilst some of them can eventually be solved, others are insoluble. It is likely that Nietzsche read both essays, but no examination of this probable reading and its possible relevance for him has been published. Some sort of response is likely, for Nietzsche sent *On the Genealogy of Morals* to Du Bois-Reymond in November 1887, and he seems to have been influenced by the *ignorabimus* debate (see his letter to Constantin Georg Naumann, *KGB* III/5, 8 November 1887).

That Nietzsche's reception of critical positivism (empiriocriticism) has been largely ignored by scholars is surprising, because he read books by the two founders of the school, Richard Avenarius and Ernst Mach (although, typically, he never mentions or discusses either of them explicitly).[29] An understanding of the historical context and the influence these figures exerted on Nietzsche will help us understand his often paradoxical statements regarding positivism, empiricism, truth and science. Contrary to received opinion, Nietzsche's reading of Avenarius and Mach shows that he was familiar with, and even stimulated by, some of the principal ideas of what was to become logical positivism, the philosophy of the Vienna Circle and analytical philosophy.

Richard Avenarius (1843–96) began to argue his empiriocritical views – that is, that it is the task of philosophy to develop a concept of the world based on 'pure experience' and the principle of 'economy of thought' – in his *Philosophie als*

Denken der Welt gemäss dem Princip des kleinsten Kraftmasses: Prolegomena zu einer Kritik der reinen Erfahrung (*Philosophy as Thinking the World in Accordance With the Principle of the Least Amount of Energy: Prolegomena to a Critique of Pure Experience*, 1876), which in many respects is similar to Mach's later, but independently conceived, *Beiträge zur Analyse der Empfindungen* (1886). According to a bill in Weimar, Nietzsche's bookseller sent him the book as early as September 1876, but it seems as if he then returned it without buying it: there is no copy of the book in Nietzsche's library today. However, he certainly read it for the first, if not the second, time during the winter of 1883–84. There is no conclusive evidence of his reading Avenarius before this date, but it is at least possible: three notes written between early 1880 and early 1881 discuss sense perception and the relation between thinking and energy which may well have been inspired by Avenarius (or Mach) (*KSA* 9, 10[F97], [F100], [F101]). In 1883–84 Nietzsche excerpted passages from, and took issue with, Avenarius's book in several longer notes (*KSA* 10, 24[9], [10], [13]). Shortly thereafter he wrote to Overbeck stating that he needed to revise his views on epistemology and metaphysics (*KGB* III/1, 7 April 1884). It is thus likely that Avenarius was of great importance for his thinking at this time.

The excerpts from and discussions of Avenarius's *Philosophie als Denken der Welt* that Nietzsche made during the winter 1883–84 concern important questions such as energy (relevant for Nietzsche's thinking about will to power) and the critique of causality, will, purpose and the concept of substance. Sections 14 and 15 of *Beyond Good and Evil*, with their critique of positivism and physiologists who uphold the 'principle of the "least possible energy"' (*Princip der 'kleinstmöglichen Kraft'*), were probably written in response to his reading of Avenarius.

Two other things connected Nietzsche to Avenarius. He knew and corresponded with Richard Avenarius's brother, Ferdinand, who in 1887 founded the cultural journal *Der Kunstwart*, to which Nietzsche contributed. More importantly, in 1877 Richard Avenarius himself founded a journal, *Vierteljahrsschrift für wissenschaftliche Philosophie*, which he thereafter edited, with the assistance of Max Heinze and Wilhelm Wundt, until 1889. Avenarius wrote several articles in the first issues of the journal, which Nietzsche may have read, including 'Über die Stellung der Psychologie zur Philosophie' and 'In Sachen der wissenschaftlichen Philosophie'. The journal was marked by a clear anti-speculative and anti-metaphysical tendency and published articles about 'scientific philosophy', including many about epistemology and the relation between psychology, natural science and philosophy. If Nietzsche sometimes read *Vierteljahrsschrift für wissenschaftliche Philosophie*, as seems likely, he could have acquired a reasonably detailed knowledge of critical positivism, but I have been unable so far to find any conclusive evidence that he did so. However, there are several reasons why it is probable that he was familiar with the journal, and this therefore makes further investigation worthwhile. For example, Nietzsche twice refers to 'wissenschaftliche Philosophie' in a manner that suggests that it could have been a response to the *Vierteljahrsschrift für wissenschaftliche Philosophie*, but that is far from assured (*HH* 131, *BGE* 211). Furthermore, the contents of the journal were advertised in several other periodicals and it is almost certain that Nietzsche would have at least secondhand knowledge of the material from these. In addition, he may have known about the journal and been inspired to read it through his acquaintance with Max Heinze, one of the co-editors. Heinze had not only been Nietzsche's teacher and private tutor at Pforta, but had remained a friend of the Nietzsche family ever since. Heinze was called to the chair of philosophy in Basle in early 1874 and thus became a colleague of Nietzsche's. They met frequently during 1874, even celebrated Christmas together, and remained in close contact afterwards as well.

The other great representative of critical positivism or empiriocriticism, Ernst Mach (1838–1916), who was indebted to Fechner's *Elemente de Psychophysik* (1860) and worked as a researcher on sense perception, argued that all empirical knowledge was the result of simple sense experience (*Empfindungen*). His ideas were extremely influential and constituted an important starting-point for the logical positivists in the early twentieth century. Nietzsche may have read several works by Mach. In July or August 1882 he wrote down the title of Mach's lecture *Die Geschichte und die Wurzel des Satzes von der Erhaltung des Arbeit* (*History and Root of the Principle of the Conservation of Energy*, 1872), presumably as intended reading relating to his thinking about eternal recurrence, but we have no certain evidence that he actually read it – or indeed, any of the other titles on the list.[30] In this essay Mach discusses some of the views he would later present in his *Beiträge zur Analyse der Empfindungen* (*Contributions to the Analysis of the Sensations*, 1886), including ideas about causality, space and time. According to Max Oehler, Nietzsche read texts by Ernst Mach in a public reading-room in Zürich in 1884.[31] It is also possible that Nietzsche read essays by Mach, or reviews and discussions of his work, in philosophical journals. Later, probably in 1886 or 1887, Nietzsche purchased and read one of Mach's major works, *Contributions to the Analysis of the Sensations*; two pages of his copy of the book contain annotations. Mach argues here that all sense experience ultimately consists of simple elements (like colour, sound, pressure, heat, and so on) and calls into question the concept of 'substance' and the belief in *a priori* concepts. Our notions of discrete objects and of the self, the 'I', are only a stopgap, Mach argues, for the purpose of acquiring a preliminary sense of orientation. The 'I' is nothing but a complex of memories, impulses and sense perceptions. By challenging the concepts of matter and soul Mach is able to suggest a solution to the mind–body problem, and thereby makes the connection between physiology and psychology more probable. The goal of science is not explanations through determining causes, but the description of the simple elements.

As Nietzsche never explicitly discusses Mach, further investigation is necessary to determine his response to his work.[32] Certainly he would have approved of, and shared, Mach's critique of concepts, especially of the concept of the self. *Twilight of the Idols*, and to a lesser extent, *On the Genealogy of Morals* and *Beyond Good and Evil*, contain a number of passages which may well bear the influence of Nietzsche's reading of Mach (and Avenarius). An example might be the prominence Nietzsche attaches to empiricism and sense perception in *Twilight of the Idols* (see, for example, *TI* 'Reason' 2), his constructivist view of how we conceive the world (with an emphasis on biological evolutionism), his belief that fundamental concepts and science are useful and necessary fictions, the emphatic anti-metaphysical strain of his thought, his conviction that a connection exists between physiology (and physics) and psychology, and his pragmatic understanding of truth. Another similarity between Nietzsche's thinking and empiriocriticism is the former's vigorously expressed idea that psychology can be employed as a way of solving epistemological and scientific problems (see, for example, *BGE* 23). It is also not impossible that Mach and Avenarius were the inspiration for Nietzsche's plans for a physiology of aesthetics and his claim that atoms with extension do not exist, but are only useful fictions (*KGB* III/1, end of August 1882).[33] And yet there is truth in the conventional view that in the late 1880s Nietzsche was by and large explicitly critical of positivism. He denies the possibility of pure facts (they always involve interpretation) and emphasizes the importance of *values*, something that the positivists more or less ignored. Nietzsche even seems to disparage the critical positivists in one section of *Beyond Good and Evil* (*BGE* 10).

Nevertheless, Nietzsche sent Mach a copy of *On the Genealogy of Morals* in November 1887, and this presumably indicates that overall he was favourably disposed towards him and his work. Nietzsche's library also contains a reprint of an article in *Repertorium der Physik* written by Mach (together with P. Salcher), 'Photographische Fixierung der durch Projectile in der Luft eingeleiteten Vorgänge', with the following authorial dedication: 'Herrn Prof. Dr. Nietzsche hochachtungsvoll E. M.'. The paper was published in the summer of 1887 and Mach must have sent it to Nietzsche in return for his *On the Genealogy of Morals*.[34]

But what other science books did Nietzsche read between 1883 and 1888? For a start, he continued to work through several works that he had first read in 1880–81. The most important of these were: Roux's *The Struggle of the Parts in the Organism*, Vogt's *Force*, Semper's *The Natural Conditions of Existence as they Affect Animal Life* and Espinas's *Animal Societies*. Nietzsche's copies of all four books are annotated throughout.[35]

During the period 1883–88 Nietzsche seems to have read few works in physics, chemistry or mathematics. The only known examples are Vogt and an article by Werner Siemens entitled 'Über die Zulässigkeit der Annahme eines elektrischen Sonnenpotentials und dessen Bedeutung zur Erklärung terrestrischer Phänomene', which Heinrich Köselitz sent him in 1883. As I mentioned earlier, Nietzsche was primarily interested in books in the fields of physiology, biology and psychology. Probably the most important books included in this category – certainly the most heavily annotated (indeed, they are among the most annotated books in Nietzsche's library) – were by Dumont, Rolph and Nägeli. Léon Dumont's *Vergnügen und Schmerz: Zur Lehre von den Gefühlen* (*Pleasure and Pain: On the Theory of the Feelings*, Leipzig, 1876), which Nietzsche read and took notes on in 1883 (but possibly also once or twice earlier, in the late 1870s and in 1881) contains philosophical and physiological discussions of responses to the biological imperatives of pleasure and pain, as well as an appraisal of Darwin's importance. The Anglo-German anti-Darwinist biologist W.H. Rolph's *Biologische Probleme, zugleich als Versuch zur Entwicklung einer rationellen Ethik* (*Biological Problems, also an Attempt at an Elaboration of a Rational Ethics*, 1884), was a significant influence on his understanding of Darwinism and evolutionary ethics, as well as on his conception of the will to power. Carl von Nägeli's *Mechanisch-physiologische Theorie der Abstammungslehre* (*Mechanico-Physiological Theory of Descent*, 1884), which Nietzsche acquired in 1886, is another anti-Darwinian work but one which also contains extensive general and philosophical discussions of natural science in general. Nietzsche annotated his copy of this massive book extensively and toyed with some of Nägeli's ideas in his notebooks (see, for example, *KSA* 12, 7[9], [25]).[36]

Other relevant books which Nietzsche read during this period include the following. In 1883 he read E.H. Weber's *Untersuchungen über den Erregungsprozess im Muskel- und Nervensystem* (*Investigations of the Process of Stimulation in the Muscular and Nervous System*, 1870), James Bell Pettigrew's *Animal Locomotion* (1872, published in German as *Die Ortsbewegung der Tiere*, 1875) – which seems to have mainly served to furnish Nietzsche with apt metaphors for flying – and, most importantly, two books by Georg Heinrich Schneider: *Der thierische Wille: System, Darstellung, Erklärung der thierischen Triebe* (*The Animal Will: System, Exposition and Explanation of the Animal Drives*, 1880) and *Der menschliche Wille vom Standpunkte der neueren Entwicklungstheorien (des 'Darwinismus')* (*The Human Will from the Standpoint of Recent Theories of Evolution (of 'Darwinism')*, 1882). Nietzsche heavily annotated the former in 1883 (his notebooks of the period also show evidence of his reading of Schneider; see, for example, *KSA* 10, 7[197], [198], [199], [211]). It is not known when he scrutinized and lightly annotated the latter, but it is likely to have been in or near the same year, and his reading of these two early examples of evolutionary psychology is likely to have been important for his concept of the will to power. In 1884, in addition to Rolph's book, Nietzsche also read Francis Galton's *Inquiries into Human Faculty and its Development* (1883) and an article entitled 'Heredity' by H.W. Holland in the *Atlantic Monthly* about Galton. In April 1886 Overbeck sent Nietzsche a print of *Vitalismus und Mechanismus* (1886), the inaugural lecture of the new professor of physiology at Basle, Gustav von Bunge, a leading vitalist. Nietzsche marked almost every page of this short text. Finally, Nietzsche also purchased Joseph Delboeuf's *La matière brute et la matière vivante* (1887), but it is unclear whether or not he actually read it.

In 1887–88 Nietzsche carefully read three specialist neurophysiological and psychiatric works: Alexandre Herzen's *Le cervau et l'activité cérébrale au point de vue psycho-physiologie* (*The Brain and Cerebral Activity from a Psycho-Physiological Viewpoint*, 1887) and Charles Féré's *Sensation et mouvement* (1887) and *Dégénérescence et criminalité: Essai physiologique* (*Degeneration and Criminality: A Physiological Essay*, 1888). Nietzsche's copy of the latter book is heavily annotated; Herzen's also contains some markings.[37] All three of these works played a role in the expansion of his extraordinary range of medicalized vocabulary in the late 1880s and in the elaboration of his concept of *décadence*. In the same vein is August Krauss's *Die Psychologie des Verbrechens* (*The Psychology of Crime*, 1884), extant in Nietzsche's library, but unfortunately there are no clues as to whether or when he read it. In 1887 he also read and annotated three further works on psychology. Henri Joly's *Psychologie des grands hommes* (*Psychology of Great Men*, 1883), treats the influence of environment and heredity on genius, a theme that Nietzsche discusses in *Twilight of the Idols*, and includes discussions of the theories of Darwin, Galton and William James. The Danish philosopher Harald Höffding's *Psychologie in Umrissen auf Grundlage der Erfahrung* (*Outlines of Psychology on the Basis of Experience*, 1887) contains detailed discussions of psychology ranging from classical philosophy to modern British and German theories. Traces of Nietzsche's reading of Höffding found their way into the text of *On the Genealogy of Morals* – ranging from his borrowing of certain concepts and examples to his remarks on 'English' philosophy and the psychology of forgetting. In 1887 Nietzsche also read and annotated *L'homme et l'intelligence: Fragments de physiologie et de psychologie* (1884) by the future Nobel laureate Charles Richet.[38] He also possessed Richet's *Essai de psychologie générale* (1887), although there is no evidence when or if he read the book. Nietzsche also owned and read at some point after 1883 two works in French translation by the English philosopher and psychologist, James Sully (1842–1923), who was an adherent of the associationist school of psychology and held views similar to those of Alexander Bain: these were *Le pessimisme: histoire et critique* (*Pessimism: A History and a Criticism*, 1882), and *Les illusions des sens et de l'esprit* (*Illusions: A Psychological Study*, 1883). Other psychological treatises in his library include: Henry Houssaye's *Les hommes et les idées* (1886) and Frédéric Paulhan's *Les phénomènes affectifs et les lois de leur apparition: Essai de psychologie générale* (*The Laws of Feeling*, 1887).

Conclusion

In this chapter I have shown that Nietzsche read a fairly large number of books concerning natural science, including some detailed specialist works by experts. I have attempted to give a general account of what he read, and when he read it. This reading suggests that Nietzsche was better informed and more engaged in questions relating to natural science than has generally been assumed. However, most of the detailed analysis and discussion concerning exactly how this material influenced his thinking could not be done here. This will be the task of the following chapters.

Notes

1 The Bank of Sweden Tercentenary Foundation has financially supported my work on Nietzsche and science.

2 Thomas H. Brobjer, 'Nietzsche's Reading and Private Library, 1885–1889', *Journal of the History of Ideas*, 58 (1997), 663–93.

3 See, for example, I.M. Zeitlin, *Nietzsche: A Re-Examination* (Cambridge: Cambridge University Press, 1994). This view is frequently held by those who take a hostile view of Nietzsche.

4 There are other books in Nietzsche's library which deal with the topic of Darwinism, but there is no evidence that Nietzsche actually read them. These include Eugen Dreher's *Der Darwinismus und seine Konsequenzen in wissenschaftlicher und sozialer Beziehung* (*Darwinism and its Scientific and Social Consequences*, 1882), which Nietzsche listed in the summer of 1883 and probably bought shortly thereafter; and Oscar Schmidt, *Descendenzlehre und Darwinismus* (*The Theory of Descent and Darwinism*, 1873).

5 An English translation of the essay can be found in G.J. Stack's article 'Nietzsche's Earliest Essays' in *Philosophy Today*, 37 (1993), 154–56. Much of the essay is inspired by Emerson, but it seems to me unlikely that the emphasis on natural science was inspired by his reading of Emerson.

6 There is little direct evidence of Schopenhauer's importance for Nietzsche's view and knowledge of science, but it is likely to have been considerable. We know that Nietzsche

read Johann Czermak's article about Schopenhauer's theory of colour, published in *Berichten der Wiener Akademie der Wissenschaft* in 1870 (it was also republished in Zöllner's *Die Natur der Kometen* (1873)) and approved of its conclusion that Schopenhauer's theory was correct and anticipated the work of Young and Helmholtz. For a more detailed discussion of Schopenhauer's importance for Nietzsche in this regard, see Maudemarie Clark's 'On Knowledge, Truth, and Value: Nietzsche's Debt to Schopenhauer and the Development of his Empiricism', in Christopher Janaway (ed.), *Willing and Nothingness: Schopenhauer as Nietzsche's Educator* (Oxford: Clarendon Press, 1998), pp. 37–78.

7 The most important are: 'Die *pinakes* der Democritea' from July–September 1867 (*HKW* 3, pp. 246–79), 'Notizen zum Democritaufsatz' (*HKW* 3, pp. 327–28), 'Zu Demokrit' (*HKW* 3, pp. 348–50), 'Democritea' (*HKW* 3, pp. 362–68), 'De Democriti librorum indice a Thrasyllo confecto: Prolegomenon in Democriti vitam' (*HKW* 5, pp. 58–61) and 'Zur Demokrit-vita' (*HKW* 5, pp. 148–51). Other notes have been collected by the editors of the *HKW* under the title 'Zu Demokrit' (*HKW* 3, pp. 332–35, 345–48) and general notes without titles referring to Democritus can be found in *HKW* 4, pp. 42–67 and pp. 72–87 and *HKW* 5, pp. 172–75.

8 See his letters to Rohde (*KGB* I/2, 3 April 1868) and to Paul Deussen (*KGB* I/2, end of April/early May 1868). In the latter Nietzsche claims that his dissertation will be one half philosophy and one half natural science, and that the preliminary work is already almost finished. Nietzsche's notes for his work (written in April and May 1868) have been published in *HKW* 3, pp. 371–94, and in *KGW* I/4, section 62, pp. 548–78.

9 Otto Kohl studied classical philology with Nietzsche in Leipzig, and was one of the founders, together with Nietzsche, of the classical society there. He was critical of Kohl's dissertation, partly because it contains a critique of Schopenhauer. But it is likely that it acted as a stimulus for him to write something similar or better.

10 We know that Nietzsche read both Strauss's *Voltaire* and *Lessing* in 1873 – at least there are references to these works and their contents in *David Strauss: The Writer and Confessor*. The third chapter of *The Old Faith and the New* forms the core of Strauss's book and contains sections with titles such as 'The development of our solar system according to Kant'; 'The origin of life on earth', 'Goethe as Darwin's predecessor', 'The genesis of Darwin's theory', 'Ape and man. Darwin on the descent of man from the apes', 'Materialism and idealism. Dualism and monism'.

11 Because Spir and his works received little critical attention in his day, it is unclear how Nietzsche came to discover him. One possible source is Friedrich Überweg's *Outline of the History of Philosophy*, which briefly mentions Spir as having 'arrived at a doctrine close to that of Parmenides'. Nietzsche discusses Spir in relation to Parmenides in *Philosophy in the Tragic Age of the Greeks* (*PTA* 15).

12 On Nietzsche's reception of Spir see Paolo D'Iorio's exemplary essay 'La superstition des philosophes critiques: Nietzsche et Afrikan Spir', *Nietzsche-Studien*, 22 (1993), 257–94; Robin Small, *Nietzsche in Context* (Aldershot: Ashgate, 2001), pp. 1–20; Michael S. Green, *Nietzsche and the Transcendental Tradition* (Urbana: University of Illinois Press, 2002).

13 Elisabeth Förster-Nietzsche, *Das Leben Friedrich Nietzsches* (Leipzig: Naumann: 1895–1904), vol. 2, p. 521.

14 Works other than those discussed below include: Liebmann's article 'Über subjective, objective und absolute Zeit' in *Philosophische Monatshefte*, 7 (1871–72); Hermann Kopp's *Geschichte der Chemie* (1844) and *Beiträge zur Geschichte der Chemie* (1869); Albert Ladenburg's *Vorträge über die Entwicklungsgeschichte der Chemie in der letzten hundert Jahren* (1869); Friedrich Mohr's *Allgemeine Theorie der Bewegung und Kraft als Grundlage der Physik und Chemie* (1869); Johann Heinrich von Maedler's *Der Wunderbau des Weltalls oder populäre Astronomie* (1867); Johann Müller's *Pouillet's*

Lehrbuch der Physik und Meteorologie (1847); and Helmholtz's *Handbuch der physiologischen Optik* (1867) in the *Allgemeine Enzyklopädie der Physik*. Everything except the Liebmann article (which he read in 1872) Nietzsche borrowed from the library of Basle University in 1873. There is also a short excerpt from Büchner's *Geist und Körper* in *KSA 7*, 30[20].

15 Paolo D'Iorio has discussed and identified a number of the sources to the notes for 'The Pre-Platonic Philosophers'; see his essay 'L'image des philosophes préplatoniciens chez le jeune Nietzsche', in Tilman Borsche, Federico Gerratana and Aldo Venturelli (eds), '*Centauren-Geburten*' (Berlin: de Gruyter, 1994), pp. 383–417. See also his detailed commentary (together with Francesco Fronterotta) in the French translation of the notes: *Friedrich Nietzsche: les philosophes préplatoniciens* (Paris: editions de l'échat, 1994).

16 See Andrea Orsucci, 'Unbewußte Schlüsse, Anticipationen, Übertragungen: Über Nietzsches Verhältnis zu Karl Friedrich Zöllner und Gustav Gerber', in '*Centauren-Geburten*', pp. 193–207; Small, *Nietzsche in Context*, pp. 59–71.

17 Nietzsche borrowed both these texts on the 28 March 1873, 2 October 1873 and 13 April 1874. On 14 November 1874 he only borrowed the latter work, which has been missing from the Basle library since then. The work consists mainly of a reproduction of Mendelssohn's critical discussion of Boscovich, with critical commentaries by the anonymous author in footnotes. I believe it unlikely that this work could have strongly influenced Nietzsche's views. Karl Schlechta and Anni Anders discuss the work in their study *Friedrich Nietzsche: Von den verborgenen Anfängen seines Philosophierens* (Stuttgart: Fromann, 1962), pp. 130–40.

18 Schlechta and Anders have suggested Fechner as the source of Nietzsche's knowledge of Boscovich. However, Fechner discusses Boscovich only to a very limited extent, and we have no evidence that Nietzsche even read the relevant work by Fechner. Several commentators have claimed that Lange was responsible for Nietzsche's interest in Boscovich, but he is not mentioned in the first edition of Lange's work. Another, unidentified source seems to me more likely.

19 The *Journal of Nietzsche Studies*, volume 20 (Fall 2000), contains a translation of the note into English, and three long articles discussing the note.

20 Schlechta and Anders, *Friedrich Nietzsche*, pp. 148–53.

21 Alwin Mittasch, *Nietzsche als Naturphilosoph* (Stuttgart: Kröner, 1952), p. 82.

22 It is important not to lose sight of the fact, when discussing the different phases of Nietzsche thought, that many of his views remained fundamentally unchanged – for example, his high regard for Goethe and the Greeks, his interest in culture and values and so on. Also, many of the views that are thought to be typical of the 'middle period' are anticipated in notes written during his youth and 'early period'.

23 See, for example: Ronald Hayman, *Nietzsche: A Critical Life* (London: Weidenfeld, 1980), pp. 188–89, and Ruth Abbey's *Nietzsche's Middle Period* (Oxford: Oxford University Press, 2000).

24 Paolo D'Iorio has discussed Nietzsche's reading of Caspari in his 'Cosmologie de l'Éternel Retour', *Nietzsche-Studien*, 24 (1995), 62–123. I have briefly examined his reading of another work by Caspari, *Die Urgeschichte der Menschheit* (*The Prehistory of Humanity*), in 'Women as Predatory Animals or Why Nietzsche Philosophized with a Whip', in Christa Davis Acampora and Ralph R. Acampora (eds), *A Nietzschean Bestiary: Animality Beyond Docile and Brutal* (Lanham, MD: Rowman & Littlefield, 2003).

25 Nietzsche's reading of Mayer, Vogt and Roux has been discussed in some detail in the secondary literature, but that of Proctor and Schmitz-Dumont has received almost no attention. A note from 1881 is based on Proctor (*KSA* 9, 11[24]). I will discuss his reading of these and other authors mentioned here in more detail in a forthcoming study.

26 Lou Andreas-Salomé, *Nietzsche in seinen Werken* (Frankfurt am Main: Insel, 2000), p. 172.

27 Nietzsche's discussion with the Wagners is mentioned in Cosima's *Diaries*; see *KSA* 14, p. 78.

28 These essays were first published in 1872 and 1880, respectively.

29 See Nadeem Hussain's essay in Chapter 5. Vivette Vivarelli, who is also working on Nietzsche and Mach, presented a paper entitled 'Nietzsche e le "spiegazioni rassicuranti" (tra Mach e Avenarius)' at a Nietzsche conference in Lecce, Italy, on March 2003, which will be published in the proceedings.

30 *KGW* VII 4/2, 1[112], p. 67. The other titles are: Kurd Lasswitz, *Atomistik und Kriticismus: Ein Beitrag zur erkenntnisstheoretischen Grundlegung der Physik* (Braunschweig, 1878); Carl Neumann, *Über die Principien der Galilei-Newton'schen Theorie* (Leipzig, 1870); Gustav Fechner, *Die physikalische und philosophische Atomlehre* (Leipzig, 1864).

31 See Alwin Mittasch, *Friedrich Nietzsches Naturbeflissenheit* (Heidelberg: Springer, 1950), p. 21n. The reliability of this information is open to conjecture.

32 Mach may have influenced Nietzsche's discussion of 'imaginary causes' in *TI* 'Errors' 4. Compare what he says there with the following remarks by Mach:

> Wiederholt habe ich ein interessantes hierher gehöriges Phänomen beobachtet. Ich sass in die Arbeit vertieft in meinem Zimmer, während in einem Nebenzimmer Versuche über Explosionen angestellt wurden. Regelmässig geschah es nun, dass ich *zuerst* erschreckt zusammenzuckte, und *nachher* erst den knall hörte. Da im Traum die Aufmerksamkeit besonders träge ist, so kommen in diesem Fall die sonderbarsten Anachronismen vor, und jeder hat wohl solche Träume erlebt. Wir träumen z.B. von einem Mann, der auf uns losstürzt und schiesst, erwachen plötzlich, und bemerken den Gegenstand, der durch seinen Fall den ganzen Traum erzeugt hat. Es hat nun nichts Widersinniges anzunehmen, dass der akustische Reiz verschiedene Nervenbahnen zugleich einschlägt, und hier in beliebiger verkehter Ordnung von der Aufmerksamkeit angetroffen wird, so wie ich bei der obigen Beobachtung *zuerst* die allgemeine Erregung, und *dann* den Explosionsknall bemerkte. Freilich wird es in manchen Fällen zur Erklärung auch ausreichen, ein Verweben einer Sinnesempfindung in ein vorher schon vorhandenes Traumbild anzunehmen. (Mach, pp. 107–108)

33 However, Nietzsche and Mach are probably making different points: Nietzsche rejects atoms with extension, while Mach denies the existence of atoms as such since we have no sense experience of them.

34 *Repertorium der Physik* 23 (1887), 587–600. It seems to me unlikely that Nietzsche read the article; its presence in his library confirms only that Mach received Nietzsche's book. I am not aware of any examination of possible influences of Nietzsche on Mach.

35 In fact, the Roux volume in Nietzsche's library contains no annotations, but according to a note by Rudolf Steiner, who worked in the Nietzsche Archive under the guidance of Elisabeth Förster-Nietzsche, Nietzsche's copy then contained annotations in the form of marginal lines. Most probably Nietzsche's copy has been lost and replaced by a new one. For this information I am grateful to an early version of the detailed listing of the content of Nietzsche's library, *Nietzsches Bibliothek*, prepared by Giuliano Campioni, Paolo D'Iorio and Andrea Orsucci.

36 On Dumont, see Maria Christina Berti, 'Beiträge zur Quellenforschung', *Nietzsche-Studien*, 26 (1997), 580–81. On Rolph, see Greg Moore, *Nietzsche, Biology and Metaphor* (Cambridge: Cambridge University Press, 2002), pp. 46–54. On Nägeli, see Andrea Orsucci, 'Beiträge zur Quellenforschung', *Nietzsche-Studien*, 22 (1992), 371–88.

37 For a general discussion of Nietzsche's reading of these works, see Bettina Wahrig-Schmidt's article '"Irgendwie–jedenfalls physiologisch". Friedrich Nietzsche, Alexandre Herzen (fils) und Charles Féré 1888', *Nietzsche-Studien*, 17 (1988), 434–64. Another recent work which emphasizes the biological influences on Nietzsche's view of decadence is Anette Horn's *Nietzsches Begriff der décadence: Kritik and Analyse der Moderne* (Frankfurt am Main: Peter Lang, 2000).

38 For examples of Nietzsche's reception of Joly, see Giuliano Campioni, 'Beiträge zur Quellenforschung', *Nietzsche-Studien*, 21 (1992), 401–407 (402–403). On Höffding, see Marco Brusotti, 'Beiträge zur Quellenforschung', *Nietzsche-Studien*, 21 (1992), 390–97 (390–91) and Thomas H. Brobjer, 'Beiträge zur Quellenforschung', *Nietzsche-Studien*, 30 (2001), 418–21. On Richet, see Greg Moore, 'Beiträge zur Quellenforschung', *Nietzsche-Studien*, 27 (1998), 535–51 (546).

Chapter 2

Nietzsche: 'That Profound Physiologist'[1]

Richard S.G. Brown

In *Ecce Homo* Nietzsche refers to the Buddha as 'that profound physiologist' (*EH* 'Wise' 6). Why would Nietzsche say such a thing? And would the Buddha, or anyone else for that matter, return Nietzsche's compliment, assuming that it was, in fact, meant to be a compliment? And if someone were to return the compliment, would they do so for the very same reasons? Who might have already thought of Nietzsche, in one way or another, as a physiologist of sorts, profound or otherwise? And lastly, what exactly is the role that 'physiology', and its strict cognates, plays in Nietzsche's writings (both published and unpublished) that might lead someone to say of him: 'Nietzsche: that profound physiologist'?

Buddha: 'That Profound Physiologist'

Although it is in *Ecce Homo* that Nietzsche refers to the Buddha as 'that profound physiologist', his reasons for doing so are given in considerably greater detail in several sections of *The Antichrist*, where he compares, and sharply contrasts, Buddhism and Christianity (*A* 20–23). While Nietzsche considers both to be kindred, nihilistic and decadent religions, only Christianity is worthy of his full condemnation. The main charge that Nietzsche levels at Christianity is that it is deceived when it mistakenly considers that human suffering has a moral or religious cause – namely, as a result of a 'struggle against sin'. Christianity's major physiological error consists therefore in being ignorant of the real cause of suffering by considering the individual's moral struggle against sin to be the actual state of affairs – when it most assuredly is not.

Buddhism, Nietzsche says, is one hundred times colder, more truthful, more realistic and more objective than Christianity. While philosophy sprang up after the advent of Christianity to serve as its handmaiden, Buddhism arose in India after a long philosophical tradition which had already abolished the concept of 'God'. As preconditions for its development, Buddhism required a mild climate, liberal social traditions and a higher and learned class of the cultured and overly intellectual. However, Buddhism also required a race of peoples who had already grown fatigued and weary in spirit along with a civilization as late, and as close to its spiritual end, as its people. This is important because, according to Nietzsche, this spiritual exhaustion or weariness is the most fundamental fact in the causal sequence he will develop.

Apart from being the most positivistic religion, Nietzsche describes Buddhism as a strict phenomenalism. Buddhism is 'too positivistic' according to Nietzsche, because it is *strictly* empirical. 'It [Buddhism] merely says what it feels' (*A* 23). It is extra-moral – that is, beyond 'good and evil' – insofar as it no longer misunderstands itself by means of moral concepts. A case in point: Buddhism considers the 'struggle against suffering', not 'sin', to be the actual state of affairs which, Nietzsche says, is in accordance with what is actual. In other words, for the Buddha, suffering was a matter of fact. It did not require interpretation, justification or compensation; it only required removal. This insight was testimony to the Buddha's physiological acumen. Therefore Nietzsche says of Buddhism, 'It is *not* morality that speaks thus; thus speaks physiology' (*EH* 'Wise' 6).

But in what decisive way, for Nietzsche, is physiology speaking? It is Nietzsche's contention that Buddhism recognized two basic physiological facts or conditions which Nietzsche develops by means of a complicated causal sequence. The most fundamental physiological condition, however, which serves as the trigger for the entire sequence, is the fact that the peoples of India at the time of the Buddha were spiritually exhausted. This spiritual exhaustion manifested itself in two ways corresponding to Nietzsche's two physiological facts. First, Nietzsche diagnoses an 'excessive excitability of sensibility', which in turn reveals itself in a 'refined capacity for [feeling] pain' and suffering (*A* 20). Buddha's First Noble Truth, after all, is that 'All is suffering'. Second, Nietzsche detects an 'over-intelligibility' or an 'excessive objectivity' (*A* 22) that is expressed in an immoderate preoccupation with the use of logic and abstract concepts, culminating in a dangerous preoccupation with the 'impersonal' and therefore a general weakening of the instinct of self-interest or egoism (the no-self doctrine of *anatmanvada*). Together, these two physiological conditions, or their symptoms, result in a state of depression. According to Nietzsche's analysis, Buddhism was designed precisely to overcome (to oppose, to combat) this state. However, because the Buddha, that profound physiologist, understood the entire causal sequence which ended in this depression, he was able to prescribe the appropriate 'hygienic' measures to alleviate it. These measures, however, were not designed to remedy the spiritual exhaustion which lies at the root of the depression (accepting that state as a given); rather, the hygienic measures were designed merely to relieve it by minimizing the secondary manifestations of the spiritual weariness which brought it about (namely, pain and impersonalism) by minimizing in turn the primary manifestations of the spiritual weariness (namely, excessive sensitivity and objectivity).

Nietzsche's list of hygienic measures prescribed by the Buddha is rather extensive. Anything and everything which either increases the excessive excitability of the senses or exacerbates the immoderate capacity for feeling pain and suffering must be avoided. This is what Nietzsche refers to as 'the main dietetic objective' of Buddhism (*A* 20), an 'ordered diet' (*A* 22). The Buddha prescribed doing certain things because they were essentially health-promoting: a life of wandering out in the open air (why the mild climate is imperative); relaxation and stillness which, along with cheerfulness (defined as the complete absence of desire or *nirvana*), are thought by Nietzsche to be Buddhism's 'supreme goal' (*A* 21). Additional prescriptions include cultivating an attitude of benevolence and kindness in a concerted effort to put the focus back on the individual person by making egoism and selfishness everyone's relative duty (to become the 'selfish' *arhat* from the *bodhisattva*'s point of view). The Buddha also condemned certain practices because they are essentially unhealthy: namely, prayer and asceticism; the mental states of resentment, enmity, vengefulness and antipathy; being both moderate and fussy in one's diet and the consumption of alcohol and any gall-producing, blood-heating emotions such as anxiety. Buddhists were also to avoid opposition, resistance and struggle at all costs and were to avoid compulsion or any (absolute) moral obligations or requirements, including remaining a Buddhist.

Who Thought What about Nietzsche and Physiology?

Answers given by various interpreters to the question as to what extent Nietzsche was a physiological thinker range from not at all – in virtue of the fact that no mention is made whatsoever of physiology in connection with any aspect of Nietzsche's thought (I shall refer to such critics collectively as the Unmentionables) – to the opposite extreme where Nietzsche's thought is read in its essence, and even in its entirety, through the lens of physiology. There is also a wide range of modern commentators who fall somewhere between these two extremes, though for quite different reasons. I think it is illuminating to have a look at a representative, but by no means exhaustive, sample from across the range of physiological interpretations of Nietzsche.

There is little sense in mentioning the Unmentionables. Just above the Unmentionables, however, at least in terms of pausing to reflect seriously on Nietzsche's use of physiology, albeit negatively, would be someone like Dan Breazeale, who opines that Nietzsche merely 'toyed' with 'speculations about the physiological determinants of thought and behaviour, though he had nothing very profound to say on this subject and most of these speculations remained in his notebooks'.[2] Walter Kaufmann likewise dismisses Nietzsche's comments on physiology as relatively unimportant, off-the-cuff or later disavowed by Nietzsche himself. Next in line are Claudia Crawford, Babette Babich, Alistair Kee and Robert Morrison who all recognize, each in their own way, that physiology is an important and positive element in Nietzsche's thought.

Walter Kaufmann

In his classic *Nietzsche: Philosopher, Psychologist, Antichrist*, Walter Kaufmann rejects what he refers to as Nietzsche's late 'physiologism'.[3] As if to rescue Nietzsche from the charge of physiologism, Kaufmann suggests that a three-level theory of values might be constructed on the basis of a threefold conception of interests which, though rooted in Nietzsche's philosophy, are nevertheless not explicitly developed by Nietzsche himself. According to Kaufmann, human beings have certain physiological interests in those things which are commonly required instinctually or unconsciously because of our animal natures – that is, because we have bodies (Kaufmann lists such physiological requirements as water, air, calcium, thyroid, and certain minimal climatic conditions). But over and above our physiological interests, we also have certain psychological interests which require some level of consciousness, that would separate us from those animals that do not have consciousness, and these psychological interests may differ considerably in individuals. But, as human beings, we also have certain 'ontological interests' which, Kaufmann says, are interests that we all share (just as we all share physiological interests), but are occasioned by the fact that human beings qua human beings experience the same 'ontological predicament' because we are creatures who are constituted by both a body and a spirit.[4] It is Nietzsche's interest in the spirit vis-à-vis the body that, Kaufmann believes, saves him from the charge of vulgar 'physiologism' – that is, an irrational and simplistic reduction of the human being to the body as if consciousness and spirit (*Geist*) did not exist or had no value. To buttress his case, Kaufmann refers to Nietzsche's criticism of Darwin in *Twilight of the Idols* for having 'forgotten the spirit' (*TI* 'Expeditions' 4). 'It is apparent', Kaufmann says, that 'even in his last period ... Nietzsche repudiated "physiologism"'.[5] According to Kaufmann, Nietzsche's later philosophy in particular has been misconstrued as an irrationalistic physiologism primarily because of his attack – especially vehement in *Twilight of the Idols* – on so-called mentalistic psychology and rejection of all conscious motivation: that is, Nietzsche's claim that consciousness lacks efficacy if his epiphenomenalism is to be taken seriously.[6] It therefore seems rather ironic, for Kaufmann, that Nietzsche's 'insistence that man's spiritual life cannot be understood completely apart from his body has been emphasized out of all proportion and thoroughly misunderstood because his concomitant insistence that man's physical life cannot be understood apart from the spirit was ignored'.[7]

Claudia Crawford

Claudia Crawford's paper entitled 'Nietzsche's Physiology of Ideological Criticism' is designed to demonstrate, in her words, that 'Ideology is physiology translated through forgetfulness by the seduction of language'.[8] While Crawford's paper is essentially an elaborate vehicle for exposing the ideological assumptions at work behind any act of cultural criticism, it is nevertheless predicated on a finely detailed analysis of Nietzsche's rhetorical model of unconscious perception which is itself a complex series of physiological operations involving at least five specific steps and including at least three distinct metaphors: displacement, transposition, and marking (this rhetorical model of perception is examined further by Christian Emden in Chapter 4 of this volume).

Crawford begins her analysis with an overview of Nietzsche's threefold use of the term 'physiology', which she takes from Wolfgang Müller-Lauter. 'Physiology' is used variably as 'biology, [secondly] as the prerequisite for consciousness and [thirdly] as a process of struggle for power in connection with individuals, societies, and the human species'.[9] While I generally applaud Crawford's exposing of the 'unconscious physiology of power relations' in which ideological criticisms such as Marxist art critic John Berger's *Ways of Seeing* necessarily participate, I have a serious reservation about her claim that 'Nietzsche's application of physiology and the various definitions of its operations function with the thematics of the will to power'.[10] Nietzsche's use of the term 'physiology' certainly pre-dates the invention of the concept of the 'will to power', which did not make its first published appearance until *Thus Spoke Zarathustra*. A great deal of what Crawford has to say about Nietzsche concerning the physiology of unconscious perception is based on

Nietzsche's notebooks of the early 1870s. The assumption would be that these early notebooks cannot be read in light of the 'will to power' without some kind of argument – whatever that might turn out to be. But when Crawford utilizes such phrases as 'the agonistics of power', she seems to me to be simply conflating or collapsing the chronology and the distinction between such concepts as *agon* (contest) or *eris* (strife) from the early 1870s and the later concept of the 'will to power', supporting her claims with copious quotes from the eponymous 'book'.

A second difficulty I have with Crawford's analysis has to do with the ambiguity surrounding the question of whether or not Nietzsche's views on unconscious perception make him a subjective relativist. Crawford writes: 'The character of the image which figures the excitation is determined only by the relationship that the perceiver has with the excitation, that is, it is determined by the "proper persuasion" of the perceiver at the moment of awareness of excitation.'[11] Is it determined the same way for each and every perceiver in virtue of the fact they belong to the same species? This is not at all clear. One might think so following Crawford's quotation from section 299 of *The Gay Science*, where Nietzsche writes that 'consciousness does not really belong to man's individual existence but rather to his social and herd nature ... each of us will always succeed in becoming conscious only of what is not individual but "average"'. In the notebooks of the 1870s Nietzsche argues that 'All that philosophy can do now is to emphasize the relativity and anthropomorphic character of all knowledge' (*KSA* 7, 19 [37]). But because it is described as anthropomorphic, the relativity in question would not seem to mean relative to me as opposed to you, but rather relative to us as human beings as opposed to some other species that 'perceives' things in a different way using different perceptual apparatus. Nietzsche, I think, says as much himself when he distinguishes between human beings and plants: 'For the plant the world is thus and such; for us the world is thus and such .... For the plant, the whole world is a plant; for us, it is human' (*KSA* 7, 19[158]). Both human beings and plants are the measure of all things except for the fact that they measure in radically distinct ways. For example, Nietzsche says: 'Nature is acquainted with neither shape nor size; only to the knower do things appear to be large or small' (*KSA* 7, 19[133]). 'Large' and 'small', 'hot' and 'cold', 'light' and 'heavy' are qualities that are relative to the entity who happens to do the measuring. 'A quality exists *for us*', Nietzsche says, 'i.e., it is measured by us. If we take away the measure, what remains of the quality?' (*KSA* 7, 19[156])

Crawford claims that we share ideological knowledge because we share comparable social relations, but we must share something much more fundamental than social relations at the unconscious physiological level if we perceive the same things. After all, Crawford says, both Berger and the art historian see the same thing when they look at paintings by Franz Hals; they only interpret what they see differently. And she adds that we, too, see the same thing that they see when we look at the Hals paintings. But because of Berger's 'persuasive technique', we are seduced, as it were, into accepting his interpretation over other interpretations competing with it. Kant makes the distinction quite succinctly in the *Logic* which I think is important to quote here along with his illuminating example.

> In every cognition there is to be distinguished matter, i.e., the object, and form, i.e. the manner how we cognize the object. For example, when a savage sees a house in the

distance, the use of which he does not know, he has the same object before him as another who knows it as a dwelling furnished for men. But as to form, this cognition of one and the same object is different in both. In the one it is mere intuition, in the other intuition and concept at the same time.[12]

Because the savage and the non-savage are both human beings and share the same basic physiology and perceptual apparatus, both perceive the same house. According to Kant, the savage only lacks the appropriate empirical concept under which to subsume the intuition and therefore cannot say what it is that he sees. If they shared the same basic perceptual apparatus and saw different things, then human perception would be arbitrary and the preservation of the species would be in serious doubt – even if the entire process is considered to be 'untruthful' from start to finish. As Nietzsche says:

> There is a twofold artistic power here: that which produces images and that which chooses among them ... it is obvious that here too there is no totally free artistic invention – for that would be something arbitrary and hence impossible. Instead, these images are the finest emanations of nervous activity as it is viewed on a surface. The images are related to the underlying nervous activity which agitates them in the same way that Chladni's acoustical figures are to the sound itself .... Considered physiologically, the artistic process is absolutely determined and necessary. (*KSA* 7, 19[79])

> ...we produce these representations in and from ourselves with the same necessity with which the spider spins. If we are forced to comprehend all things only under these forms, then it ceases to be amazing that in all things we actually comprehend nothing but these forms. (*TL* 1, *KSA* 1, pp. 885–86)

Like Feuerbach, Nietzsche writes that 'Knowing is only possible, as a process of mirroring and measuring oneself against one standard (sensation)'. (*KSA* 7, 19[146])

Thomas Nagel

If my interpretation of Nietzsche is correct, then Nietzsche's epistemological position seems to be very close to that of Thomas Nagel. In his article, 'What is it Like to be a Bat?', Nagel suggests, as a matter of principle, that an 'organism has conscious mental states if and only if there is something that it is like to be that organism – something it is like for the organism'.[13] He then asks: what does it mean for a bat, for example, to be conscious, to experience the world as a bat? Nagel distinguishes between two distinct perspectives. On the one hand, there is the subjective point of view – or what it is like to experience the world as I alone experience it – which seems to be singularly and uniquely my own and therefore both personal and private. On the other hand, there is the objective point of view, which is universal and public. The subjective experience is qualitative – that is, an internally described account of experience – whilst the objective experience is quantitative – that is, an externally described account of the same experience. According to Nagel, you simply cannot reduce subjective qualia to objective quanta. Since the subjectivity of experience is connected to one single point of view, namely my own, it is tantamount to the question: what is it like to be me? It must therefore address what my physiology is like. We assume as well that bats have subjective experiences – that is, that there is something like what it is to be a bat. But bats have a different sensory apparatus from our own, namely sonar or echo location, which requires that bat brains be different from ours, translating high-frequency noise into information about the distance, shape, size, location, motion and texture of objects. We know objectively what it is like to be a bat – but not subjectively. As human beings we are subjectively different from bats in our perceptual experiences. So the question can be asked: is there a way for us as human beings, to know, experience or even to imagine what it is like for the bat to be a bat? In other words, is it possible to move out of our subjective framework, our own uniquely human perspective, into the bat's subjective framework or the bat's perspective? Nagel says that when we try to imagine what it is like within the bat's framework, we unavoidably bring our own human framework along. We try to imagine 'seeing' with our eyes closed, or 'seeing' blindly as bats do, or making sounds so high pitched that we can't hear them (though we could if we were bats). But, Nagel says, imaginative descriptions are not descriptions of what it is like to be a bat but only fanciful descriptions of what it is like for me as a human being to imagine what it is like to be a bat. In sum, we have no access to the bat's perspective.

As human beings, we are restricted to our own framework precisely because of our particular sensory apparatus, our own fundamental structure and physiological organization. Since we are unable to change our sensory apparatus, and, without changing it, we cannot understand what it is like to be a bat, then we are restricted, as it were, to knowing things only from the human perspective. Knowledge is anthropomorphic. Like Nietzsche, Nagel argues that our conceptual schemes or systems of representation – that is, how we represent things to ourselves – are dictated by our fundamental structures, our physiology. And, like Nietzsche, Nagel sees this raising a rather basic epistemological question: what is the object of perception? The answer for both obviously depends on a point of view. For both, to speak about the object, apart from any and all points of view (Nagel's view from nowhere!) would be ridiculous. For Nagel, it would be asking what is it like to be a bat apart from the bat's point of view. For Nietzsche, it would be asking what the world would be like for the human being who cut off his own head or who attempted to peer down through the crack of consciousness. Nevertheless, both concur that an object can be viewed from any number of different perspectives.

Babette Babich

The third chapter of Babette Babich's *Nietzsche's Philosophy of Science*, 'On the Ecophysiological Ground of Knowledge', is subtitled 'Nietzsche's Epistemology', although her further description of it as 'Nietzsche's anticpistemology' would probably be, to speak tongue in cheek, much closer to the 'truth'.[14] Nietzsche's anti-epistemology concerns itself with our human interaction with the world or nature as Real, and this Reality qua will to power is variously described by Babich as irresolvable, ambiguous, complex, indeterminate, chaotic, in constant flux, effectively unknowable and universally virtuous in its interpretative, better, artistic function, which includes human knowledge, but is not confined to the human or even to the organic.

Babich characterizes Nietzsche's philosophy as a 'hyperrealism' because she wants to emphasize the scrupulous attention which Nietzsche paid in his observations and descriptions to acute differences and fine details. 'One wants in Nietzsche's spirit', she says, 'to count in all the little mistakes, the marginalities, the fuzzy details previously disregarded ("annihilated" or "damned") by science' – that is, to pay heed to 'the fuzziness of perceptual/conceptual achievement'.[15] Unlike science, but quite like Nietzsche, Babich does not want to clarify the fuzzy, to gloss over the fuzzy, or to avoid the fuzzy altogether; rather, she wishes to rejoice in the fuzzy and to embrace it seriously. Using Nietzsche's example of the 'leaf' from *On Truth and Lie in an Extra-Moral Sense* as a point of departure, Babich ventures out on another, not so trivial, limb by asking the seemingly simple question: is it raining? True or false? Yes or no? Nietzsche rejects the two-valued logic (a given for traditional science) which would demand either a true or false answer because it 'falsely' assumes that the question itself is straightforward and obvious. Instead, Babich argues, Nietzsche adopts a multivalued logic; dare I say, a 'fuzzy' logic?

But what exactly does Babich mean when she characterizes Nietzsche's epistemology as an 'eco-physiology' or, at times, a 'physio-ecology'? Babich defines 'ecology' as a 'relative world circumstance' which is 'reciprocally determined' with physiology.[16] In other words, when Babich refers to ecology *and* physiology, she is emphasizing that they are inseparably connected. On the side of physiology, Babich includes the sensual, empirical and organic whereas on the side of the ecological, she places the will to power as world-interpretive expression, a 'power-expressive interpretation' which seems to be in perpetual response to the physiological or to be a perpetual reflex in conjunction with the physiological.[17] Although – by my reading at least – it is not said explicitly, the conceptual element in what she has called 'the fuzziness of perceptual/conceptual achievement' would appear to fall on the physiological side right along with the perceptual – given their inseparability. In sum, according to Babich, Nietzsche's epistemology has to be understood as basically a 'product of body and world', once again with very special emphasis on the 'and'. Both body and world together serve as the 'ground for knowledge'. As Babich writes: 'to be true to Nietzsche's thought, an ecophysiological focus on the interpretive becoming (*physis*: understood as self-generating, self-manifesting) of the world and the body (the last understood as the expenditure of organic force expressing a particular perspective) is needed'. She continues: 'The plausibility of Nietzsche's perspectivalist possibility is ecophysiological, that is, determined by the physiological or physical constitution of the interpreting perspective and its relative (ecological) position in the world or cosmos.'[18]

In several passages in this chapter, Babich makes reference to the linkage between conceptual and perceptual knowledge and specifically how 'raw sense perception [is] "informed" by concepts'.[19] Nietzsche does seem to distinguish between two so-called prejudices or forms of falsification both of which arise unconsciously and unçontrollably: '*Twofold falsification*, on the part of the senses and of the spirit, to preserve a world of that which is, which abides, which is equivalent, etc.' (*KSA* 12, 7[54]). These two prejudices constitute what Nietzsche calls a 'sensual-spiritual appropriation' and jointly determine our 'perspectival kinds of outlook' (*KSA* 12, 5[11]). To have a perspective entails that our perceptions, as well as our conceptions, are altered in some significant fashion. 'Positing of equality' at the level of sensation serves as the foundation for our subsequent thinking or conceiving about the world which we perceive. 'The development of reason is adjustment, invention, with the aim of making similar, equal – the same process that every sense impression goes through' (*KSA* 13, 14[152]).

What, then, would Babich's assessment of Nietzsche's leaf say about Kant's assessment of the house as seen by both a savage and non-savage? It seems that although Kant would say that we can distinguish between what we see (as mere rather than pure intuition) from what we interpret that something to be (empirical concept), Nietzsche would argue that there is as much interpretation in seeing as there is in interpreting in Kant's sense. Indeed, how we interpret at the level of sense (intuition) will determine how we interpret at the level of concept and word. In other words, for Nietzsche, perceptions are already interpretations. To quote Babich: 'The reality reported by our senses structures what we know and the very same filtering event that characterizes perception streamlines conception.' And again: 'When we subsume perceptions under concepts (which is what we do in every act of human understanding), we achieve an effective estimation of similarities, rather than an affirmation of difference as difference or ambiguity.'[20] There is only one particular, one individual leaf. It is not even fair to say that there can be two leaves since the word 'leaf', to be accurate, should only apply to a single particular. There cannot be more than a single token of any type. Indeed, in reality, there are only unique tokens and no types. It becomes problematic, bordering on the nonsensical, to have a plethora of words for various kinds of snow to the point of requiring a different word for every flake, let alone just different words in the enriched vocabularies of skiers and indigenous populations. But even if these distinctions are considered existentially legitimate for self-preservation or for human need and interest, having a different word for each and every snowflake, let alone raindrop or oxygen atom, would certainly demand a 'god-like perspective' and whilst not strictly speaking a private language, it would have precisely the same negative effect on our ability to 'think', let alone communicate.

In conclusion, Babich, like Nagel, also draws a distinction for Nietzsche between human and animal perspectives. The difference, she says, 'lies in the superior plasticity of human interpretations: human perspectival fictions are proof against error. Human knowledge perspectives are not merely life-preservative errors; they are themselves further preserved in metaphysical, religious or scientific concepts'.[21] While it is undoubtedly true in general that human fictions are proof against error, not all human fictions are created equal. The Christian 'struggle against sin', for example, is the kind of fiction which Nietzsche thinks will ultimately prove itself to be an insurmountable error, especially when it is viewed against the Buddhist understanding of the 'struggle against suffering'.

Alistair Kee

In his book *Nietzsche against the Crucified*, Kee not only rehearses the well-known fact that Nietzsche had read F.A. Lange's *History of Materialism* at a very young age (1866), but also suggests that Nietzsche knew the works of Ludwig Feuerbach.[22] However, whilst Nietzsche is said to follow Feuerbach in seeing 'the value in morality to be its embodiment of human self-understanding', Kee points out, without mentioning someone like Jakob Moleschott explicitly, that 'the most radical materialism in Germany at that time was not left-wing Hegelianism, but medical materialism, and it is of some significance that Nietzsche attempted to trace such phenomena as morality and religion back not to anthropology but *physiology*'.[23] According to Kee, Nietzsche's genealogy of morality, as well as his aesthetics, rested entirely on the assumption that all human values express human needs and that these needs are biological, chemical and physiological. Hence Kee is confident in making the blanket claim that Nietzsche's philosophy is, in fact, 'built upon physiology' insofar as 'Nietzsche has traced everything back to physiology: morality, ethics, ascending life and ... will to power'.[24] For the most part, Kee's claims about the relationship between values and physiology in Nietzsche's philosophy remain general, although he does offer one telling example, where Nietzsche is said to 'mix' 'physiological and moral judgements'. To Kee's way of seeing, '"unhealthy" looks like a physiological category, while "degenerate" is a moral category'.[25] Apart from suggesting that Nietzsche conflates physiological and moral categories, Kee never says whether this is a category mistake on Nietzsche's part, a simple oversight, a slip of the pen, an illegitimate move, or what one should expect given Nietzsche's reduction of everything to the physiological. But, given the central importance which Kee has already ascribed to physiology in Nietzsche's thought, I think it would be safe to say that, for Kee, Nietzsche eventually came to regard them as simply two different ways of referring to the very same thing: moral values are simply signs and symptoms of underlying physiological conditions. In the notebooks of 1888 Nietzsche wrote the following: 'Today we know moral degeneration is no longer to be thought as something separable from the physiological: it is a mere symptom-complex of the latter; one is necessarily bad, just as one is necessarily sick' (*KSA* 13, 14[113]).

While Kee considers Nietzsche, like Feuerbach, to be both a thoroughgoing materialist and reductionist, he also believes that both Nietzsche and Feuerbach took great pains to despiritualize morality and religion in order to be able to humanize them again. There is a need, first, to 'despiritualize the earth', to borrow a phrase from Nietzsche. But although Kee never develops the relationship between Nietzsche and Feuerbach beyond just the few points of contact to which I have referred, I think it is nevertheless important to see how Nietzsche and Feuerbach are representatives of what might be called the 'you are what you eat' school.

Ludwig Feuerbach

Feuerbach himself understood his own philosophy to be a materialist anthropology or naturalism, writing: 'We only translate and interpret the book of nature.'[26] Feuerbach's words foreshadow those of Nietzsche; after we have despiritualized the earth, Nietzsche proclaims, we need 'to translate man back into nature' (*BGE* 230). In making their respective claims, both philosophers were prescribing realistic, objective and, therefore, like the Buddha, atheistic philosophies that would unite philosophy on a practical basis with the empirical sciences – something that Nietzsche had called for, even as a schoolboy, in his essay 'Fate and History'. Both Nietzsche and Feuerbach consider the human being to be the ground of philosophy.

Everything is ultimately reducible to the human being because philosophy must reflect reality through human needs and experience. Both Nietzsche and Feuerbach think they have overcome the dualism of mind (spirit, soul) and matter. They both make, as Feuerbach puts it, 'anthropology, with the inclusion of physiology, the universal science'.[27]

In 1837, even before writing the *Principles of the Philosophy of the Future,* Feuerbach was studying anatomy and physiology. In 1850 the physiologist Jakob Moleschott, who was both a student and a close friend of Feuerbach, sent him a copy of his recently published *Theory of Nutrition* (*Lehre der Nahrungsmittel*). Feuerbach adopted the central slogan of Moleschott's work, an untranslatable pun that has become synonymous with Feuerbach: 'Der Mensch ist, was er ißt' ('Man is what he eats'). In 1850 Feuerbach reviewed Moleschott's text very favourably in an essay entitled 'Natural Science and the Revolution'. Despite the fact that Moleschott's text only dealt with eating and drinking, Feuerbach claimed it had the highest philosophical significance, 'the true *Fundamental Principles of the Philosophy of the Future* and of the present'. In Moleschott's work Feuerbach thought he had discovered the solution to the mind–body problem:

> How philosophers have racked their brains over the problem of the link between body and soul! Now we know from science, what the common people long knew from experience, that eating and drinking hold body and soul together, that food is the link we have been searching for .... Only sustenance is substance. Sustenance is the identity of spirit and nature. Where there is no fat, there is no flesh, no brain, no spirit. But fat comes only from sustenance .... Everything depends on eating and drinking. Difference in Nature is but difference in food. The brain cannot be formed without phosphorous-bearing fat .... Without phosphorus, no thought ....[28]

Marx W. Wartofsky has succinctly summarized Feuerbach's position as follows: 'If life is the precondition of thought and food is the precondition of life, then food is the precondition of thought.' But beyond solving ancient philosophical puzzles, what Wartofsky caricatures as the Moleschottian 'gastrointestinal vision of the world',[29] also had, Feuerbach thought, far-reaching moral and political significance:

> Food becomes blood, blood becomes heart and brain, the stuff of thoughts and attitudes. Human sustenance is the basis of human education and attitudes. If you want to better the people give them better food instead of exhortations against sin. *Man is what he eats.*[30]

Feuerbach suggests, for example, that the failure of the 1848 revolution in Germany was due to the sluggish blood produced by a diet of potato and cabbage. Revolution and revolutionary thinking are not possible, it seems, if that is all you eat:

> Shall we then therefore despair? Is there no other food which can replace potatoes even among the poorer classes and at the same time nurture them to manly vigour, and a manly disposition? Yes, there is such a food, a food which is the pledge of a better future, which contains the seeds of a slower and more gradual, but also of a more thorough revolution: it is beans.[31]

The practical ramifications of adopting this medical materialism, then, are intriguing, especially when we see that Nietzsche, rather like Feuerbach, took precisely this kind of reductionism very seriously.

Morrison on Nietzsche and the Buddha

In his book, *Nietzsche and Buddhism: A Study in Nihilism and Ironic Affinities*, Robert G. Morrison claims that recognizing otherworldly values as human inventions pulled the value rug out from under Europe's feet,[32] just as it had before the rise of Buddhism. In both cases the consequence was psychological feelings of melancholy and depression which, Morrison says (confusing what is cause and what is effect) brought about certain psychosomatic illness such as weariness of spirit, physiological exhaustion and a general decline in health, strength and power. Nietzsche regarded 'passive nihilism' as a sign or symptom which reflected the underlying fatigue of life and body. The Buddha never attempted actively to overcome nihilism but simply advocated a way of accepting it stoically. Morrison argues that Nietzsche did not think that the Buddha was profound enough as a physiologist to recognize that passive nihilism was itself an interpretation of existence that reflected the psychological state of depression. In a phrase, the Buddha failed to understand nihilism and its implications. Nevertheless, according to Nietzsche, the Buddha's advocacy of retreat from the world was both an enlightened and appropriate response to believing that life is worthless. Such world withdrawal was the Buddha's desire for personal extinction (*nirvana*), which was achieved by attaining a condition without desire. The Buddha therefore offered a set of rules for moving from a state of depression to one of cheerfulness in the here and now.

Where Babich refers to Nietzsche's 'little errors', Morrison devotes an entire chapter to what Nietzsche called 'little things', a phrase taken from *Ecce Homo* (*EH* 'Clever' 2). Nietzsche's 'little things' include food, diet, climate (hours of sunlight, dryness of the air) and metabolic rate – issues that are discussed in Chapter 3 of this volume by Gregory Moore. Although Nietzsche may refer to such matters as 'little things', he understood them to be very important, though overlooked by most, if not all philosophers (the Buddha being an obvious exception), since these apparent trivialities have a profound determining effect. Nietzsche, Morrison claims, demonstrates an ironic affinity to early Buddhism because both accept a reciprocal interaction between two influences which, when taken together, determine our values and world-view. There are 'external' influences (nature), which include climate, land, sky, one's neighbours, parents, country of origin and education, and 'internal' influences (nurture), which include our human drives, our basic capacity for power (that is, whether fundamentally we are of the ascending or descending type). According to Morrison, the greater one's capacity for power, the more internal influences (nurture) count over external influences (nature), so that noble or ascending types presumably have a much greater ability to alter their course towards becoming the *Übermensch*. Eventually, Morrison claims, for both Nietzsche and Buddhism, the external forces (nature) become negligible and only the internal forces (nurture) become decisive in their respective processes of self-overcoming. It is Morrison's contention that, for Nietzsche as well as for Buddhism, we can undo and counteract our 'conditioning'; that we can 'learn anew' and 'free ourselves' from these effects. However, Morrison assumes, incorrectly, I would argue, that, according to Nietzsche, we do not know the underlying workings of our bodies (that is, the physiological causes). And yet, Morrison does say that such knowledge is really not that important for Nietzsche since we do know the symptoms or signs of the body which manifest themselves as affects which are defined by Morrison as 'the whole gamut of a person's subjective experience'.[33] So self-overcoming therefore depends exclusively on 'working with the affects'. In conclusion, Morrison suggests that Nietzsche's view of these 'little things' is basically Nietzsche's own personal account of his 'unsuccessful' attempts to find an environment which would be conducive to his own well-being and that Nietzsche's 'comments on the relation between the German diet and the German spirit are not offered as serious physiology, just Nietzsche grumbling'.[34] Morrison's final word is that Buddhism understands the so-called 'little things' much better than Nietzsche did, and therefore Buddhism makes much more sense than Nietzsche. Indeed, if Nietzsche had really understood the little things better, he would not have been sick all of the time and undoubtedly would have lived to be 80 years of age – just like the Buddha!

Nietzsche: That Profound Physiologist

I think it can be demonstrated that rather than simply toy with physiological speculations, as Breazeale thinks he did, Nietzsche actually took such musings very seriously – even when, from the vantage-point of our own time, many of them appear to be both far-fetched and comical.[35] Nor are these physiological speculations just tucked away in Nietzsche's notebooks, as Breazeale suggests; they can be found liberally scattered throughout his published works as well.

A brief accounting. There is but a single reference to 'physiology' (or to one of its strict cognates – for example, 'physiological', 'physiologist' and so on) in *The Birth of Tragedy* (published in early 1872) and an additional ten references in the notebooks between 1870 and 1872 inclusive, when Nietzsche was at work on such unpublished material as *On Truth and Lie in an Extra-Moral Sense* (1872) – material which has a decidedly different flavour than *The Birth of Tragedy*.[36] Similarly, there is only a single reference to physiology in the four *Untimely Meditations*. There is one reference to physiology in the notebooks of 1873; none in 1874; three in 1875; and none in 1876. In fact, there are no references to physiology whatsoever in Nietzsche's notebooks from 1876 to 1879. There is but a single reference to physiology in *Human, All Too Human*, *Assorted Opinions and Maxims*, and *The Wanderer and His Shadow* (1878–1880) and none in the notebooks from 1877 to 1879. After this, physiology begins to make an increasingly steady appearance. It crops up a dozen times in 1880 as Nietzsche started to work on *Daybreak* (1881), in which there are some eight references to physiology and 13 more in the notebooks of the same period (1881). *The Gay Science* (1882) makes six references; the corresponding notebooks three. The four parts of *Thus Spoke Zarathustra* (1883–85), as might be expected, make no reference, although in the notebooks of the same period, there are 26. In *Beyond Good and Evil* (1886), there are 16 references, with 17 in the 1886 notebooks. Then follows Nietzsche's published work which has the greatest number of references to physiology, namely: *On the Genealogy of Morals* (1887) has 33 references with another 23 in the 1887 notebooks. Then, last, but certainly not least, there are the works from 1888: *Twilight of the Idols* (26); *The Antichrist* (13); *Ecce Homo* (17); and *The Case of Wagner* (9). In the 1888 notebooks, there are 104 references to physiology. In total, there are 133 references to physiology in Nietzsche's published writings (excluding *Nietzsche contra Wagner*, a compilation of earlier material), of which 114 (a full 85 per cent) are to be found in the texts starting with *Beyond Good and Evil* (1886). There are 212 references to physiology in the notebooks as a whole, with 144 (68 per cent) falling in the same timeframe. However, the notebooks of 1888 alone account for almost 50 per cent of those 212 references.

Nietzsche, it seems, was a very well-informed amateur in matters pertaining to natural science in general and in physiology in particular. His interest in physiology began shortly after his reading of the first edition of Lange's *History of Materialism* in 1866[37] but, as Thomas Brobjer shows in Chapter 1 of this book, he began to study physiology in earnest around 1880. He describes this period of his life in *Ecce Homo*: 'A truly burning thirst took hold of me: henceforth I really pursued nothing more than physiology, medicine, and natural sciences' (*EH* 'Human' 3).

Morrison argues that Nietzsche was consistently in the dark about matters physiological through to 1888. No doubt Nietzsche quite appreciated his lack of such knowledge during his early writings up to and including *Daybreak*. In *On Truth and Lie in an Extra-Moral Sense*, for example, he expresses this ignorance about the body:

> What does man actually know about himself? Is he, indeed, ever able to perceive himself completely, as if laid out in a lighted display? Does nature not conceal most things from him – even concerning his own body – in order to confine and lock him within in a proud, deceptive consciousness, aloof from the coils of the bowels, the rapid flow of the blood stream, and the intricate quivering of the fibres! She threw away the key. (*TL* 1, *KSA* 1, p. 877)

In the 1875 notebooks Nietzsche was already describing morality as starting from 'mistaken physiological assumptions' (*KSA* 8, 3[76]). In *Daybreak* he railed against 'Christian interpreters of the body' who regard whatever proceeds 'from the stomach, the intestines, the beating of the heart, the nerves, the bile, the semen' as moral and religious phenomena (*D* 86). This becomes a recurrent theme in his notes. In later years, he referred to 'ancient physiological *errors*' (*KSA* 9, 11[173]) and 'physiological confusions' (*KSA* 13, 14[168], p. 354); the ascetic ideal was 'a physiological contradiction' (*KSA* 12, 8[3]). In fact, Nietzsche regarded the last two millennia – that is, the time since the advent of Christianity – as representative of 'physiological absurdity' (*KSA* 13, 25[1]). 'How impotent has all physiological *knowledge* hitherto been!' he grumbled in 1881 (*KSA* 9, 11[173]).

Sometime around the writing of *Daybreak*, Nietzsche decided to look for the key to physiology which nature had thrown away. When physiology spoke, particularly about morality, Nietzsche started to listen. In the notebooks of 1886, Nietzsche traces his philosophical genealogy and suggests that he has something in common with the 'mechanistic movement': the aim of reducing all moral and aesthetic questions to physiological ones, all physiological questions to chemical ones, all chemical questions to mechanical ones, 'but with the difference that I do not believe in "matter" and consider Boscovich to be one of the great turning-points, like Copernicus' (*KSA* 11, 26[432]). He would soon speak of advancing philosophy towards a 'genuine physio-psychology' (*BGE* 23), the result of which would be a 'physiology of morality' (*KSA* 11, 27[14]) and a 'physiology of art' (*KSA* 12, 7[7]). By the time of *The Antichrist*, Nietzsche was in a position to make his strongest claims for philosophers to acquire knowledge of physiology.

> As regards the animals, Descartes was the first to have dared, with admirable boldness, to understand the animal as *machina*: the whole of our physiology endeavours to prove this claim. And we are consistent enough not to except man, as Descartes still did: our knowledge of man today goes just as far as we understand him mechanistically. (*A* 14)

Nietzsche came to regard 'the *body* and physiology [as] the starting-point' (*KSA* 11, 40[21]) because, in the body, we find the origin of all '*physiological* value judgements and racial conditions' (*BGE* 20): in other words, 'physiological demands for the preservation of a certain species of life' (*BGE* 3); even the 'physiological capacity for life' (*GM* iii 11). He says that '[b]efore we learned to understand physiological conditions physiologically', with such interpretations as 'the struggle against sin' in mind, 'man had to deal with moralistic conditions. Consequently, the area of morality shrank extraordinarily – and is becoming smaller still' (*KSA* 9, 3[10]). Morality gradually disappears as physiological knowledge increases because 'moral conditions are *physiological* conditions' (*KSA* 9, 6[445]). Or more precisely, they are pathological conditions: as early as 1880 Nietzsche is convinced that 'morality has hitherto been a means of disturbing the development of mankind in its physiological foundation' (*KSA* 9, 3[97]). Morality is a disease, he would later say, and 'we Europeans are the invalids' (*KSA* 12, 7[6], p. 282). And what primarily is to blame? In a word, ignorance. Ignorance about physiology has been the cause of human misery: 'Ignorance *in physiologicis* – the Christian has no nervous system –; contempt and the deliberate neglect of the demands of the body, of the discovery of the body ...' (*KSA* 13, 15[89]). At the same time, knowledge of physiology is the means to man's salvation: 'Man did not know himself physiologically down the long chain of the millennia: he does not know himself even today. To know, for example, that one has a nervous system (but no "soul") is still the privilege of the best informed' (*KSA* 13, 14[179]).

Moreover, in *Twilight of the Idols* (*TI* 'Expeditions' 33), Nietzsche continues to delve deeper into physiological matters and indeed, I would argue, understands physiology to be the key to solving the most fundamental problems and questions posed by philosophy. In fact, Nietzsche goes so far as to state, as a matter of principle, that '[e]very individual may be scrutinized to see whether he represents the ascending or the descending line of life'. It is not just individuals who may, indeed *should* be placed against this standard to see how they measure up; the same thing can be done for all values, moralities, beliefs, political institutions and all cultures, past, present, and future. What this standard is measuring is essentially the degree to which something is healthy or sick, strong or weak, potent or impotent, physiologically well or pathologically in decline.

Physiology, therefore, is no doubt one perspective on Nietzsche's philosophy, although some who are less than generous might argue that even Nietzsche would have regarded it as partial, foreground, a nook, corner, or narrow perspective. Might a case be made for its being a privileged perspective or even regarded by Nietzsche as the ground for perspectivism itself? One might think that the theme of physiology would have no claim at this deep level of explanation, that his account of the will to power would be helpful in explaining his use of physiology but not vice versa, especially since the will to power, being universal, includes the inorganic, while physiology, one would think, would necessarily exclude it. However, is there a way of making the will to power parasitic on physiology? Is it possible, in other words, that perspectivism and even the will to power are all reducible to physiology? In his notebooks, Nietzsche does refer to the will to power as the 'physiology of the order of rank' (*KSA* 12, 2[74]) and to the 'physiology of power' (*KSA* 12, 2[76]), and although he does make one reference to physiology (or at least the physiological process of cell division) 'as a consequence of will to power' (*KSA* 13, 12[1]), thereby reducing physiology to the will to power, he also refers to the 'physiology of the will to power' (*KSA* 13, 13[2]), which seems to reduce the will to power to physiology. Nietzsche himself reduced what there is universally (be it organic or inorganic) to the Greek word *physis*, using the term in much the same way as we would use the word 'physics', though not the mechanistic Newtonian physics of inert atoms but rather the Boscovichian physics of dynamic force-points which interact universally *as if they were organic*. As Nietzsche earlier wrote, although in a rather different context, '[n]one of us knows how deep and high *physis* extends' (*KSA* 7, 16[42]).

I think a case could be made that the will to power is, in fact, parasitic upon physiology. For, as we have already seen, along Nietzsche's epistemic axis, human physiology dictates what will be of value to *us*. The will to power manifests itself through *our* perspective valuations and these, in turn, can be reduced to purely physiological valuations (*BGE* 20). In addition, Nietzsche thinks that all categories or general horizons of meaning, which include the Kantian categories of the understanding, (and indeed even the will to power), are sensual–physiological, regardless of their claim to apriority. The Kantian categories are therefore mistakenly thought to be *a priori* – that is, necessary and strictly universal (*BGE* 4) – simply because, for a time, they really are structures which are necessary for human existence.

To make this a little clearer, let us examine Nietzsche's answer to the question 'Why is it that we have "permanence" and "causality" as the two most fundamental categories?' Nietzsche's response, I think, would be twofold: we have these particular categories (1) because of general *human* need and usefulness and (2) because this general need and usefulness at a *physiological* level has been translated through the senses to a conscious level. Epistemologically speaking, Nietzsche argues that thinking is impossible without first having 'things' to think about. It is therefore necessary, if we are going to make any judgements about objects, that there be 'things' which are self-identical (*KSA* 12, 9[89]). Hence, to be able to judge that 'the book is red' or 'the book is on the table' is to presume that there are such self-identical objects as 'books' and 'tables' and that something can be asserted about them. In fact, Nietzsche claims that belief in subject–attribute and cause–effect is to be found in every judgement (*KSA* 12, 2[83]). Since the actual state of affairs existing in the world independently of the knowing subject is deemed by Nietzsche to be utterly chaotic and disunited – that is, without 'things' – he argues that we must first undertake a construction at the level of the senses – *that is, from our own unique, physiological, human qua human perspective.* It is therefore only in terms of this 'pre-given' perspective that we are able to understand 'things' and how these 'things' can then subsequently serve as the basis for our judgements. Thus, our senses are falsified by our unique human situatedness, and it is this situtatedness, this human physiological perspective, that organizes the chaotic multiplicity of sense impressions so that we apprehend at least the *semblance* of 'identical things'. The fiction of permanent objects at the level of the senses then serves as the basis for implicitly believing and explicitly judging through a general, conceptual framework which regards 'thinghood' as a necessary category:

> The perspective of all organic functions, all the strongest instincts of life: the force in all life that *wills* error; error as the precondition even of thought. Before there is 'thought' there must have been 'invention'; the *construction* of identical cases, of the appearance of sameness, is more primitive than the *knowledge* of sameness. (*KSA* 12, 10[159])

In sum, Nietzsche's argument is that we judge or think that there are permanent entities in the world because there *seem* to be permanent entities. Nevertheless, such belief, thought or judgement is based directly on the *way* that the world appears to the knowing subject. Becoming ('change' on the ontological level) is thereby falsified by the senses, by our human physiology, to appear to consciousness as being ('permanence' on the believed-to-be ontological level). It is therefore necessary within Nietzsche's epistemological framework for an equalization or manufacturing process at the level of sense to precede thought. Thus, as Nietzsche writes in *Daybreak*: 'The habits of our senses have wrapped us up in a tissue of lying sensations *which in turn* lie at the base of all our judgements and our "knowledge"' (*D* 117; emphasis added). Subsequently, the semblance or appearance of being is categorized epistemologically under 'predicates of being in general' – that is, unity, thinghood, permanence, substance (*TI* 'Reason' 5).

It should be apparent that, given these particular epistemological parameters – namely, physiological, sensory and even psychological prejudice – it is neither possible nor necessary for Nietzsche to escape the subject–object dichotomy in order to ascertain the actual state of affairs existing in the world independently of the knowing subject. Nietzsche's epistemology might therefore be described as a 'misrepresentational realism' in which two intermediaries prevent us from securing direct knowledge of the way things are. According to Nietzsche, 'the character of existence is to be *misunderstood*' (*KSA* 13, 17[3]). Man, by his very nature, 'conceals *reality* from himself, he falsifies it' (*KSA* 13, 15[91]; emphasis added). Although some might assume that 'men, who are products of an exacting process of development would not be likely to survive if their minds were falsifying agents',[38] Nietzsche argues that the very opposite is true. The 'exacting process of development' through which we have evolved has found it necessary for us to be such falsifying agents because, for Nietzsche, the bottom line is the survival of the human species. Both subjective modes for representing objects – that is, intuition and concepts – precisely in so far as they represent objects, must misrepresent reality. Since there must always be this human physiological perspective first, there must always be a misrepresentation of the Heraclitean flux which constitutes empirical reality as it exists independently of the knowing subject – that is, of psycho-physiological human beings.

The Nietzschean categories are therefore not understood to be true, in terms of correspondence, nor could they ever be so understood. They are not *a priori* (*KSA* 13, 14[152]) because they have what Nietzsche calls a 'sensual origin' (*KSA* 12, 9[98]) – that is, as 'sensual-intellectual appropriation' (*KSA* 12, 5[11]). The categories of permanence and causality are discovered in the world because they are imposed by reason on products already sensibly determined, already physiologically formed. Thus, Nietzsche accepts that activities, at both the intuitive and the conceptual levels, strictly determine the ways in which the world of the senses (*Sinnenwelt*) becomes the product of our organization. Reality does not, and cannot, exist in the form in which it presents itself to us because we, as doubly active subjects, must inevitably misrepresent it. There is in fact no empirical world of objects at all beyond the subject, beyond our subjective modes of cognizing. 'We can *comprehend* only a world that we ourselves have *made*' (*KSA* 11, 25[470]). That is to say, we, as human beings, can only grasp 'a world' for us because we already possess a species-specific physiological starting-point. 'The overwhelming consensus among men regarding things proves the complete homogeneity of their perceptual apparatus' (*KSA* 7, 19[157]).

Conclusion

In conclusion, I think there is little doubt that the Unmentionables commit a serious error of omission when they completely ignore the role of physiology in Nietzsche's thought. The more considered route, as we have seen, is to recognize that physiology does play an important, a positive and even an indispensable role. Whilst he is not perhaps exactly a 'profound' physiologist, especially from our temporal vantage-point, I think there is little question that Nietzsche must be understood in light of what he had to say about the physiological question.

Notes

1 I should like to thank the editors of this volume for their insightful suggestions and also my colleague Brian Lightbody for his help in preparing this manuscript for publication.
2 Friedrich Nietzsche, *Philosophy and Truth: Selections from Nietzsche's Notebooks of the Early 1870s*, trans. and ed. Daniel Breazeale (Atlantic Highlands, NJ: Humanities Press, 1979), p. 25, n.56.
3 Walter Kaufmann, *Nietzsche: Philosopher, Psychologist, Antichrist* (Princeton, NJ: Princeton University Press, 1974).
4 Ibid., pp. 254–55.
5 Ibid., p. 294.
6 A way of overcoming this misinterpretation of Nietzsche, Kaufmann says, can be found in what he calls 'Nietzsche's *Physics*' (sections 618–39 in *The Will to Power*) which he suggests might even be called 'Nietzsche's *Monadology*' (ibid., pp. 262–63).

7 Ibid., p. 295.
8 Claudia Crawford, 'Nietzsche's Physiology of Ideological Criticism', in Clayton Koelb (ed.), *Nietzsche as Postmodernist: Essays Pro and Con* (Albany, NY: SUNY Press, 1990), pp. 161–86 (p. 170).
9 Ibid., pp. 162–63. Daniel R. Ahern expands Crawford's third point into an entire book which examines Nietzsche from his 'clinical standpoint': *Nietzsche as Cultural Physician* (University Park, PA: Pennsylvania State University Press, 1995).
10 Crawford, 'Nietzsche's Physiology', pp. 161, 163.
11 Ibid., p. 164.
12 Immanuel Kant, *Logic*, trans. by Robert Hartman and Wolfgang Schwarz (New York: Bobbs-Merrill, 1974), p. 37.
13 Thomas Nagel, 'What is it Like to be a Bat?', in Fred D. Miller and Nicholas D. Smith (eds), *Thought Probes: Philosophy through Science Fiction* (New Jersey: Prentice-Hall, 1981), pp. 179–91 (p. 160).
14 Babette E. Babich, *Nietzsche's Philosophy of Science: Reflecting Science on the Ground of Art and Life* (Albany, NY: SUNY Press, 1994), pp. 77–134 (p. 113).
15 Ibid., pp. 115, 114.
16 Ibid., p. 120 n.11.
17 Ibid., pp. 107, 94.
18 Ibid., pp. 6, 80, 84, 82.
19 Ibid., p. 99.
20 Ibid., pp. 92, 99.
21 Ibid., p. 97.
22 Alistair Kee, *Nietzsche Against the Crucified* (London: SCM Press, 1999), p. 41.
23 Ibid., pp. 55, 41.
24 Ibid., pp. 188, 97.
25 Ibid., p. 89.
26 Ludwig Feuerbach, *Sämtliche Werke*, ed. Wilhelm Bolin and Friedrich Jodl, 10 vols (Stuttgart: Fromann, 1903–11), vol. 10, p. 50.
27 Ludwig Feuerbach, *Principles of the Philosophy of the Future* (Indianapolis: Hackett, 1986), p. 70. There is only one explicit reference to Feuerbach in Nietzsche's published writings, namely, in *GM* iii 4. There are some five references to Feuerbach in the notebooks, none of which is flattering to him: *KSA* 8, 9[1], p. 131 (in the context of extensive notes on Dühring's *The Value of Life*); *KSA* 11, 26[8], 26[377], 26[412]; *KSA* 12, 7[4], p. 261.
28 Feuerbach, *Sämtliche Werke*, vol. 10, pp. 12–15.
29 Marx W. Wartofsky, *Feuerbach* (Cambridge: Cambridge University Press, 1977), pp. 415–16.
30 Feuerbach, *Sämtliche Werke*, vol. 10, p. 22.
31 Ibid., p. 23.
32 Robert G. Morrison, *Nietzsche and Buddhism: A Study in Nihilism and Ironic Affinities* (Oxford: Oxford University Press, 1997).
33 Ibid., p. 109.
34 Ibid., pp. 107, 108.
35 For example, Nietzsche discusses the stupidity of allowing women, who know nothing about the physiological after-effects of their cooking, into the kitchen (*BGE* 234); how fakirs abstain from salt in a deliberate hygienic measure designed to abolish their will and desire (*GM* iii 17); how a diet of newspapers, politics, beer and Wagnerian music may be connected with overt nationalism or Parkinson's disease (*GM* iii 26).
36 I am, of course, only referring to the term 'physiology' and its strict cognates. A wide variety of other related medical and physiological terms can also be found in Nietzsche's writings, even in earlier material. Still, I would argue that in the later works – that is,

after *Human, All Too Human* – these related concepts can be seen with increasing *frequency.* A by no means exhaustive selection is: anaemia, cretinism, depression, dyspepsia, epilepsy, erotic precocity, *folie circulaire*, hemiplegia, hypersensibility, hypochondria, *idée fixe*, *induction psycho-motrice*, inherited melancholy, monomania, neurosis, neurasthenia, rhachitis, syphilis and tuberculosis.

37 See George J. Stack, *Lange and Nietzsche* (Berlin: de Gruyter, 1983) and Curt Paul Janz, *Friedrich Nietzsche. Biographie* (Munich: Hanser, 1978), vol. 1, pp. 239, 407, 504–507.

38 Marvin Farber, *Phenomenology and Existence: Toward a Philosophy within Nature* (New York: Harper, 1967), p. 66.

Chapter 3

Nietzsche, Medicine and Meteorology

Gregory Moore

From the 1860s onwards Nietzsche's letters to his friends and family recite a litany of sickness. His eyes pain him appallingly. Migraines reduce him to acute agonies. His stomach is utterly disordered. He succumbs to nervous exhaustion. Plagued by unremitting bad health, he would spend a decade wandering around Europe, moving from doctor to doctor, from health spa to the mountains. Like many patients searching for an acceptable diagnosis, he sought the opinions of different medical men, trying their remedies, before giving up and turning in despair to another expert and another treatment. Yet, although Nietzsche would later transform his condition into a self-consciously heroic struggle through his writings, his distress was ultimately little different from the experiences of his contemporaries. A great number of nineteenth-century intellectuals were unwell: Charles Darwin, Thomas Carlyle, Herbert Spencer, and Eduard von Hartmann, to name just a few prominent writers with whose ideas Nietzsche took issue, were all chronically debilitated. Many of Nietzsche's friends, too, endured recurrent bouts of illness: Franz Overbeck suffered from rheumatism; Heinrich Köselitz complained of hysteria; Paul Rée was struck down by nerve fever on more than one occasion. With illness a part of everyday life in the nineteenth century it is unsurprising that, as Bruce Haley has argued, 'No topic more occupied the Victorian mind than Health'[1] – neither religion, nor Empire, nor progress, nor Darwinism. And just like Nietzsche, his friends and contemporaries flocked to coastal resorts, retreated to the Alps, experimented with various dietary regimens, swallowed pills of dubious medicinal value, took the waters, endured electrotherapy and nasal douches, stretched their limbs in gymnastic exercises – all in the name of health.

Despite the perfectly ordinary nature of his illnesses, Nietzsche's spectacular intellectual collapse in Turin in January 1889 ensured that an extensive literature on his pathology grew up in the late nineteenth and early twentieth centuries – written by psychiatrists such as Max Nordau, Paul Möbius, Paul Cohen, Wilhelm Lange-Eichbaum and later catalogued by Pia Volz in her own exhaustive treatment of the subject.[2] Nevertheless, much of the work on Nietzsche's medical state tends to limit itself, for one reason or another, to identifying or discussing the conditions from which he may have been suffering, or to reveal the complex interrelations between ill-health and his extraordinary fertile inner life. But it seems to me that there are other, more interesting questions that might be asked about Nietzsche's ailing body and which address not Nietzsche's *medical history*, but the *cultural history of his illness*. How did Nietzsche himself perceive his own illnesses? What did it mean to be healthy and sick in the nineteenth century? And to what extent did these discourses impact upon his writings? This chapter will attempt to shed some light on these questions.

One way of approaching these issues is to examine Nietzsche's enthusiastic reading of medical works. What distinguishes his long-standing fascination with medicine from his interest in other scientific disciplines – in physics and biology, say – is that the former had little to do with the acquisition of abstract knowledge or extending the metaphorical resources of his writing. Rather, it was of a much more pragmatic nature; Nietzsche wanted to learn about his own sick body and to alleviate his suffering, if not cure his illnesses altogether. In other words, Nietzsche did not only read about medicine; in a rather misguided fashion he tried to practise it, too. In an early note from the 1870s Nietzsche writes, presumably planning a programme of study, 'Every day for one hour: hygiene' (*KSA* 8, 41[75]). Some years later, he would admit: 'the little that I can manage with my eyes now almost exclusively belongs to physiological and medical studies' (*KGB* III/1, 20–21 August 1881). He certainly read widely. He acquired works on physiology such as *Lehrbuch der Physiologie* (1881) by the eminent Cambridge scientist Michael Foster, a textbook aimed at medical students and general practitioners, C.F. Kunze's *Kompendium der praktischen Medizin* (*Compendium of Practical Medicine*, 1881), as well as a range of more popular books: the Danish social reformer Emil Hornemann's *Beiträge zur praktischen Gesundheitspflege* (*Contributions to Practical Hygiene*, 1881); the bestselling guide *Das Buch vom gesunden und kranken Menschen* (*The Book of the Healthy and Sick Person*, 1870) by Leipzig professor of anatomy, Carl Ernst Bock; M. Beande's *Dictionnaire de médecine usuelle* (1837); Gustav Münter's *Geschichtliche Grundlagen zur Geisteslehre des Menschen oder die Lebensäußerungen des menschlichen Geistes im gesunden und krankhaften Zustande* (*Historical Foundations of Psychology, or the Human Mind's Expressions of Life in its Healthy and Sick State*, 1850). He also bought books detailing particular cures: Daniel Gottlob Moritz Schreber's *Ärtzliche Zimmergymnastik* (*Medical In-door Gymnastics*, 1877); W.H. Schüssler's *Eine abgekürzte Therapie gegründet auf Histologie und Cellular-Pathologie* (*A Short Therapy Based on Histology and Cellular Pathology*, 1881); and Josef Wiel's *Diätetisches Kochbuch* (*Dietetic Cookbook*, 1873). Nietzsche's derogatory remarks in *Untimely Meditations* about 'the universally admired "popularization" … of science, that is to say the infamous trimming of the coat of science to fit the body of the "general public"' (*HL* 7, p. 99) is rather ironic given his own dilettantish dabbling in medicine, as well as his reading of such popular books.

Nietzsche's own habit of self-doctoring mirrors wider cultural developments in the nineteenth century – the enormous explosion of medical and biological knowledge, and especially the increasing popularity and availability of self-help books and faddish do-it-yourself cures. These books, primarily polite physiologies and guides to dietetic medicine aimed at the swelling ranks of middle-class society, offered cogent discussions of the structures and functions of the body and the rational regimen necessary for maintaining health.[3] Hygiene meant, above all, a systematic code of behaviour with respect to diet, exercise and evacuation, sleep and sex – all the habits that govern the well-being of the body. Echoing the injunction of the Delphic Oracle, popular hygienists urged their readers to 'know thyself'. This same imperative was issued by Carl Bock in the introduction to his *Book of the Healthy and Sick Person*, a phenomenally successful work which sold 100 000 copies between 1855 and 1876, a copy of which was owned by Nietzsche. Bock railed against the physical and mental degeneracy of modern humanity, evidenced in such phenomena as the profusion of digestive disorders, nervous irritability, epidemics of piles and gout, the lack of subjects fit for military service, the increase of sedentary habits and the concomitant decline of willpower. He laid the blame for this physical decline squarely at the door of those 'artificial practices' caused by ignorance of basic hygiene and the workings of the body. The regeneration of humanity could lie only in 'a treatment of the healthy and sick body that is natural and based on a knowledge of the human organism'. 'Every sensible person', he concluded, 'is a doctor.'[4] Nietzsche could not agree more with these sentiments; he himself sought to overcome his own medical naiveté and railed against the ignorance of others. 'But from now on I want to be absolutely my own physician, and people shall say of me that I was a *good* physician – and not just for myself' (*KGB* III/1, 9 July 1881). His preoccupation with the body, and with physical and cultural health, should be seen in this light.

I

To understand how Nietzsche understood his own body, both in sickness and in health, we should begin by exploring how the body in general was conceived in the nineteenth century. The dominant model used to conceptualize the human organism was that of the machine or engine. Ever since Hermann von Helmholtz had formulated the law of the conservation of energy in 1847 – according to which the forces of nature (mechanical, electrical, chemical, and so on) are forms of a single, universal energy, or *Kraft*, that cannot be created or destroyed – hygienists and physiologists had enlisted the new language of thermodynamics to describe bodily processes. By extending the concept of the correlation of forces to include vital as well as physical force, scientists laid the foundation of an experimental physiology based on the study of metabolism. Organisms were conceived as animate engines which transformed food into a definite quantity of force, including heat, muscle or labour power (*Arbeitskraft*), and, most importantly of all, nerve force (*Nervenkraft*). The concept of nerve force was an old one; arguments for the existence of some kind of invisible, subtle fluid channelled through the nerve fibres stretch back at least to the Enlightenment. But, by the mid-nineteenth century, nerve force was beginning to be associated with electricity, and, although its precise nature was often undetermined, it was generally held that it was relayed from one point to another in the body, and then transformed into information, feeling and other elements in the cerebral hemispheres. According to the Scottish psychologist and educationalist Alexander Bain, the German translation of whose *Mind and Body* Nietzsche owned, nerve force was 'an agent with various powers – mechanical agency, heat agency, chemical agency; all of which are due to the molecular alteration of the nerve-substance, the complement of the changes being a supply of blood in proportion to the force set free'. It was supposed, then, to exert some kind of mediating influence between mind and the physical forces of nature to which the body was subject. Thought, Bain declared, 'can never transcend the physical pace of the nerve-force', a speed he estimated at 90 feet per second.[5]

Bain and his contemporaries viewed nervous organizations as an energy system which, like a battery, expended energy only in proportion to its capacity to store it via a dynamo. Energy was released when the body received stimuli from outside. But if energy had been overspent – that is, if the organism had been excessively stimulated – there would be signs of 'dimming', and the system would be unable to recharge, with disastrous consequences. As Carl Bock put it, 'all powerful stimulants, when they are applied often, [leave behind] an irritable weakness of the cerebral nervous system, which, with continued irritation, ultimately leads not infrequently to mental illness, even to imbecility'.[6] Medical texts were obsessed with maintaining stable quantities of vital energy or *Kraft* – and particularly its most precious form, nerve force – for every person possessed only limited amounts. A balance had to be struck between activity and repose, between the expenditure of energy and recuperation. Health consisted precisely in maintaining this delicate equilibrium.

Heedless overexertion, whether mental or physical, could drain an individual's supply of nerve force, leaving an exhausted nervous system incapable of all endeavour. With internal harmony so crucial to health, people had to manage their lives very carefully indeed. Health writers agreed on a number of general rules: daily routines had to be thorough and carefully structured; the principle of moderation and temperance was paramount; the extremes of abstinence and excess were to be avoided. For the nineteenth-century body was an extraordinarily fragile structure, under threat from the countless dangers everyday life might throw up: overwork, coffee, sexual promiscuity and even excessively animated conversation at the dinner table. As one of Nietzsche's authors expressed it, ill-health arose from a 'want of balance between our powers and the demands we make upon them, and the want of harmonious physical and moral energy'.[7]

Failure of nervous power meant utter incapacitation. It was not even necessary to exhaust the whole body's supply of nerve force before nervous collapse ensued; it sufficed to overwork only part of the body, for the interdependence of the entire nervous system meant that depletion anywhere drained away the nerve force needed for other physical functions. Bain, for example, spoke of 'the very great falling off' in the 'processes of organic life … when the nerve force is monopolised for intense mental or muscular exertion'.[8] Overworking the brain could damage digestion or hinder reproduction. Even worse, the will was fatally weakened, allowing victims to succumb to all manner of vicious, even criminal, behaviour. Exhaustion – and especially nervous exhaustion or neurasthenia – was the great fear of the nineteenth century,[9] a fear which Nietzsche shared. 'Where does our modern world belong,' he asked in 1888, 'to exhaustion or ascent?' (*KSA* 13, 15[20]). And as he becomes preoccupied with the problem of decadence and sharpens his attack on 'our nervous age' (*KSA* 9, 6[129]), his late notebooks obsessively list the pathologies he sees afflicting European civilization: not only crime, prostitution and alcoholism, but also democracy, pessimism and Christianity are all social symptoms of nervous debility, weakness of will and hyper-irritability, of that 'certain inability to "master" oneself (*not* to react to stimuli, even to faint sexual stimuli)' that is 'one of the most frequent consequences of total exhaustion' (*KSA* 13, 23[1]).

Perhaps unsurprising in the great age of capital, another master metaphor employed with regard to the body as a whole, and to the nerves in particular, was drawn from the world of economics. As the term 'animal economy' implied the entire range of physiological functions, so the nineteenth century viewed the nervous system in terms of a nervous economy. Nerve force was a precious resource that could be spent wisely, hoarded for future investment or squandered recklessly to the point of bankruptcy. Nietzsche himself was more than familiar with this conception of the human body as a stockpile of nervous energy, as an economy of vital forces. He himself frequently resorted to it: 'The doctrine of moderation derives from observation of nature. That which is to become strong and tall must always increase its energy like capital and must not wish to live off it' (*KSA* 9, 6[183]). The same model also underpins the arguments of a number of the books on hygiene which Nietzsche owned. This is how one of his authors outlines it:

> The physical man, during the second half of life, lives upon the interest of the capital of those powers he amassed during the first half. Well for him if he has not in youth consumed that capital, but by careful stewardship of the revenues, has rather increased it. The revenues are those organic powers which should be carefully husbanded, and be used in the further development of our capacities. The capital is that ability to continually reproduce and perfect fresh power, – the innermost spring of vigorous life. The time for increasing our capital is past. We can only continue to enjoy our revenues by careful husbandry, by suitable usage of our strength, by careful spending of our stored-up materials. By this means alone can we preserve our capital of vigorous life untouched ... so long as possible ....[10]

The author of those remarks was Daniel Gottlob Schreber, whose phenomenally successful illustrated handbook *Ärztliche Zimmergymnastik* (*Medical In-Door Gymnastics*) Nietzsche acquired in its 15th edition. It was in September 1878 that Nietzsche embarked – 'enthusiastically', as he wrote to his publisher from Interlaken (*KGB* II/5, 10 September 1878) – on Schreber's course of gentle, rather undemanding exercises aimed in particular at 'relieving or drawing off the causes of nervous overstrain, and refreshing the mind by gently strengthening and stimulating the system while it soothes the over-tense nerves'.[11] Even before this time, however, Nietzsche appreciated the vital importance of physical culture. For, like many nineteenth-century observers, Nietzsche despaired of the way in which the body was increasingly neglected at the expense of mental exertion, with too much nerve force wasted in cerebration, leaving insufficient quantities for other bodily functions. He was not alone in believing that modern civilization was breeding a race of pallid, emaciated and enfeebled neurasthenics – the scholars and metaphysicians against whom he so often rails. Herbert Spencer, incapable of prolonged mental effort for nearly the last 50 years of his life after suffering a nervous breakdown, believed that intense brain work involved a waste in nervous tissues, which in its turn would lead to a 'consumption of nervous matter'. In his *Education: Intellectual, Moral, and Physical*, Spencer typically resorts to the metaphor of the bodily economy when he asserts that 'the amount of vital energy which the body at any moment possesses, is limited'. Nature, he continued, 'is a strict accountant; and if you demand of her in one direction more than she is prepared to lay out, she balances the account by making a deduction elsewhere'.[12]

Nietzsche himself thought along similar lines. In a note written in 1877, he claimed that, because 'modern education expects of people much greater brain activity, humanity must struggle more energetically for *health* in order not to produce nervously overwrought, even insane offspring'. To avoid a future populated by the mad and overtaxed, he demands not only that parents themselves be healthy in order to ensure robust offspring, but also that a variety of hygienic measures be universally implemented: daily gymnastic exercises, 'which must be as habitual and as coveted as one's daily bread', a balanced and sensible diet, sanitary living quarters, instruction in anatomy and so on (*KSA* 8, 23[151]). Like Spencer before him, Nietzsche insists that the cultivation of the mind must be complemented by a *physical education*.

As Nietzsche began to turn his back on the idealism of his youth, so otherworldly philosophers and theologians also came under fire. In *Daybreak*, one of the crimes which Nietzsche lays at the door of idealism and 'the teaching of *pure spirituality*' is that it has 'destroyed nerve force with its excesses: it has taught deprecation, neglect or tormenting of the body'. This depletion in nerve force has predictable results: a 'general chronic over-excitability was finally the lot of these virtuous pure-spirits: the only *pleasure* they could still recognise was in the form of ecstasy and other precursors of madness' (*D* 39). But then again, thinking was a luxury in an age suffering from what, in *Schopenhauer as Educator*, he called 'the sickness of this undignified haste' (*SE* 6, p. 168). In *Daybreak* once again, Nietzsche inveighs against the nineteenth-century ideology of the 'dignity of work', for the daily grind 'uses up an extraordinary amount of nerve force, which is thus denied to reflection, brooding, dreaming, worrying, loving, hating' (*D* 173). Nietzsche returns to the concept of nerve force once more in *Daybreak*, echoing these same sentiments in what, this time, is perhaps an autobiographical remark:

> A thinker can for years on end force himself to think against the grain: that is to say, to pursue not the thoughts which offer themselves from within him but those to which an office, a prescribed schedule, an arbitrary kind of industriousness seem to oblige him. In the end, however, he will fall sick: for this apparently moral overcoming of himself will ruin his nerve force just as thoroughly as any regularly indulged in excess could do. (*D* 500)

But for all Nietzsche's concern about the neglect of the body, his disgust at scholars 'crippled and enervated by pedantry and antiquated procedures' (*HH* 250), he himself was hardly innocent. According to the diagnosis of one of the many physicians he consulted, Otto Eiser, Nietzsche's chronic, debilitating headaches were due partly to a 'predisposition in the irritability of the central organ' (*KGB* II/4, 6 October 1877), originating in excessive mental activity (notoriously, Eiser would later agree with Richard Wagner's malicious suggestion that Nietzsche had become unbalanced by overtaxing his nervous system in a different way – through habitual masturbation). Nietzsche was prescribed the usual treatment: reading and writing were banned for several years; methodical alternation between work and relaxation was recommended; bright light and extreme exertion were to be avoided, as were spicy or indigestible foods, strong coffee and heavy wines. Nietzsche ignored this advice; such intellectual abstinence was intolerable. But whether or not he agreed with Eiser's diagnosis at the time, in *Ecce Homo* Nietzsche later conceded – with the benefit of hindsight – that he had indeed overtaxed himself as a young academic. His admission is couched in the language of the bodily economy: 'In my time at

Basle my entire spiritual diet, the division of the day included, was a perfectly senseless abuse of extraordinary powers without any kind of provision for covering this consumption, without even reflection on consumption and replacement' (*EH* 'Clever' 2).

II

If Nietzsche complained that his spiritual diet was hopelessly flawed, then this was also true of his diet proper. Throughout his life Nietzsche suffered from chronic stomach ailments and felt that he was fighting a losing battle with his gut for control of his body, sighing: 'If only I were master of my stomach once more!' (*KGB* II/5, 30 March 1879). But this was nothing unusual: gastrointestinal disorders seem to have reached almost epidemic proportions in the nineteenth century. 'There is no complaint,' wrote F.W. Pavy in 1867, 'that the physician is so frequently called upon to treat, in one shape or another, as that of indigestion'.[13] Like nervousness, dyspepsia was regarded as one of the diseases of civilization, an inevitable byproduct of the unnatural practices of modern life. Yet dyspepsia was not the only problem caused by a disordered stomach, for this delicate and capricious organ was granted a singularly important place in the bodily economy, which, as the author of most of man's miseries, it frequently abused. 'The stomach,' observed Lionel Beale, 'is the cauldron in which most of the diseases which are preventable by our own knowledge and efforts are brewed.... It is commonly in the stomach that the primary derangements occur, which lay the foundation for all varieties of ill health.'[14] Nietzsche, too, points to those 'acts of roguery' whose 'revolting source is so often the belly' (*HH* 243) and has Zarathustra declare 'an aching stomach' to be the 'father of affliction' (Z iii 'Law-Tables'). The digestive system – its physiology, pathology and hygiene – exercised, then, a peculiar grip on the Victorian imagination, with innumerable treatises, therapeutic guides and public lectures devoted to the subject.[15] One Sydney Whiting even wrote a bestselling book called *Memoirs of a Stomach*, which purported to be the personal account of the trials and tribulations experienced by a beleaguered digestive organ forced to consume an immoderate diet of unchewed, poorly cooked and spicy food.

What lies behind this nineteenth-century obsession with digestive regularity and inner hygiene? The digestive system – through the metabolic process, which had been recently discovered in the 1840s – was the means of maintaining the thermodynamic equilibrium, of transforming one form of *Kraft* into another. The human being was a heat engine (*Wärmemaschine*) that converted the energy contained in the raw materials with which it was supplied into nerve force and socially productive labour. The body was typically viewed as a throughput system, requiring efficient digestion and speedy waste disposal. Any form of blockage or improper functioning of the 'machinery' that resulted in a failure to evacuate noxious substances was ruinous. In the words of Daniel Schreber: 'one of the most common causes of obstructions in the development and continuity of the chemical process of life is insufficient combustion of material and insufficient evacuation of the used-up, and therefore useless, particles of matter; for when these are retained in the body, there must be a disturbance of the equilibrium between their consumption and their assimilation'.[16] A good stomach animated the system as a whole, but in particular it replaced and nourished the blood. Maintaining the proper quality and quantity of the blood supply was essential to health. This was brought home to Nietzsche in 1875, when, after years of suffering, he spent a month at a health resort being treated by the renowned stomach specialist Josef Wiel.[17] Wiel's diagnosis was that he suffered from chronic 'gastric catarrh' with significant dilation of the stomach, causing vascular engorgements which prevented the brain from being supplied with sufficient blood. Nietzsche began a regime of regular small meals, self-administered enemas and a spoonful of Carlsbad mineral salt to relieve his distended stomach; the abnormal acidity, which Wiel attributed to mental and nervous pressures, was treated by applying leeches to Nietzsche's head. Wiel's diagnosis of a 'nervous affection' played on Nietzsche's mind. He increasingly began to suspect that his stomach pains were merely symptomatic. By 1887 Nietzsche was certain that 'with me the stomach is only indirectly a factor and never the *causa prima*' (*KGB* III/5, 30 June 1887).

He also frequently suffered from constipation, which followed his bouts of illness, and was vulnerable to the obsession with regularity that stalked most Victorians. If Nietzsche's bowels were in order, then he was able to think clearly: 'Today, after a strong laxative, a good day', he observed in 1881. Such was his relief, he reports, that afterwards he was able to settle on the final arrangement of the aphorisms for *Daybreak* (*KGB* III/1, 24 February 1881). Nietzsche often bemoaned his painfully slow metabolism, another hint at chronic constipation: 'I have the most sluggish intestines in the world' (*KGB* III/3, end of July 1885). A strong stomach and rapid metabolism were prized assets in the nineteenth century. Nietzsche's difficulties with digestion – the assimilation of nutrients and evacuation of waste matter – was ominous. As he would have known from his reading of Bock, among others, constipation was caused by a lack of vital energy.

As a result of his chronic stomach complaints, Nietzsche spent a lifetime experimenting with the kinds of food he could and could not tolerate. Sometimes he would survive on just milk and eggs; he tried eating very little, in accordance with Luigi Cornaro's *A Treatise of Temperance and Sobriety*. At various times he eliminated from his diet meat, potatoes, cabbage, cauliflower, vinegar, mustard, pepper, black bread, onions, cheese and replaced them with artificial foodstuffs such as Liebig's meat extract and malto-leguminose.[18] Liebig's meat extract was a thick dark liquid which, when mixed with water, instantly formed a beef broth. It was marketed as a medicine, a useful but non-essential stimulant to the nervous system, and was credited with aiding digestion and the assimilation of food, as well as increasing mental alertness. Malto-leguminose was a vegetable equivalent. Both were supposedly highly concentrated forms of food, but in reality were nutritionally worthless.[19]

For all his woes, Nietzsche was well informed about dietary matters, for works of popular hygiene in the late nineteenth century are unthinkable without nutritional advice. In addition to the more general handbooks already mentioned, Nietzsche owned books and pamphlets such as Edward Smith's two-volume *Die Nahrungsmittel* (*Foods*, 1874); C.A. Meinert's *Wie nährt man sich gut und billig?* (*How Does One Eat Well and Cheaply?*, n.d.); Friedrich Kinkelin's *Über Ernährung* (*On Nutrition*, 1872). He was also well informed about the mechanics of metabolism, having not only annotated the relevant sections of Foster's *A Text Book of Physiology*, but also having read a copy of *Die Ernährungsgesetze Liebigs in neuester Fassung und das neue Nährmittel Malto-Leguminose* (*Liebig's Laws of Nutrition and the New Foodstuff Malto-Leguminose*, 1878) by the pioneering organic chemist who, by 1840, had laid the foundations for the study of metabolism and established that energy input and output formed a balance sheet. In addition, he read *Die organische Bewegung in ihrem Zusammenhange mit dem Stoffwechsel* (*The Motions of Organisms and their Relation to Metabolism*, 1845) by Liebig's student, Julius Robert Mayer. The biologist William Rolph's ideas about the role nutrition plays in evolution also exerted a powerful influence on Nietzsche's thinking about physiological processes.[20]

Given Nietzsche's own problems and the ventral obsessions of his contemporaries, it is perhaps unsurprising that digestion and digestive dysfunction are a master metaphor running through his works. An organism that lacks the strength to incorporate what it has acquired, to excrete waste matter, is anathema. Thus Nietzsche often dwells on the costive social organism: Germany cannot assimilate its Jewish population, unlike the Italians, French and British, who have 'stronger digestions' (*BGE* 251); modern society can no longer be construed as a 'body', for it 'no longer has the power to *excrete*' (*KSA* 13, 16[53]). It is perhaps no coincidence that what Nietzsche terms the 'liberated man' (*freigewordene Mensch*), the forerunner of the *Übermensch* who appears in notes from 1881, is envisaged as a perfectly functioning, self-regulating organism, with particular emphasis being placed on his capacity to assimilate, digest, metabolize and excrete (*KSA* 9, 11[182]). Furthermore, Nietzsche understands the intellect as 'a kind of stomach of all affects (which require nourishment)' (*KSA* 11 25[185]).[21] In one note, he describes modernity 'from the perspective of nutrition and digestion', suggesting that the bombardment of the intellect with impressions results in 'a weakening of the power to digest' (*KSA* 12, 10[18]). The assimilative character of the will to power is even elevated to the fundamental process of the universe. And when he enumerates the supposedly physiological side-effects of listening to Wagner, Nietzsche cites disturbances not only to the nervous system, but also to digestion. Wagner is said to obstruct 'the stomach with phlegm' – that is to say , his music causes gastric catarrh, the very complaint from which Nietzsche suffered (*CW* 'Postscript'). Wagner also causes constipation. Referring to a brand of laxatives, he declares: 'To listen to Wagner, I need pastilles Gérandel' (*NCW* 'Where I Offer Objections').

III

During the 1880s Nietzsche began to realize that experimenting with diet was not the only way to improve his poor digestion and headaches. Climate and the weather played an increasingly important role in his thinking about his body and his various illnesses. Towards the end of his professorship in Basle it seems that Nietzsche began to suspect that certain climates and atmospheric conditions exacerbated, and some relieved, the symptoms of his ailments. His retirement from his teaching post in 1879 was at least partly motivated by his 'Basileophobia' (*KGB* II/5, 3 April

1879), his realization that Basle's dismal climate itself was inimical to his well-being. Leaving his family home in 1880, he spent the rest of his intellectual life wandering Europe in search of the perfect climate, engaging – just as he had done and would continue to do with his diet and his amateur pharmacology – in 'pure experimentation' to find the milieu most appropriate to his specific needs. Most of the locations he visited proved detrimental to his health: there was too much rain; it was excessively hot or cold or too humid; the sky was overcast. 'Basle, Naumburg, Geneva, Baden-Baden, almost all the mountain resorts with which I am familiar, Marienbad, the Italian lakes, etc. – these are my ruin. Winter by the sea is tolerable, spring (Sorrent and Genoa) incessant suffering (because of the unsettled cloud cover)' (*KGB* III/1, 7 July 1881). He knew what he was looking for, but struggled to find it: 'Where is the land with plenty of shade, eternally clear sky, an *equally* strong sea breeze from morning to evening, without changes in the weather? It is there, there that I want to go! Even if it lies outside Europe!' (*KGB* III/1, 23 June 1881). A little more than a week after this outburst, he found the place he would thereafter spend every summer: Sils-Maria, high up in the Alps. Two years later, in December 1883, Nietzsche discovered Nice, a place he later described as a '"paradise for the sick and the old"' (*KGB* III/3, 2 January 1886). In 1888, he stumbled upon what he called his 'third residence', Turin. He planned to divide his year between Sils, Nice and Turin, enjoying 'the dry air, which is the same in all three places', and, together with clear skies and sunshine, crucial, he believed, for his welfare (*KGB* III/5, 13 May 1888).

Nietzsche's letters are filled with detailed, almost obsessive, descriptions of the weather. Yet his fascination with meteorology is not frivolous, and, at a time when this new science overlapped with several other dominant discourses in nineteenth-century European culture – physics, biology, medicine, hygiene – many of his contemporaries shared his curiosity. The diaries of John Ruskin, for example, are haunted by the black 'plague-wind' he thought responsible for the deterioration of the British climate. But where Ruskin was primarily concerned with the effects of industrialization on weather conditions, Nietzsche's interest is part of the discourse of 'medical climatology' or 'climatotherapy', a form of treatment that gained huge popularity in the late nineteenth century and, in the words of one expert, was predicated on 'the utilization of climate to promote comfort or recovery of the sick, or to prevent the development of disease'. The ideal climate for such purposes, one that favoured 'the development of races of men both vigorous in body and capable in mind', was generally mild, but characterized by moderate variations in the weather: 'Such a climate exercises the powers of adaptation and resistance of the different organs without subjecting them to excessive strain, and thus keeps the body in proper working condition.'[22] Once again, the principle of moderation was paramount. Nietzsche's lifestyle mirrored that of many of his contemporaries; he joined the increasing numbers of middle-class invalids who, in their restless search for the perfect mild climate in which to convalesce, headed for the Mediterranean coast to spend the winter months and retreated to the mountains during the summer. Nice, Sils, Venice, Genoa – those places to which Nietzsche would return again and again from 1880 onwards – were all climatic health resorts.

'One is punished good and proper for one's ignorance,' Nietzsche once complained to his mother and sister. 'If I had occupied myself *at the right time* with medical, climatological and similar problems, instead of with Suidas and Laertius Diogenes, I wouldn't be a half-ruined man' (*KGB* III/5, 14 March 1885). In order belatedly to rectify these shortcomings, he purchased and read a number of books which dealt either principally or partly with meteorology: Paul Reis's *Lehrbuch der Physik* (*Textbook of Physics*, 1872); Friedrich Schoedler's much-reprinted *Das Buch der Natur* (*The Book of Nature*, 1875); John Tyndall's *Das Wasser in seinen Formen als Wolken und Flüsse, Eis und Gletscher* (*The Forms of Water*, 1873).[23] In 1881 he asked Franz Overbeck to procure for him Pierre Foissac's *On Meteorology, and its Relations to the Science of Man, principally medicine and public hygiene* (1859) because he had begun to suspect that his various ailments were exacerbated by the deleterious influence of atmospheric electricity, and he wanted to understand this phenomenon better and take preventive measures (28 October 1881). Yet Foissac's rather outdated book disappointed him. This 'medical meteorology', he wrote disconsolately, is 'unfortunately a science still in its infancy and, with respect to my personal requirements, just a dozen question marks more' (*KGB* III/1, 14 November 1881). He was left to wonder whether scientists knew more than they had done in Foissac's day, but in fact, although most meteorologists of the age conceded that atmospheric conditions directly influenced the human organism, they were at a loss to explain how and to what extent this was the case. Nietzsche consequently felt it necessary to develop his own theories, to take every opportunity to discuss with doctors the importance of weather, to compile detailed meteorological tables (see, for example, *KGB* III/5, 17 June 1888), and speculate with friends and fellow-patients about the relationship between climate and health (*KGB* III/5, 1 February 1888). In 1883 he complained that the agonies he had suffered the previous winter were caused by 'electrical "storms"' which coincided with 'the appearance of *large* sun spots' (*KGB* III/1, 10 May 1883). A year later, in 1884, he claimed that his best work – *The Birth of Tragedy* and *Zarathustra* – was conditioned by the periodic increases in solar magnetism and that, on the other hand, his decision to become a philologist and disciple of Schopenhauer (which he described as a form of madness), as well as the worst crisis in his health around the time he was composing *Human, All Too Human*, was concurrent with the lowest ebb of the sun's influence (*KGB* III/1, 20 September 1884).[24]

Meteorological conditions *did* undoubtedly have an impact upon Nietzsche's work, though perhaps not in the insidious way he sometimes assumed. His writings not only bear the imprint of the location in which they were composed – he describes, for example, the first two parts of *Thus Spoke Zarathustra* and Book IV of *The Gay Science*, as 'the products of light and clear skies' (*KGB* III/1, mid-August 1883). His books also possess, as it were, an imaginary climate and can, consequently, be described in meteorological terms. In the preface to the second edition of *The Gay Science*, Nietzsche relates how that book is the expression of a certain joyfulness, of pleasure taken in one's strength that has lain dormant after long illness. It enacts, in other words, the rites of spring, and hence seems 'to be written in the language of the wind that thaws ice and snow: high spirits, unrest, contradiction, and April weather are present in it, and one is instantly reminded no less of the proximity of winter than of the triumph over the winter' (*GS* 'Preface' 1). Nietzsche returns to this theme in *Ecce Homo*: 'He who knows how to breathe the air of my writings knows that it is an air of the heights, a *robust* air. … The ice is near, the solitude is terrible – but how peacefully all things lie in the light! how freely one breathes! how much one feels *beneath* one!' (*EH* 'Foreword' 3). Just as Nietzsche and his contemporaries searched for the most healthy climate in order to restore their ailing bodies, so Nietzsche's works constitute a self-contained, independent milieu that allows the *intellect* to recuperate and become strong. Both Sils-Maria and Nietzsche's books boast the same pure air. Nietzsche once described himself as a 'weather prophet', whose sensitivity to minute changes in atmospheric pressure made him a human barometer (Letter to Overbeck, *KGB* III/1, 14 November 1881).[25] But through the process of writing, he transcends this role as a passive instrument at the mercy of the elements. He becomes godlike; he *makes* weather. In a letter to his mother – after complaining, as usual, about the rain and cold – he reports that he has received messages from admirers, saying: 'Your son is positively a force: he strengthens and refreshes; he creates "good weather" for other people' (*KGB* III/5, 3 December 1887).

Furthermore, Nietzsche construes not only his own writings, but all cultural artefacts as imaginary climates: German politics, like all things from 'Europe's *flatland*' (*TI* 'Germans' 3), 'is merely another kind of permanent winter and bad weather' (*KGB* III/5, 24 February 1887). And it is not just Nietzsche who thought in these terms. His friend Reinhart von Seydlitz, who was himself undergoing a cure in an Egyptian climatic health resort, wrote to Nietzsche complaining of the effects of the Chamsin wind, 'which is like a meteorological version of a Brahms symphony: ruthless, sandy, dry, mysterious, nerve-shattering, like a ten-fold scirocco [*sic*]' (*KGB* III/6, 21 March 1888). Nietzsche, too, possibly influenced by Seydlitz's words, subsequently likens the 'Wagneran orchestral tone' to the effects of the scirocco, which he had once described as his 'deadly enemy' (*KGB* III/1, end of March 1884): 'I break into a disagreeable sweat. *My* good weather is gone' (*CW* 1). Wagner's music has the same physiological effect on Nietzsche as a detrimental climate. Just as Nietzsche reports that 'all animal functions are depressed' after a spell of bad weather (*KGB* III/5, 13 May 1887), so, when hearing Wagner, he can no longer breathe easy; his stomach, heart, circulation and intestines are affected. Music, though, like a favourable climate, ought to bring blessed relief, 'as if all animal functions should be quickened by easy, bold, exuberant, self-assured rhythms' (*GS* 368). It is precisely such respite from inclement weather which, in *The Case of Wagner*, Nietzsche claims to find in the work of Bizet: 'With this work one takes leave of the damp north, of all the steam of the Wagnerian ideal … it has what goes with the torrid zone: the dryness of the air, the limpidezza in the air. In every respect the climate is changed' (*CW* 2).[26]

Nietzsche had first suggested that the concept of 'climate' could be expanded to cover something more than the mere sum of various atmospheric and terrestrial conditions within a particular region in an intriguing aphorism in *The Wanderer and his Shadow*. Each culture, he argues there, constitutes a 'spiritual climate', and, like a 'physical climate', might be either injurious or invigorating to a given organism, depending on its individual needs and constitution. Nietzsche envisages a cultural physician whose mastery of world history would be such that, like a medical doctor, he would have the authority to send each person to the precise 'climate' favourable to him, either temporarily or permanently: 'To live in the present, within a single culture, does not suffice as a universal prescription: too many people of the highest utility who cannot breathe properly in it would die out. With the aid of history one can give them *air* and try to preserve them.' But this is not all. This spiritual regimen would be accompanied by the establishment of a 'medicinal geography', whose task would be to discover to what illnesses each region of the earth gives rise and, conversely, what remedial qualities it offers. Then nations, families and individuals would be transplanted for as long and continuously as would be needed for inherited physical infirmities to be conquered. One day Nietzsche prophesied, 'the whole earth will be a collection of health resorts' (*WS* 188).

He was not wrong, as he was about to discover on his travels. But what Nietzsche calls 'medicinal geography' in *Wanderer* would become known as 'medical climatology'. By the end of the nineteenth century, climatic health resorts had sprung up throughout the Mediterranean and North Africa, as well as in North America and Oceania. Countless books were published containing detailed weather charts and rainfall tables of the areas in question. Nietzsche eventually read and owned several such therapeutic guides – presumably in order to move his 'experimentation' with self-prescribed climatological treatment on from mere trial and error – and he used these to gather information about the various resorts he planned to visit. The pamphlets he consulted included *Südliche klimatische Curorte* (*Southern Climatic Health Resorts*) by Professor Carl Sigmund of the University of Vienna; Amédée Goubet's *Les stations sanitaires de la France (Alpes Maritimes). Résidences d'été* (*The Health Resorts of France. Summer Residences*, 1884); and Rudolf Gerber's *Ajaccio als Winterkurort und die Insel Corsica* (*Ajaccio as Winter Health Resort, and the Island of Corsica*, n.d.).

As Nietzsche knew from his reading of these guides, different climates and meteorological conditions were held to have a different influence on what both he and his contemporaries called the 'animal functions' – that is, on the nerves and muscles – either stimulating or depressing them. Exposure to one climate or another was designed to allow patients to recoup their depleted stocks of vital and nerve force. Those with exhausted nerves – like Nietzsche himself – went to dry, invigorating climates such as Nice (in the winter) and Sils (in the summer). Those with irritable nerves travelled to humid climates such as Pau, Venice and Tenerife. 'The *salutary influence* of the Venetian *climate*,' wrote Carl Sigmund, 'consists primarily in its mild, humid, dust-free and relatively windless atmosphere. As with Pau, its depressive, calming, relaxing effect on the circulation, the respiratory organs and the whole nervous system can be observed even after a short stay.'[27] As Nietzsche would discover to his own cost, choosing an inappropriate climate could have disastrous consequences. In March 1880, when he first began his peripatetic life drifting from one southern health resort to another, he moved to Venice, believing that the city's climate 'exerts a favourable influence on many who suffer from headaches'. 'I am performing,' he declared, 'the *very* necessary experiment to establish whether a decisively "*depressive*" climate (in medical terms) is not better for my head than an excitative one, which I have exclusively applied hitherto' (*KGB* III/1, March 22, 1880). It was not; three months later he retreated to Marienbad. Yet, though he would on occasion return, he later recognized that Nice and Venice were 'climatic antitheses'. Though some might benefit from the enervating effects of the Venetian climate, he could expect only 'headaches and melancholy' (*KGB* III/1, 25 February 1884).

Health resorts were finely graded on the basis of 'indication', or presenting complaint – that is, on the basis of what malady they most suited. If he had read Sigmund sooner, he would have known that Venice favoured 'those suffering from chest and heart complaints with extremely irritable nervous and circulatory systems, those afflicted with laryngeal and bronchial catarrh; and feverless, chronic consumptives in general'.[28] But what ailments was Nice particularly good for, according to the medical writers of the time? Sigmund recommends the resort to scrofulous and rhachitic youths; chlorotic and anaemic inmates of approved schools; those suffering from chronic rheumatism and arthritis; and, most importantly, for this is the category to which Nietzsche presumably belongs, 'those who, as a result of mental and physical strain, are exhausted, dissipated, degenerate and crippled'. Its 'predominantly dry' atmosphere, as well as the environment of Nice itself, 'have an excitative influence on the nerves and circulation of the blood'. It has a 'climate which intensifies absolutely all functions of the organism'.[29] Nietzsche agreed. When he discovered Nice in 1883 – he would spend subsequent winters there until he discovered Turin in 1888 – it was a revelation. He wrote ecstatically of the 220 clear days which Nice was said to enjoy annually, a figure plucked from Sigmund's detailed meteorological description of the resort (*KGB* III/3, 2 January 1886).[30] He had found, he told friends, his 'promised land' (*KGB* III/1, 28 March 1884): 'I cannot give you an idea of the vitalizing, indeed downright electrifying effect this abundance of light has on my entire system'. What improved most was his appetite and digestion. 'A brain like mine,' Nietzsche wrote, 'requires a *very powerful nutrition*: – and I have for years suffered from insufficient nourishment because an unfavourable climate (such as that of Basle) increased my difficulties.' In Venice he found it difficult to digest even the lightest meal. But Nice, on the other hand, 'is the only place that *stimulates* my metabolism to such a degree that my head feels *free*. The opposite happens in places which are humid and very cloudy' (*KGB* III/3, 26 June 1885). Such was the invigorating effect of the climate that he found it impossible to eat 'à la Cornaro' – that is, according to the strictures of the Renaissance merchant who advocated eating only small amounts and who, perhaps not insignificantly, was Venetian (*KGB* III/1, 28 March 1884). Even beer, something Nietzsche usually normally avoided, seemed to have medicinal qualities in Nice.

It is here, then, that Nietzsche's preoccupation with digestion *and* climate converge: a favourable, dry climate has an energizing, vivifying effect on the entire organism because it quickens the metabolism, the fundamental process that transforms raw materials into vital force, which in turn powers the mind. In *Ecce Homo*, Nietzsche goes as far as to suggest that genius is determined by such climates:

> The influence of climate on *metabolism*, its slowing down, its speeding up, extends so far that a blunder in regard to place and climate can not only estrange anyone from his task but withhold it from him altogether: he never catches sight of it. His animal *vigor* never grows sufficiently great for him to attain to that freedom overflowing into the most spiritual domain where he knows: *that* I alone can do.... A never so infinitesimal sluggishness of the intestines [*Eingeweiden-Trägheit*] grown into something mediocre, something 'German'; the German climate alone is enough to discourage strong and even heroic intestines. The *tempo* of the metabolism strands in an exact relationship to the mobility or lameness of the *feet* of the spirit; the 'spirit' itself is indeed only a species of

> this metabolism. Make a list of the places where there have been gifted men, where wit, refinement, malice are a part of happiness, where genius has almost necessarily made its home; they all possess an excellent dry air. Paris, Provence, Florence, Jerusalem, Athens – these names prove something: that genius is *conditioned* by dry air, clear sky – that is to say by rapid metabolism, by the possibility of again and again supplying oneself with great energy [*Kraft*]. (*EH* 'Clever' 2)

It was one thing to be sensitive to climatic and meteorological change. It was another to be *over*sensitive. Nietzsche knew – at the very least since his reading of Leopold Löwenfeld's *Moderne Behandlung der Nervenschwäche* (*The Modern Treatment of Nervous Illness*) in 1887 – that his professed ability in *Ecce Homo* to be able, for example, to 'read climatic and meteorological effects off from myself as from a very delicate and reliable instrument', to detect minute changes in humidity between Turin and Milan was a sign of chronic nervousness (*EH* 'Clever' 2). 'It is a familiar experience for doctors as well as laymen,' wrote Löwenfeld, 'that the nervous are influenced by weather conditions to an unusual degree.... [The neurasthenic] can be brought from a state of relative well-being to one of the most extreme discomfort through a change in the weather, with dizziness, headaches, aches and parasthesia, weakness of the limbs, sleeplessness and various other disorders arising as a result.'[31] These were all symptoms with which Nietzsche was familiar, and he himself certainly situated his chronic illness firmly within the contemporary discourse of nervous exhaustion. He wrote to Franz Overbeck in 1881: 'I have experienced terrible conditions; under the influence of the downright malignant and wild weather a general *décadence* has set in' (*KGB* III/1, 5 September 1881). He regarded his periodic bouts of illness as symptomatic of a '*thorough* nervosity' (*KGB* III/1, 22 February 1883). A few years later he wrote to the same friend, delivering an extremely detailed description of the state of his health. This, then, is how he finally came to think of his body, resorting to the metaphors of both machine and economy:

> Eternal headaches, eternal vomiting; a recrudescence of my old pains; masking profound nervous exhaustion when the whole machine is no good for anything. ... It is not just health that is lacking, but also the prerequisite for recovering my health – the vital energy is no longer intact. The losses incurred during at least 10 years can no longer be made good: during this time I have always lived off 'capital' and acquired nothing, absolutely nothing to replace it. But that makes one *poor*.... One does not catch up *in physiologicis*, every bad day *counts* ... I have ended up in an irksome and unsettled winter weather, which takes it out of me like a February in Basle. – This extreme irritability in response to meteorological impressions is *not* a good sign: it characterizes a certain general exhaustion, which is actually my real illness. Everything else, like headaches and so on, is only a *consequence* of this and relatively symptomatic.... I am certainly *not* suffering from a head complaint, *not* from a stomach complaint: but under pressure of nervous exhaustion (which is partly *hereditary*, – from my father, who also only died of *consequences* of a total lack of vitality – and partly acquired) the consequences appear in all forms. (*KGB* III/5, 4 July 1888)

Nietzsche's understanding of his illness is embedded in nineteenth-century discourses about the body. The combination of acquired and hereditary nervous exhaustion, an irritability exacerbated by bad weather, the decline in vital energy and the conviction that his health is irrecoverable – all these factors, as Nietzsche well knew, point to his degeneracy. This is all a far cry from his ideal of the genius, of the Dionysian artist, the polar opposite of the bourgeois body, given to excess and prodigality, with a total disregard for the miserly hoarding of life's energies.

IV

Nineteenth-century self-help books were not merely concerned with mental and physical well-being. What they attached most importance to – indeed, what ensured the soundness of mind and body – was *moral* hygiene. H.L. Mencken once wrote that hygiene 'is the corruption of medicine by morality. It is impossible to find a hygienist who does not debase his theory of the healthful with a theory of the virtuous. The whole hygienic art, indeed, resolves itself into an ethical exhortation.'[32] This is certainly true of the nineteenth-century philosophy of health. Illness did not exist in isolation from the will, from morality and from the ensemble of social forces that comprised the nation. Physiology was both theoretical and practical, both descriptive of the human body and prescriptive about its care. Consequently, codes of hygiene had all the moral force of the Decalogue. Herbert Spencer was adamant that the 'preservation of health is a *duty* … all breaches of the laws of health are *physical sins*'.[33] C.A. Meinert insisted that it was everyone's 'important duty' to 'preserve and to increase the valuable capital' of *Arbeitskraft*, which could only be achieved through 'sober, moral conduct and a rational nutrition'.[34] Carl Bock formulated rules for every aspect of the body's functioning, from maintaining the proper quality of the blood to cutting one's toenails. Illness, then, was a moral failing, and health was as much a state of moral perfection as morality was a means to, and external sign of, health. Nietzsche was all too aware of the systematic conflation of morality and health – his two enduring preoccupations – and his critique of Judaeo-Christian values, as well as his characteristic appeal to a medicalized vocabulary of health and sickness, must, as I have argued elsewhere,[35] be seen as a reaction to this confusion: a desire to divest physiology (and nature in general) of the moral apparel in which, for millennia, it has been disguised. Once God has been abolished and metaphysics made redundant, the only unimpeachable truths remaining belong to the dietetics of health, 'which have survived all scepticism and disintegration'. Hygiene, in other words, is the only secure foundation upon which future systems of knowledge may be built; it is from the healthy body that we must create new values (*HH* 22). Nietzsche may still demand that 'practical morality' will one day be the proper domain of physicians, but only after the pernicious and illusory concepts of good and evil are revealed to be nothing more than rational justifications of physiological states, of pleasure and pain, sickness and health, 'all those distempers, debilitations, excitations, the whole chance operation of the machine' (*D* 86).

Victorian culture conceived vitality not as the absence of disease, but in positive terms that made the quest for health a privileged form of self-development. Health was a state of constitutional growth and development in which the bodily systems and mental faculties interoperated harmoniously under the direct motive power of vital energy and the indirect motive power of the moral will. From the point of view of the individual, its signs were a sense of wholeness and unencumbered capability, and, on the level of society, the production of useful, creative labour.[36] What hygiene books effectively offered was a translation of the bourgeois ideology of self-containment and self-improvement into physiological terms. Through moderation, self-control and persistence, one becomes a subject, a rational, productive member of society. Daniel Schreber makes this point very clearly:

> To live *soberly*, *actively*, and *contentedly*, are the three rules of the Philosophy of Health, and if we obey them we may hope for a contented old age. And in closest connection with them are the commands of the Ethical Philosophy of life – 'Strive after full command over thyself, over thy spiritual and bodily weaknesses and wants. Begin this warfare bravely … – and persevere unweariedly in the struggle for this true (inward) freedom, for the perfection of self. …' For in true performance of these two commands – of the hygienic and of the ethical – lies the whole secret of the most difficult, but the most noble and the most important, of all sciences, – the science of life, the science of living *well*.[37]

How is this self to be perfected? By increasing one's energy. By fortifying the body through gymnastics, Schreber hoped also to increase the 'powers of will and action', to strengthen resolve and ward off the easy distractions of sloth and immorality.

Of course, Nietzsche, too, is concerned with self-development, and nowhere more so than in the sections entitled 'Why I am so Clever' and 'Why I am so Wise' in *Ecce Homo*. Both *Ecce Homo* and the dietetic handbooks of the nineteenth century are concerned with the 'little things' that have been hitherto ignored, with questions of the 'selection of nutriment, of place and climate, of recreation' (*EH* 'Clever' 8), and offer rules for life, for maintaining and expanding vitality. But Nietzsche's ambition is rather different. For him, of course, one must increase one's strength to overcome prevailing valuations, precisely those which are codified in the self-help books. He celebrates egoism, selfishness and self-discipline; the development of the self-legislating individual, not altruism or the bourgeois work ethic. And all this is to be achieved by dividing the day between work and rest, by choosing climate, by exercise or through a change in diet. The question he asks in *Ecce Homo* is: 'how to nourish yourself so as to attain your maximum of strength [*Kraft*], of *virtù* in the Renaissance style, of moraline-free virtue?' (*EH* 'Clever' 1). How does one overcome *ressentiment*, which is dangerous, he says, because it 'causes a rapid expenditure of nerve force, a morbid accretion of excretions, for example of gall into the stomach' (*EH* 'Wise' 6)?

If morality is, as he believes, merely an abstract and misleading interpretation of organic processes, a semiotics of unconscious bodily impulses, then any changes in diet and climate, in physical exercise, will bring with it a change in morals. Certain regimes are more beneficial than others in this regard. Just as overeating causes, according to Nietzsche, 'dissoluteness and over-excitability' (*D* 203), so undereating – or eating the wrong things – is equally dangerous. It leads to depression, to that 'fatigue and inertia' to which all religions are a rational response (*GM* iii 17). Christianity is the expression of malnourishment, and its ascetic rituals serve merely to reinforce and exacerbate the consequences of this restrictive intake: the 'diet is selected so as to encourage the morbid phenomena and to over-excite the nerves' (*A* 21). Similarly, though Nietzsche may on occasion praise Buddhist practices because they betray a superior knowledge of hygiene (as Richard S.G. Brown discusses in Chapter 2), he nevertheless attributes Eastern pessimism to the 'after-effects of some great dietary mistake', with Buddhism largely due to the excessive and almost exclusive rice diet of the Indians, which 'led to a general loss of vigour' (*GS* 134). Similarly, Nietzsche pours scorn on the attempt by his vegetarian brother-in-law, Bernhard Förster, to establish a German colony in Paraguay. Vegetarianism makes German intellectuals 'more irritable and moody'. 'Look at the "carnivorous" English: they have hitherto been the race, which has *best* founded colonies. *Phlegm* and roast beef – that has hitherto been the recipe for such "enterprises"' (*KGB* III/3, end of May 1885). The problem with vegetarianism – or the insubstantial diet of Christians – is that it does not increase one's vital force, at least not for the tasks which Nietzsche has in mind. Similarly, the German climate – cloudy, cold, Wagnerian – and the tropical heat of South America also sap one's energy. 'What is known of the moral effects of different foods?' asks Nietzsche in *The Gay Science*. 'Is there any philosophy of nutrition?' (*GS* 7). He was determined to provide the answers to these questions himself. As Nietzsche learned from Liebig and Meinert, input must match output in the human economy. The great tasks of the future would require large quantities of *Kraft*. Just as Feuerbach and Moleschott had already urged in 1850 that moral reform of the workers could be achieved by improving their diet, so Nietzsche, it seems, expected that Judaeo-Christian values could be overcome by abolishing the bad eating habits common in modern Europe.

In *Daybreak*, Nietzsche distinguishes between two different kinds of morality: a 'joyful and restless' one that accompanies an increase in nerve force and a gloomy one which is symptomatic of a decrease in nerve force (*D* 368). This first kind, then, is Nietzsche's new, post-moral ethics, one free of Judaeo-Christian *ressentiment* because it seeks to overcome the physiological preconditions for it to arise; it involves nothing more than the hygienic cleansing of the body – a '*physiological* purification and strengthening' (*KSA* 12, 5[61]) – and the concomitant increase in vital force. That the 'salvation' of humanity can be effected by attention to such trivial things suits Nietzsche's sense of the ridiculous. But of course, it is precisely because they have been ignored that they seem frivolous. What seems strange, though, is that Nietzsche's 'morality', his recommendations in *Ecce Homo*, differ so little from the conventional medical wisdom of the time. A sensible diet will promote a healthy digestion. Don't eat between meals. Abstain from alcohol and coffee. Live in a mild, dry climate. Don't overtax yourself. But can self-overcoming – the 'desire to become master, to become more, to become stronger' that is the essence of life as will to power (*KSA* 13, 14[81]) – be achieved through such moderation, the very principle of moderation that governs nineteenth-century dietetics?

Notes

1 Bruce Haley, *The Healthy Body and Victorian Culture* (Cambridge, MA: Harvard University Press, 1978), p. 3.

2 Pia Volz, *Nietzsche im Labyrinth seiner Krankheit: Eine medizinisch-biographische Untersuchung* (Würzburg: Königshausen und Neumann, 1990).

3 For a history of hygiene, especially in Germany, see Philipp Sarasin, *Reizbare Maschinen. Eine Geschichte des Körpers, 1765–1914* (Frankfurt am Main: Suhrkamp, 2001).

4 Carl Ernst Bock, *Das Buch vom gesunden und kranken Menschen*, 7th edn (Leipzig: Keil, 1866), p. viii.

5 Alexander Bain, *Mind and Body. The Theories of their Relation*, 2nd edn (London: Henry S. King, 1873), pp. 33, 38. Bain was not, of course, the only author with whom Nietzsche was familiar to employ the concept of nerve force. Georg Schneider (*Der thierische Wille* (Leipzig, 1880)) and Alfred Espinas (*Die thierischen Gesellschaften* (Braunschweig, 1879)), among others, also made use of it. The concept of nerve force crops up in several places in Nietzsche's notebooks (see, for example, *KSA* 9, 2[5], *KSA* 10, 7[239], *KSA* 11, 34[211]). The most interesting instance is Nietzsche's use of it to explain paranormal phenomena (*KSA* 10, 1[31]). On the history of the nineteenth-century preoccupation with nerves and nervous illness, see Janet Oppenheim, *'Shattered Nerves': Doctors, Patients, and Depression in Victorian England* (New York: Oxford University Press, 1991); Joachim Radkau, *Das Zeitalter der Nervosität: Deutschland zwischen Bismarck und Hitler* (Munich: Hanser, 1998).

6 Bock, *Das Buch vom gesunden und kranken Menschen*, pp. 479–80.

7 Daniel Schreber, *Medical In-Door Gymnastics* (London: Williams and Norgate, 1899), p. 16n. Compare also Bain's remarks on alcohol, tea, tobacco, opium and other narcotic stimulants: 'they are the large consumers, not the producers, of vitality; they expend our stock of power in nerve-electricity in a higher degree, and with more dangerous license, than the ordinary stimulants of the senses' (*Mind and Body*, p. 73).

8 Bain, *Mind and Body*, p. 21.

9 See Anson Rabinach, *The Human Motor. Energy, Fatigue and the Origins of Modernity* (Berkeley, CA: University of California Press, 1990).

10 Schreber, *Medical In-Door Gymnastics*, p. 16n.

11 Ibid., p. 9.

12 Herbert Spencer, *Education: Intellectual, Moral, and Physical* (London: Williams and Norgate, 1861), pp. 187 and 271.

13 F. W. Pavy, *A Treatise on the Function of Digestion, its Disorders, and their Treatment* (London: Churchill, 1867), p. 5.

14 Lionel Beale, *The Laws of Health, in Relation to Mind and Body* (London: Churchill, 1851), pp. 31–32.

15 For a cultural history of constipation and other digestive disorders in the nineteenth century, see James C. Whorton, *Inner Hygiene. Constipation and the Pursuit of Health in Modern Society* (Oxford: Oxford University Press, 2000).

16 Schreber, *Medical In-Door Gymnastics*, p. 6.

17 Wiel was the author of several dietetic cookbooks, which Nietzsche subsequently purchased: *Diätetisches Kochbuch mit besonderer Rücksicht auf den Tisch für Magenkranke*, 2nd edn (Freiburg, 1873); *Tisch für Magenkranke* (Karlsbad, 1875).

18 For a full list of the foods which Nietzsche found indigestible, see *KSA* 11, 34[13].

19 Mark R. Finlay, 'Quackery and Cookery: Justus von Liebig's Extract of Meat and the Theory of Nutrition in the Victorian Age', *Bulletin of the History of Medicine*, 66 (1992), 404–18.

20 On the influence of Rolph on Nietzsche's thought, see my *Nietzsche, Biology and Metaphor* (Cambridge: Cambridge University Press, 2002), pp. 46–54. Thanks to Thomas Brobjer for providing me with photocopies of Nietzsche's annotations to Foster's book.

21 Compare Feuerbach: 'Man eats … with the senses, too … The brain is the stomach, the digestive organ of the senses' ('Das Geheimnis des Opfers oder Der Mensch ist, was er ißt', in *Gesammelte Werke*, 14 vols (Berlin: Akademie-Verlag, 1967–2001), vol. 11 (1972), pp. 26–52 (pp. 44–45).

22 F. Parkes Weber, *Climatology, Health Resorts, Mineral Springs* (London: Rebman, 1901), pp. 17, 18.

23 In 1873 Nietzsche had borrowed Johann Müller's *Pouillet's Lehrbuch der Physik und Meteorologie* (Braunschweig, 1847) from the library of Basle University.

24 These ideas may have been influenced by Nietzsche's reading of an article by Werner Siemens, sent to him by Köselitz in 1883. Siemens alludes to the fact that the frequency of sun spots increases and decreases over a cyclical period of 10–11 years, and suggests that the consequent rise in electrical potential is responsible for such phenomena as the Northern Lights and terrestrial magnetism. See Werner Siemens, 'Ueber die Zulässigkeit der Annahme eines elektrischen Sonnenpotentials und dessen Bedeutung zur Erklärung terrestischer Phänomene', *Annalen der Physik und Chemie*, 20: 9 (1883), 108–36.

25 The concept of the 'weather prophet' which Nietzsche develops here appears in more elaborate form in *GS* 316. See also *WS* 182.

26 For a fuller discussion of the role of the weather in Nietzsche's self-understanding and his writings, see Sebastian Posth, 'Der meteorologische Komplex bei Nietzsche', unpublished MA thesis, University of Bochum, 2003.

27 Carl Sigmund, *Südliche klimatische Curorte*, 3rd edn (Vienna: Braumüller, 1875), p. 195. Nietzsche partially quotes this passage in a letter to Köselitz (*KGB* III/1, 25 February, 1884). The fruits of his reading of Sigmund can also be seen in a later letter to Overbeck: 'I am well informed about Tenerife. It has a drawback, despite many merits of the first rank: it is unequalled in its capacity to enervate and depress, by dint of its excessively humid air. In that respect it is the very opposite of Nice. Nevertheless, there are sick people on whom this composition of the air has a *calming* influence (Pau, Pisa, Palermo are related in this respect.)' (*KGB* III/5, 4 January 1887). Nietzsche also hints at his reading of Gerber in an earlier, similar remark to Paul Lanzky: 'I am very well informed about Ajaccio – humidity similar to Pisa and Corfu' (*KGB* III/3, beginning of August 1886). Lanzky, co-owner of a hotel in Vallombrosa and an admirer of Nietzsche who would later write a book in imitation of *Daybreak* called *Sunset* (*Abendröte*), was an important source of information about the climates of various European resorts.

28 Sigmund, *Südliche klimatische Curorte*, p. 195. In a letter to his sister, Nietzsche complained that 'all the famous "winter health resorts" are intended for those suffering from chest complaints: nothing for *me*!' (*KGB* II/5, 24 July, 1879).

29 Sigmund, *Südliche klimatische Curorte*, pp. 298–99.

30 Compare Sigmund, *Südliche klimatische Curorte*, p. 281.

31 Leopold Löwenfeld, *Moderne Behandlung der Nervenschwäche* (Wiesbaden: Bergmann, 1887), pp. 45–46. See also Willy Hellpach, *Geopsyche*, 5th edn (Leipzig: Engelmann, 1939), p. 271.

32 H.L. Mencken, *Prejudices. Third Series* (London: Jonathan Cape, 1923), p. 269.

33 Spencer, *Education*, pp. 229–30.

34 C.A. Meinert, *Wie nährt man sich gut und billig? Ein Beitrag zur Ernahrungsfrage* (Mainz: Mittler, n.d.).

35 See my *Nietzsche, Biology, and Metaphor*, chapter 2.

36 Haley, *The Healthy Body*, p. 21.

37 Schreber, *Medical In-Door Gymnastics*, p. 16n.

Chapter 4

Metaphor, Perception and Consciousness: Nietzsche on Rhetoric and Neurophysiology

Christian J. Emden

On several occasions throughout his writings Nietzsche emphasizes that knowledge is ultimately dependent on metaphor. His numerous reflections on the epistemological status of metaphor, and on the complex relationship between language and knowledge, have generated a considerable amount of philosophical criticism. Much of this criticism concentrates on Nietzsche's views on metaphor in terms of a philosophy of language, interpretation or rhetoric. I wish to argue, however, that Nietzsche's understanding of metaphor is far more complex than is generally assumed and that it leads in fact to a rhetorical model of the mind: Nietzsche – perhaps surprisingly at first glance – links the epistemological implications of metaphoricity to nineteenth-century debates in the field of neurophysiology.

For Nietzsche, metaphor is a rather broad concept with a range of different meanings, and he is able to adopt this position because all attempts to limit the meaning of 'metaphor' tend to be flawed and often tautological. It is easy to see that Aristotle's definition of *metaphorá* as *epiphorá* is itself of a metaphorical nature,[1] and a similarly tautological argument appears in Eva Feder Kittay's otherwise very rigorous study on the subject: 'metaphorical expressions can be used to refer metaphorically'.[2] Nietzsche is aware of the confusion which tends to bedevil most philosophical accounts of rhetoric, and he is therefore willing to abandon all rigid designations of the term 'metaphor' in order to examine the epistemological implications of this concept. He shifts metaphor from specific speech acts to a broader epistemological category at the beginning of his *Darstellung der Antiken Rhetorik* (*Account of Classical Rhetoric*, 1872–74) when he turns his attention to the relationship between nerve stimulus and linguistic utterance (*KGW* II/4, p. 426). But although this may seem a surprising direction for his lectures and notes on rhetoric to take, his interest in the physiological origin of speech acts has far-reaching consequences for his understanding of perception and consciousness. According to Nietzsche, mental activity is directly dependent on manifold forms of signification. Such an understanding of mental activity is hardly new, since Aristotle already notes in *De anima* that thinking is essentially based on the use of signs and images,[3] and recent approaches in philosophical psychology and cognitive science have characterized 'thinking' as a form of 'inner verbalization', or at least as the formation of concepts.[4]

At first sight, it seems that Nietzsche follows these commonplaces, but upon a closer look we cannot but realize that he has a very specific understanding of mental activity, one that is based on rhetoric. He enriches this rhetorical dimension with concepts from the fields of (neuro)physiology and psychology. Already in Arthur Schopenhauer's *Die Welt als Wille und Vorstellung* (*The World as Will and Representation*, 1819), Friedrich Albert Lange's *Geschichte des Materialismus* (*History of Materialism*, 1866) and Eduard von Hartmann's *Philosophie des Unbewussten* (*Philosophy of the Unconscious*, 1869), he had encountered numerous and, at times, very detailed expositions of physiology and psychology, and in Johann Carl Friedrich Zöllner's *Ueber die Natur der Cometen* (*On the Nature of Comets*, 1872) he could furthermore find a general exposition of Hermann von Helmholtz's understanding of perception, which dominated debates in physiology in nineteenth-century Germany to a considerable extent. But so far it has gone unnoticed that Nietzsche relates these discussions to his reflections on language and rhetoric, and I shall argue that this leads to a rhetorical understanding of the mind. In order to understand this relation between rhetoric and neurophysiology, it will also be necessary to consider some of Nietzsche's own sources as well as the wider intellectual constellations within which he moves.

Metaphor, Translation and the Physiology of Perception

Both physiology and psychology emerged as independent fields of research between the late eighteenth and the mid-nineteenth centuries. Whilst neurophysiology developed out of the discourse of medicine, psychology detached itself from philosophical thought but, despite their different origins, in the nineteenth century they moved along a common trajectory: the functional description of mental activity.[5] In his programmatic *Psychologie als Wissenschaft* (*Psychology as Science*) of 1824–25, Johann Friedrich Herbart demanded a model of psychological analysis along the lines of the experimental paradigms within the natural sciences, and in his equally important *Elemente der Psychophysik* (*Elements of Psychophyics*) of 1860, Gustav Theodor Fechner followed Herbart's suggestions and proposed a scientific reassessment of traditional psychological commonplaces in terms of the functional relationship between physical and mental factors.[6] This attempt by physiology and psychology to establish themselves as empirical sciences did not go unnoticed by contemporary philosophers, and Nietzsche incorporated many aspects of these emerging scientific disciplines into his own philosophical enterprise. His lively interest in physiological issues has attracted a considerable amount of scholarly attention. Most of these discussions, however, limit Nietzsche's philosophical appropriation of physiology to the period between 1878 and 1888 and focus either on his reflections on aesthetics and morality or on his conceptions of decadence and the human body.[7] In contrast, I wish to emphasize that Nietzsche's interest in physiology – especially the physiology of perception – begins much earlier, namely in the context of his studies on language and rhetoric. This early interest has, in fact, a profound influence on his understanding of the mind and on his account of the relationship between the self and its environment.

In his *Account of Ancient Rhetoric*, Nietzsche presents us with a detailed description of the relation between reality and language. He argues that linguistic reference must fail because it does not refer to reality or knowledge but only to stimuli and beliefs:

> In using language, man does not perceive things or occurrences [in nature], but only stimuli: he does not express sensations, but merely images of sensations. The sensation, triggered by a nervous stimulus, does not provide any access to the thing itself: this sensation is represented externally by an image.... (*KGW* II/4, p. 426)

In this passage, Nietzsche argues for a physiological origin of belief that is also outlined, for instance, in Rudolph Hermann Lotze's *Medicinische Psychologie oder Physiologie der Seele* (*Medical Psychology or the Physiology of the Soul*, 1855), Otto Funke's *Lehrbuch der Physiologie* (*Textbook of Physiology*, 1855–57), and especially in Lange's *History of Materialism*. For Lange, perception is already a form of abstraction and results in mental representations.[8] These representations correspond to Gustav Gerber's argument in his *Die Sprache als Kunst* (*Language as Art*) that we can think only in chains of mental images and that these images necessarily structure our linguistic abilities.[9] Gerber's book in particular, which attempts to link the aesthetic dimension of language to wider epistemological issues, had a lasting influence on Nietzsche's rhetorical ideas, although it is easy to overestimate Gerber's actual influence: his *Language as Art* is in many ways compiled from other sources, which provided Nietzsche with a great deal of guidance in the philosophy of language.[10]

Appealing to Lange, Gerber and other contemporary sources, Nietzsche is able to assert that language is a representation of abstract images produced by perception. In other words, he enriches rhetoric with physiology. This paradigm of representation (mental and linguistic) is, it seems, based on the one hand on the Aristotelian idea that perception (*aisthésis*) is related to images (*phántasmata*) and, on the other, on Lange's and Johann Carl Friedrich Zöllner's insistence that there is no such thing as pure perception.[11] Nietzsche, however, criticizes Zöllner for overlooking the rhetorical and linguistic nature of mental representations. In a series of fairly idiosyncratic essays on the history of science and the theory of knowledge, published under the somewhat misleading title *On the Nature of Comets*, Zöllner argues that perception is immediately followed by unconscious judgements.[12] In contrast, Nietzsche largely rejects any notion of unconscious reasoning as a mental automatism (*KSA* 7, 19[217]). But he also takes this cognitive and linguistic scepticism one step further and presents us with a radical description of perception in terms of rhetoric. In order to understand Nietzsche's idea of the rhetorical foundation of language in perception, it is necessary to shift our attention once more from his *Account of Ancient Rhetoric* to the essay *On Truth and Lie in an Extra-Moral Sense*. But since the essay on truth and lie is essentially based on the concept of metaphor, we must first examine the meaning of this concept.

The Rhetoric of Metaphor

In his *Account of Ancient Rhetoric*, Nietzsche quotes Gustav Gerber's *Language as Art* and declares that ordinary language is always figurative: 'Eigentlich ist alles Figuration was man gewöhnliche Rede nennt' (*KGW* II/4, p. 427).[13] He subsequently narrows down this definition and states that all tropes are, in one way or another, 'translations' (*Übertragungen*); that is, on the level of *elocutio* words are accordingly replaced by other words with a different semantic framework (*KGW* II/4, p. 449).[14] Now, the chief model for such a translation or transposition is metaphor – a model which he traces back to Isocrates and Aristotle (*KGW* II/4, p. 443).[15] But Nietzsche obviously recalls that he has to introduce his students to ancient rhetoric, and he therefore goes on to define metaphor in classical terms, without further philosophical speculation, as 'a shorter simile' (*KGW* II/4, pp. 443 and 462) and urges caution with regard to the excessive use of metaphor. A closer reading reveals, however, that he is more interested in a totally different aspect of metaphor, namely its definition as *translatio* or *Übersetzung*.

Both Quintilian and Cicero note that metaphoric speech is a necessary element of language because sometimes we have no choice but to use metaphor if a correct *verbum proprium* does not exist.[16] Certainly, Cicero goes on to argue, metaphor ultimately became a pleasing and entertaining mode of speech, but the true origin of metaphor lies in the requirement to express things which are not represented by a *verbum proprium*.[17] Nietzsche is aware of this argument (*KGW* II/4, p. 442) and he agrees with Cicero that the indeterminacy of language is the source of figurative, tropical and, ultimately, metaphoric speech. Undoubtedly, this argument is indebted to Cicero, who notes in *De oratore*: 'Nihil est enim in rerum natura cuius nos non in aliis rebus possimus uti vocabulo et nomine.'[18] That there is nothing in nature we cannot designate otherwise than with a literal use of language means, for Cicero, that we do not have to call a tree 'tree' and a snake 'snake', but that we can, and sometimes must, use different words for one and the same thing. In this sense, metaphor itself means nothing more and nothing less than an act of transference or transposition. When Nietzsche discusses metaphor in his *Account of Ancient Rhetoric*, he focuses less on its classic definition as simile (*KGW* II/4, pp. 443 and 462), but – on the basis of Richard Volkmann's and Wilhelm Wackernagel's rhetorical textbooks – he stresses the 'literal' meaning of 'meta' plus '*phérein*' as '*Übertragung*' (*KGW* II/4, pp. 443 and 449).[19] Nietzsche also encounters this description of metaphor as 'Übertragung' in Gerber's *Language as Art*,[20] but, again, his primary sources are Aristotle, Cicero and Quintilian. In his *Rhetoric*, Aristotle defines *metaphorá* as *epiphorá* which Cicero and Quintilian 'translate' into Latin as *translatio* and which becomes a commonplace in the rhetorical textbooks of the Western tradition from Augustine onwards.[21]

Signa translata, Stimuli and Transmissions

The description of language as a form of transference or translation has a further perspective which Nietzsche was aware of – namely the eighteenth-century discussion regarding the origin of language. In this debate, archaic languages are usually seen as rich in images, symbols and tropes, or, as Rousseau remarked, 'the first language must have been of a figurative nature'.[22] One prominent participant in this discussion was Johann Georg Hamann, and Nietzsche read part of Hamann's writings between January and March 1873 while he was working on his lectures on rhetoric as well as on the essay on truth and lie.[23] He will not have missed

Hamann's *Aesthetica in nuce* of 1762, in which the author criticizes the rationalistic exegesis of the scriptures and instead introduces the idea that language is a form of epiphanic translation, a passage through different spheres according to a teleological and theological hierarchy.[24] Hamann's thesis, which is to some extent prefigured in Augustine's *De doctrina Christiana* and Aquinas's *Summa Theologiae*, states that all signs are *signa translata*: speaking is translating. For Nietzsche, this is an important idea, even though he does not share Hamann's neo-Platonic tendencies. That linguistic activity is a kind of translation is an argument which runs through the Western tradition from Aristotle to Hamann and beyond, and for Nietzsche 'metaphor' is the rhetorical paradigm of 'translation' or 'transference', so that all language is, ultimately, metaphorical in nature. Stephen Houlgate has maintained that this understanding of metaphor is 'extremely loose',[25] but in what follows, I wish to argue that there are very good reasons why Nietzsche uses metaphor in such a broad sense.

As can be seen with regard to Aristotle and Augustine, ancient philosophy of language and patristic hermeneutics focused mainly on the relation between language and reality, and not so much on the more complicated connection between language and thought.[26] This situation changed in the aftermath of Antoine Arnauld's and Pierre Nicole's *Logique de Port-Royal* (1662) and John Locke's *An Essay Concerning Human Understanding* (1690) when, for instance, Johann Georg Sulzer and Johann Gottfried Herder paid considerable attention to the interdependence of language and thought. Despite the emergence of physiology, psychology and phrenology in the late eighteenth century, and despite the success of such works as Samuel Thomas Soemmering's *Über das Organ der Seele* (1796) or Franz Joseph Gall's and Gaspard Spurzheim's *Recherches sur le système nerveux en général* (1803), the connection between language and mental states continued to be a very serious problem for any reflection on language.[27] Nietzsche himself attempts to solve this problem by linking his metaphorical understanding of language to the findings of nineteenth-century neurophysiology, and we have to understand his definition of metaphor as *Übertragung* in precisely this context.

In the mid-nineteenth century, physiology experienced an unprecedented explosion of research and became one of the main fields of scientific endeavour. *Übertragung* is the term with which nineteenth-century physiology denotes the relation between nerve stimuli and mental states: thinking is the result of electromotoric transmissions. Let us remember that, when Nietzsche tries to describe the relationship between perception and language as the basis of rhetorical tropology, he assumes a break between, on the one hand, physical events and, on the other, mental states. The question now arises of how physical events, stimuli, sensory perceptions and mental representations are connected with one another. In his lectures and notes on rhetoric, Nietzsche does not answer this question, because it is sufficient for him to note here that linguistic utterances do not represent physical events but only mental states. He does not, however, ignore the problematic connection between perception and language, but, in his *On Truth and Lie in an Extra-Moral Sense* and in his notebooks of 1872–73, rephrases it in physiological terms as a form of transference, translation or transmission from nerve stimuli to linguistic utterances: the path from perception to language is a 'metaphoric' process. Thus, he is able to enrich his rhetorical findings concerning the tropical nature of language with physiological 'evidence', but in order to understand the full extent of this argument and its development, it is again necessary to return to Nietzsche's sources and the intellectual contexts of his time.

Schopenhauer and the Physiology of the Mind

When, in October–November 1865, Nietzsche began to read Schopenhauer's *The World as Will and Representation*, he found in the second volume a thorough philosophical critique of Locke's and Kant's theories of perception, understanding and knowledge. He encountered not only Schopenhauer's rejection of any privileged, direct and immediate access to reality, but also numerous passages in which Schopenhauer elaborates his critique of traditional epistemological theories with the help of examples from physiology.[28] In this context, Schopenhauer also turns his attention to the relationship between sensory perception and mental processes, and he argues for the materialist point of view that our perception of reality is the product of physiological processes, and hence that mental representations of the world are nothing but brain functions.[29] Schopenhauer goes on to exemplify this point with reference to the physiology of the brain. The central nervous system is supposed to consist of two different kinds of nerves: on the one hand, there are nerves for the perception of external stimuli and, on the other, there are nerves which serve only the 'transmission' of stimuli.[30] Schopenhauer follows the Aristotelian principle 'nihil est in intellectu, nisi quod antea fuerit in sensu', and stresses the close connection between nerve stimuli in perception and their mental representation: perception is an immediate and non-conceptual physiological act, and it is marked by a transmission between nerves and brain.

For Nietzsche, this is an important point, because in so far as perception is a translatory process between nerve stimuli and consciousness, it can be described in rhetorical terms as metaphorical. And, after his reading of Schopenhauer, Nietzsche in fact finds more 'evidence' for this argument in the nineteenth-century physiological and scientific treatises to which he turns during his years as a student in Bonn and Leipzig and as a young professor in Basle. This scientific interest persists until the late 1880s and is not unusual in nineteenth-century philosophy. He is here responding to an intellectual environment in which Hermann von Helmholtz's *Die Thatsachen der Wahrnehmung* (*The Facts of Perception*, 1879), Gustav Theodor Fechner's *Elemente der Psychophysik* (*Elements of Psychophysics*, 1860) and Franz Brentano's *Psychologie vom empirischen Standpunkt* (*Psychology from an Empirical Standpoint*, 1874) become more dominant than the seemingly more speculative philosophical systems that shaped German Idealism until its slow demise in the first half of the nineteenth century.[31] Lange's *History of Materialism* provides Nietzsche with a contrasting analysis of the philosophical implications of the natural sciences in the nineteenth century, and in the 1880s he continues to base many of his philosophical reflections on, for instance, Johannes Vogt's *Die Kraft* (*Force*, 1878), Wilhelm Roux's *Der Kampf der Theile im Organismus* (*The Struggle of the Parts in the Organism*, 1881) and Michael Foster's *Lehrbuch der Physiologie* (*A Textbook of Physiology*, 1881). In this context, Nietzsche encounters a more detailed account of the translatory relationship between physical events and mental representations, and, even more important, this relationship is almost always described in terms of *Übertragung*. This *Übertragung* is conceptually based on the phenomenon of electricity, which had been very much in vogue in the intellectual circles of German Romanticism, as well as in the psychological theories of Franz Anton Mesmer, and it became one of the main commonplaces of physiological research in the nineteenth century.[32] In order to understand Nietzsche's rhetorical concept of metaphoric *Übertragung* in the context of his physiological interpretation of mental activity, we have to turn our attention to the status of electricity in nineteenth-century physics and neurophysiology.

'Organic Electricity' and the 'Inductive Sciences'

During the second half of the eighteenth century, some of the most fundamental issues in physics focused on the phenomena of electricity and magnetism, which could not be successfully explained in terms of the dominant Newtonian mathematization of physics. As a consequence, studies such as Benjamin Franklin's widely-read *Experiments and Observations on Electricity* (1751–54), Joseph Priestley's *The History and Present State of Electricity* (1767) and Johann Christian Polycarp Erxleben's *Die Anfangsgründe der Naturlehre* (*The Rudiments of the Study of Nature*, 1772) increasingly shifted scientific attention from celestial mechanics and the properties of light to the so-called 'imponderable' phenomenon of electric flows and paved the way for the fundamental research of Alessandro Volta.[33] In particular, Volta's invention of the 'electrophorus' – a device generating static electricity – and his renowned lectures at the University of Padua from 1778 to 1784 moved the phenomenon of electric flows to the forefront of late eighteenth-century physics.[34] His research also initiated a remarkable controversy with Luigi Galvani. Whereas Volta maintained that electric flows should be understood in terms of 'metallic electricity', Galvani challenged this view in his *De viribus electricitatis in motu musculari commentarius* (1791) and argued that electric flows are mainly an organic phenomenon. Galvani attempted to prove his theory with a famous experiment carried out in 1786, which investigated the effect of an electric current on the muscles in the leg of a frog. This experiment involved an arc made of two different metals, one of which was placed in contact with a nerve and the other with a muscle. The electric current between the two metals would stimulate the nerve and thus move the muscle. However, Volta pointed out that the electric current itself is actually independent of the frog's nerve and muscle, which only act as conductors.[35] But although Volta's experimental studies and his explanations proved to be more successful, Galvani's theory of 'animal electricity' had a profound impact on the understanding of electric flows in the context of German Romantic *Naturphilosophie*, so that – much like Mesmer – Johann Wilhelm Ritter came to regard electricity as an immaterial force of nature which pervades both the inorganic and the organic world.[36]

In the aftermath of Ritter's somewhat speculative suggestions, nineteenth-century research into the phenomenon of electricity is marked by an understanding of physics as an exact and unified science of nature, and the seminal work of Hans Christian Oersted, Michael Faraday and James Clerk Maxwell aimed at an experimental, as well as a mathematical, explanation of electric currents, electromagnetism and electrodynamics.[37] This approach was especially successful in Germany, where the continuing institutionalization of the physical sciences led to an unprecedented explosion in experimental research.[38] Although Nietzsche was certainly aware of these developments through his ongoing interest in the natural sciences, we are safe in assuming that his knowledge of nineteenth-century physics was rather limited. Research into electric currents was, however, not confined to physics, for Galvani's and Ritter's ideas about 'animal electricity' – or, more precisely, 'organic electricity' – received renewed attention in the context of physiology, so that, for instance, in his voluminous *Untersuchungen über thierische Electricität* (*Investigations in Animal Electricity*, 1848–60), Emil Du Bois-Reymond attempted to apply the results of physical research into the phenomenon of electricity to the physiology of animals.[39] Under the influence of the Berlin school around Hermann von Helmholtz, which aimed at an understanding of physiology according to physical principles, physiology itself emerged as one of the central scientific disciplines in nineteenth-century Germany, and in this context, the idea of measurable electric currents in organic material, such as human nerve fibres, played an essential role.

In nineteenth-century neurophysiology, the concept of electricity was used on two different levels. On the one hand, artificially induced electric flows in the brains of animals (for example, dogs) were examined in order to explain the connection between certain mental states, such as pain, and motor actions, such as a sudden movement of the leg.[40] On the other hand, similar methods were used in order to localize mental operations in specific regions of the brain according to a topographical model, which can be found, for instance, in the work of Pierre Paul Broca, Pierre Flourens and David Ferrier.[41] Since Nietzsche himself was a keen reader of neurophysiological studies such as Eduard Hitzig's *Untersuchungen über das Gehirn* (*Investigations into the Brain*, 1874) and philosophical discussions of the emerging brain sciences in Lange's *History of Materialism* (1866) as well as the German translation of Alexander Bain's *Mind and Body* (1873), we can assume that he was quite familiar with these specific aspects of nineteenth-century neurophysiology, even if he did not have enough specialist knowledge to follow the discussions in every detail.

The translatory model of nervous processes as electric transmission in nineteenth-century physiology was not an isolated phenomenon, and it is not particularly surprising that, at the very same time that electrophysics and physiology were dealing with the problem of conductors and electric inductions, philosophers became increasingly fascinated by the concept of induction as a principle of logic. William Whewell, one of the founders of the philosophy of science at Cambridge, attempted to formulate inductive logic as the methodological foundation of the natural sciences.[42] The most famous example of this trend in the philosophy of science, however, is John Stuart Mill's *A System of Logic, Ratiocinative and Inductive* (1843). On the one hand, Mill follows Whewell and others in that he seeks to establish inductive logic as a form of scientific reasoning, but he also thematizes the close relationship between inductive and analogical judgements, which can also be found in the rhetorical treatises of Cicero, Quintilian and Aristotle.[43] Furthermore, Mill presents us with a model of the mind, which is indebted to the psychology of association in the tradition of Locke and Hume, so that induction itself is at least to some extent a fundamental feature of mental operations.[44]

Induction, association, analogy and translation build the theoretical framework of Nietzsche's intertwining of metaphor and mental processes, which itself is based on the idea that the relationship between nerve stimuli and mental activity can be explained according to the relatively new assumption that nerve fibres serve as conductors for electric flows. This background is of crucial importance for Nietzsche's rhetorical concept of a metaphoric *Übertragung* between nerve stimuli, mental representations and language. He could find this idea in a variety of publications, but in the following paragraphs I shall limit myself to some central examples, which Nietzsche either read himself or was at least aware of.

Nerves, Images and Language

Let us first consider Ludwig Büchner's *Kraft und Stoff* (*Force and Matter*, 1855) and Rudolph Hermann Lotze's *Medical Psychology* (1852). Büchner refers to the physiology of the brain in some detail, although he prefers to discuss these issues in a more general manner. He emphasizes a constant 'metamorphosis' or 'transposition' of forces in nature which he then exemplifies with reference to the nervous system: mental processes can be explained in terms of 'electric currents'.[45] In contrast to Büchner's more general account, Lotze's study is far more precise in that he aims at a very detailed understanding of nerve processes.[46] Already in his *Allgemeine Physiologie des körperlichen Lebens* (*General Physiology of Corporeal Life*, 1851), he presents an account of human physiology, and especially neurophysiology, which he then discusses in more detail in his *Medical Psychology*.[47] Although Lotze is, to some extent, quite sceptical about the concept of 'organic electricity', he regards nerve processes as transitions between different fibres, which result in a 'transmission' from an initial stimulus to an ensuing form of mental activity.[48] He finally notes that 'nerve processes' are the 'medium between the external world of stimuli and the internal world of representations', and he goes on to describe the step from an external stimulus to internal mental representations as a continuous process of transmission.[49] This physiological 'translation' and 'transformation' of nerve processes advances, according to Lotze, in six steps: from external reality to stimuli, from stimuli to the nerve endings, from there to the various nerve processes which lead to the brain and trigger an unconscious mental state, which itself turns into a specific perception.[50] Although Lotze apparently has to assume a direct physiological access to reality, he concludes that such a direct access is impossible, since any act of perception – and of mental activity in general – involves too many complicated neurophysiological transpositions.

This translatory description of sensory perception is explained in even more detail in Otto Funke's *Lehrbuch der Physiologie* (*Textbook of Physiology*, 1855–57), which Nietzsche read in 1870 – that is, shortly after he turned his attention to linguistic problems in his *Vorlesungen über lateinische Grammatik* (*Lectures on Latin Grammar*) and while he was already working on his later lectures on rhetoric. We can assume that Funke's voluminous work had some impact on Nietzsche's scientific knowledge as well as on his concept of *Übertragung*. In fact, Funke is quite explicit and describes the 'nerve fibre' as a 'conductor'.[51] In this respect, Funke concludes, perception is quite literally a 'transmission from fibre to fibre'.[52] This physiological description of sensory perception as a translatory process is a common argument in nineteenth-century thought and can also be found in the writings of Hermann von Helmholtz, in Fechner's treatise on psychophysics and in Hartmann's study on the philosophy of the unconscious.[53]

The materialist tendency of these physiological reflections is obvious, and it is in Lange's *History of Materialism* that Nietzsche finds the ultimate connection between perception, nerve processes and linguistic utterances. Lange, who concludes that the physiology of sensory perception is an extremely important field, uses almost all the descriptive terms of psychophysics that we have encountered above,[54] but in contrast to the physiological studies of, for instance, Lotze and Funke, he also stresses the connection between the nervous system and language. After deconstructing the concept of the 'soul' and replacing it with the concept of the 'brain', he notes that, in the nerve fibres, we not only find 'conductors' and 'electric currents', but that they also have a direct 'effect on the contraction of muscles'.[55] Among the muscles which Lange has in mind are certainly mouth, tongue and the laryngeal muscles as well as the 'respiratory muscles' which are set in motion by 'nerve currents'. Not only are mental states such as thinking or feeling physiological functions, but so are the foundations of language. Hence, Lange can assume a 'peculiar connection between thinking and speaking'. This is a crucial point, and Nietzsche will not have overlooked it, especially since Lange describes these nerve processes as a constant 'fluctuation of stimuli'.[56]

When Nietzsche turned to Gerber's philosophy of language, he encountered the final stage of this argument: for Gerber, linguistic activity has its origin in the physiological processes of perception.[57] Nietzsche integrates this point into his lectures on rhetoric when he argues that perception is not so much based on external reality as on stimuli which produce an image of reality. In this context, he remarks that perceptions do not refer to things but only to stimuli which produce images (*KGW* II/4, p. 426).

The Rhetoric of the Mind

At this point, one might well ask why Nietzsche speculates about such issues in lectures which are otherwise dedicated to Greek and Roman rhetoric. We have already seen that the most important aspects of rhetoric – namely, the tropical nature of language – are forms of transference and transposition, and we came to the preliminary conclusion that metaphor becomes, for Nietzsche, a master trope in so far as it accurately describes these forms of transference. We can now see that, in addition, he was able to apply the same model to the relation between perception and language. The steps from stimulus to language involve two distinct transpositions from one realm of experience into another. This argument, which Nietzsche proposes in his detailed lectures and notes on rhetoric under the wider influence of nineteenth-century physiology and on the basis of Lange and Gerber, is fully expressed in his essay *On Truth and Lie in an Extra-Moral Sense*: 'To begin with, a nerve stimulus is transferred into an image: first metaphor. The image, in turn, is imitated in a sound: second metaphor. And each time there is a complete overleaping of one sphere, right into the middle of an entirely new and different one' (*TL* 1, *KSA* 1, p. 879). It is obvious that the concept of transference or transposition plays a crucial role here. The physiological movement from stimuli to linguistic representations is described in even more detail in a fragment of 1872–73, which corresponds to the accounts presented by Lotze and Funke:

> Stimulus – mnemonic image
> connected by means of metaphor (analogical inference).
> Result: similarities are discovered and revitalized. In a mnemonic image the *repeated* stimulus occurs once again.
> *Stimulus perceived* – then *repeated*, in many metaphors, whereby related images from various categories throng together. Every perception achieves a multiple imitation of the stimulus, but transferred into different realms.
> Stimulus sensed
> transferred to related nerves
> repeated there, in transferred form, etc.
> What occurs is a translation of one sense impression into another: ... This is a wholly universal phenomenon. (*KSA* 7, 19[227])

The transition from perception to language accordingly proceeds in a 'metaphorical' manner; it is a 'transition from *image* to *image*' (*KSA* 7, 19[107]) and does not start with any kind of external reality or some pseudo-Kantian thing-in-itself. Nietzsche, and this is an important point, does not maintain that psychological states *are* metaphors: they are undoubtedly biological phenomena, and he is plainly aware of this, for he endorses a biologization of knowledge – any form of experience, from ethical feelings to our conception of reality, must be seen as the result of 'our own nervous processes' (*KSA* 9, 10[E95]).[58] Thus, he is later able to point out that the functions of the 'mind' are merely results of organic functions (*KSA* 11, 25[356]). But he also maintains that the interaction of psychological states and our awareness of them can be efficiently described in metaphorical terms and according to a rhetorical model.

This approach distinguishes Nietzsche's position very clearly from that of Lange and Gerber. Whilst Lange is certainly able to provide a relatively detailed and critical insight into the physiological dimension of perception, he is relatively unaware of its implications for any philosophy of language. At first sight, Gerber seems to take a step in the right direction in his occasional references to the question of perception, especially when he describes perceptions in general terms as 'metaphors' or 'analogies'.[59] In fact, his approach lacks any real foundation in the natural sciences, let alone physiology. In contrast, Nietzsche is much more aware of the epistemological implications of physiology for linguistic and rhetorical thought, and he realizes that rhetorical models are also applicable to a description of physiological processes.

This does not mean that Nietzsche is a kind of eliminative materialist in so far as he reduces mental events to neurological processes. The question whether he is indeed a 'materialist' or, as Maudemarie Clark has recently claimed, an 'empiricist',[60] is in fact beside the point, especially with regard to the development of his linguistic thought. He merely uses materialist arguments when they seem reasonable and rejects them if they seem problematic. What is at stake here is rather his emphasis on the tropical character of language and, by extension, the tropical or metaphorical character of perceiving and knowing. In this respect, he shares Gerber's assumption that 'pure thought' is a 'fantasy'.[61] But even though Gerber declares that we think in 'chains of images',[62] paralleling Nietzsche's statement that thought is nothing but the 'interconnection' of symbols (*KSA* 7, 5[80]), Gerber's argument moves within the boundaries of spoken and written language. In contrast, Nietzsche's vision of rhetoric is a springboard leading to questions regarding the complex relationship between physiology, thought and language. It is in this sense that Nietzsche can conclude that knowledge is dependent on metaphor (*KSA* 7, 19[228]).

And What about Consciousness?

Many philosophers before Nietzsche, especially Descartes, Locke and Kant, did not regard the self as a particularly difficult problem, since it was always seen as the main point of reference for human action. For many early modern European conceptions of the self and rationality which present 'ratio' as a constituent of the soul – such as Christian Wolff's *Psychologia rationalis* (1734) – (self-) consciousness had to be taken as a given if rationality and reason were to make any sense. Naturally, reason was located in the self, and the latter necessarily had to be constructed as unified. The groundwork for this view was, to some extent, laid by Descartes's *Meditationes de prima philosophia* (1641), in which *cogitare* as the central function of the *sensus communis* combines such diverse human features as *dubitare*, *intelligentia*, *affirmare*, *volere*, *nolere*, *imaginatio*, *sentire* and so forth.[63] In his *Einleitung zu der Vernunfft-Lehre* (*Introduction to Logic*, 1691), Christian Thomasius also adopts this model and supplies the anthropological connection between reason and the integral self. The use of mental faculties such as reasoning, imagining and reflecting, he suggests, is only possible if we already have a kind of *a priori* understanding of our own wholeness.[64] This concept of wholeness became especially fashionable in eighteenth-century anthropology. As an attempt to construct a unified science of mind and body, eighteenth-century anthropology had to presuppose that the self exists as an entity which can be explained by various methods, including theories of evolution, phrenology, morality, physiognomy and aesthetics. Such models of the self and of consciousness run into serious difficulties if we accept Nietzsche's claim that reason is dependent on the tropical nature of language and that philosophical concepts are of linguistic origin. This also implies that the self, our idea of the mind, and to a considerable extent even our idea of the body as located in space and time are all connected to language. We can accordingly expect Nietzsche to reject any notion of subjectivity and consciousness which relies on an immediate connection between 'being' and 'thinking'. During the late 1870s and in the context of his criticism of metaphysics, he becomes increasingly interested in the concepts of 'mind' and 'consciousness', and in the early 1880s he finally discusses the inadequacies of Descartes' assumption of a self-contained *cogito*. Self-awareness, Nietzsche points out, is not dependent on a privileged access to 'pure thought', but it is a result of our immersion in language and the physiological processes of the brain.

In his *Meditationes* and elsewhere, Descartes had argued that self-consciousness, and therefore selfhood, is a product of our awareness that we think.[65] Hence, *cogitare* distinguishes us from all other existing matter: thinking is self-contained and has no *extensio* in the way that trees, stones and rivers do.[66] Although he seems to stress the close relation between *res cogitans* and *res extensa* in so far as thinking takes place within the body, Descartes also emphasizes that thought and knowledge do not rest on bodily predispositions but are merely dependent on mental activity, and he concludes that there is a crucial difference between the mind and the body: objects in nature – such as the human body – can be divided into their parts, whereas the mind itself is indivisible ('magnam esse differentiam inter mentem & corpus, in eo quod corpus ex natura sua sit semper divisibile, mens autem plane indivisibilis').[67] The fact that the *cogito* cannot be divided into subdepartments in the way that human bodies can be divided into arms, legs and so forth, leads to a unified conception of self-consciousness as the foundation of all human endeavours. Even though the world of appearances may be marked by uncertainty, the statements 'I am' and 'I exist' are always 'necessarily true' ('Ego sum, ego existo … necessario esse verum'), and our self-consciousness of being is always an 'innate idea' which cannot be doubted.[68]

Obviously, for Descartes, 'being' precedes language, since the latter is not innate. But, for Nietzsche, such a point of view is simply silly, since our understanding of 'being', 'thinking' and 'self' is already formed by linguistic operations. Any immediate access to selfhood and to thinking is nothing but a mythological construction or a metaphysical fiction (*KSA* 11, 40[20], [22], [23], [24]; and *KSA* 12, 10[158]), and in a famous passage of *Beyond Good and Evil* (1886), Nietzsche notes that there is no pure thought which could serve as a foundation for Descartes' *cogito*:

> There are still harmless self-observers who believe 'immediate certainties' exist, for example, 'I think'.… But I shall reiterate a hundred times that 'immediate certainty' contains a *contradictio in adjecto*: we really ought to get free from the seduction of words! Let the people believe that knowledge is total knowledge, but the philosopher must say to himself: when I analyse the event expressed in the sentence 'I think', I acquire a series of rash observations which are difficult, perhaps impossible, to prove – for example, that it is *I* who think, that it has to be something at all which thinks, that thinking is an activity and operation on the part of an entity thought of as a cause, that an 'I' exists, finally that what is designated by 'thinking' has already been determined – that I *know* what thinking is. For if I had not already decided that matter within myself, by what standard could I determine that what is happening is not perhaps 'willing' or 'feeling'? Enough: this 'I think' presupposes that I *compare* my present state with other known states of myself in order to determine what it is: on account of this retrospective connection with other 'knowledge' at any rate it possesses no immediate certainty for me. (*BGE* 16)[69]

In order to know that we think, Nietzsche argues, we necessarily have to distinguish 'thinking' from all other kinds of mental and physical activity, and there is consequently no such thing as a unified (self-)consciousness. Christopher Janaway has argued that Nietzsche takes this 'deconstruction' of the self from

Schopenhauer, but it comes in fact from Gustav Teichmüller's *Die wirkliche und die scheinbare Welt* (*The Real World and the World of Appearance*), which Nietzsche read shortly after its publication in 1882.[70] Much like Nietzsche himself, Teichmüller remarks that we can only be aware of our thinking once we have distinguished thinking itself from other forms of mental activity.[71] In contrast to Nietzsche's assertion that this leads to a rejection of any unified form of self-consciousness, Teichmüller holds on to the Cartesian model.[72] Nevertheless, such a distinction between thinking and other activities is based on the assumption that 'thinking' *means* something different from, say, 'feeling', 'doing', 'desiring' and so forth. First of all, meaning is a linguistic phenomenon: things do not mean anything, and it is only through our use of some kind of language that we attach meaning to them, generate concepts and integrate them into our system of beliefs. Within this system, meaning can only be established by comparing certain concepts we have with other concepts, and – as Charles Sanders Peirce noted – 'the meaning of a proposition is itself a proposition'.[73] This relation between concepts, beliefs and propositions is essentially a rhetorical process, because comparison is possible only through seeing similarities and differences, through transferring concepts into different contexts, and through relating them to one another and to what we see as the 'world'. In order to be aware of anything at all – be it thinking, the world, or chairs – we need conceptual thought, and the generation of concepts is, for Nietzsche, a linguistic phenomenon. Even if we cannot express something or have no concept of it, we at least have the concept of not having concepts; any notion of non-conceptual thought or experience is itself conceptual. Without language, there is no being, no thinking, no relation between them, and thus no self. In this context, Descartes' *cogito* appears as an epiphenomenon of language, and in the fifth book of *The Gay Science* (1887), Nietzsche consequently notes that the evolution of language and the emergence of consciousness within any given individual are inextricably linked (*GS* 354).

The direction of Nietzsche's argument against Descartes is obvious and targets the self-evidence of subjectivity and consciousness as cornerstones of so-called 'metaphysical' thought: human experience has no transcendental value; thus, the human condition is a physiological and rhetorical problem which we ought to deal with in linguistic terms (*BGE* 54). Seen from this perspective, it becomes increasingly clear that Nietzsche emphasizes that 'consciousness' itself is basically nothing but a word and therefore a linguistic abstraction which rolls different mental states, experiences, perceptions and concepts into one.[74] In much the same way as external reality is not an ontological fact but the product of an interplay between perception, mental activity and linguistic concepts, 'consciousness' – or what Stephen Toulmin pointedly terms 'inwardness'[75] – is not a given fact or some kind of eternal truth (*D* 116), but rather a product of the same processes which lead us to assume that the world exists. This does not mean that Nietzsche simply says that 'consciousness' and the 'self' do not possess any value, but merely that they are what he calls, in a note of mid-1885, a 'regulative fiction' (*KSA* 11, 35[35]). Nevertheless, 'consciousness', the 'self' and the 'mind' are necessary fictions, for they enable us to relate ourselves to what we see as the world. The assumption that there is something such as phenomenal consciousness or the self makes sense for two reasons: first, concepts such as 'consciousness' and the 'self' group different bodily experiences together and enable us to possess a more or less coherent perspective on our relationship to the world; and, second, these conceptual fictions order the world according to this perspective.[76] Thus, Nietzsche does not say that consciousness and subjectivity do not exist or do not make sense; he merely reminds us that the phenomena which we label as such are ultimately based on the tropical and topical character of language. Taking this background into account, we can expect Nietzsche to be highly critical of any *Bewußtseinsphilosophie* and hence of the idealistic notion of *Geist*. Already in his lecture-series *Die vorplatonischen Philosophen* (*The Pre-Socratic Philosophers*), written in 1872, he points out that the dichotomy between 'mind' and 'body' does not really seem to hold water, since 'mind' is in any case generated by the brain and, thus, a physical function (*KGW* II/4, pp. 294 and 306; see also *D* 39). Here, he was able to follow once more Alexander Bain's *Mind and Body* (1873). The latter explicitly concluded that mental states are nothing but physical states and that our traditional understanding of the mind and of consciousness is merely a result of linguistic vagueness.[77] In this respect, it is not surprising that Nietzsche regards mental activity and thinking as a continuous association between different ideas or as a chain of images which unfolds in our mind. As a consequence, he regards mental activity as a form of 'language' (*KSA* 7, 5[89] and *KSA* 12, 1[75]). Against this background, we are now able to conclude that his argument that knowledge is ultimately dependent on metaphor only seems to make sense once we realize that it is based on a complex intertwining of rhetorical discourse and neurophysiology.

Notes

1 See Aristotle, *Poetics*, 1457b, and *Rhetoric*, 1405b, 1406bff and 1410bf, in *The Complete Works of Aristotle*, ed. Jonathan Barnes, 2 vols (Princeton, NJ: Princeton University Press, 1984).

2 See Eva Feder Kittay, *Metaphor: Its Cognitive Force and Linguistic Structure* (Oxford: Clarendon Press, 1987), p. 313.

3 See Aristotle, *De anima*, 403a9f, 431a16, 431b2 and 432a5ff.

4 Peter Carruthers, *Language, Thought and Consciousness: An Essay in Philosophical Psychology* (Cambridge University Press, 1996), p. 225.

5 On the historical development of psychology and physiology in late Enlightenment and nineteenth-century thought, see David E. Leary, 'The Philosophical Development of the Conception of Psychology in Germany, 1780–1850', *Journal of the History of the Behavioral Sciences*, 14 (1978), 113–21, and Brigitte Lohff, *Die Suche nach der Wissenschaftlichkeit der Physiologie in der Zeit der Romantik* (Stuttgart: Fischer, 1990).

6 See Johann Friedrich Herbart, *Psychologie als Wissenschaft, neu gegründet auf Erfahrung, Metaphysik und Mathematik* (Königsberg: Unzer, 1824–25), vol. 1, p. 185, and Gustav Theodor Fechner, *Elemente der Psychophysik* (Leipzig: Breitkopf & Härtel, 1860), vol. 1, p. 8. On Fechner's 'psychophysics' and its philosophical implications, see especially Michael Heidelberger, *Die innere Seite der Natur: Gustav Theodor Fechners wissenschaftlich-philosophische Weltauffassung* (Frankfurt am Main: Klostermann, 1993), pp. 217–19.

7 See, for example, Eric Blondel, *Nietzsche, le corps et la culture: La philosophie comme généaologie philologique* (Paris: PUF, 1986); Helmut Pfotenhauer, *Die Kunst als Physiologie: Nietzsches ästhetische Theorie und literarische Produktion* (Stuttgart:

Metzler, 1985); and Volker Gerhardt, 'Von der ästhetischen Metaphysik zur Physiologie der Kunst', *Nietzsche-Studien*, 13 (1984), 374–93.

8 See Friedrich Albert Lange, *Geschichte des Materialismus und Kritik seiner Bedeutung in der Gegenwart* (Iserlohn: Baedeker, 1866), pp. 484, 492 and 496.

9 See Gustav Gerber, *Die Sprache als Kunst*, 2nd edn (Berlin: Gaertner, 1885), vol. 1, pp. 260 and 326–27.

10 Anthonie Meijers, 'Gustav Gerber und Friedrich Nietzsche: Zum historischen Hintergrund der sprachphilosophischen Auffassungen des frühen Nietzsche', *Nietzsche-Studien*, 17 (1988), 369–90, has examined Gerber's influence in much detail, but ultimately tends to reduce Nietzsche's reflections on rhetoric to Gerber's stipulations.

11 See Aristotle, *De anima*, 427a19–21 and 432a12–14; Lange, *Geschichte des Materialismus*, p. 494; and Johann Carl Friedrich Zöllner, *Über die Natur der Cometen: Beiträge zur Geschichte und Theorie der Erkenntnis* (Leipzig: Engelmann, 1872), pp. 201–203 and 362.

12 See, in contrast, Zöllner, *Über die Natur der Cometen*, pp. 201–202.

13 See Gerber, *Die Sprache als Kunst*, vol. 1, pp. 225, 232 and 259.

14 The word *Übertragung* is also used by Richard Volkmann in his *Die Rhetorik der Griechen und Römer in systematischer Übersicht* (Berlin: Ebeling & Plahn, 1872), pp. 353 and 391–92.

15 See Wilhelm Wackernagel, *Poetik, Rhetorik und Stilistik: Academische Vorlesungen*, ed. Ludwig Sieber (Halle: Verl. d. Buchh. d. Waisenhauses, 1873), p. 394, who regards metaphor as the trope which 'am Ende auf alle figürlichen und tropischen Wendungen passt'. Volkmann writes: 'Der häufigste und schönste, dabei allgemeinste Tropus, so dass sich die meisten übrigen Tropen im Grunde genommen als Unterarten derselben betrachten lassen, ist die Metapher' (*Die Rhetorik der Griechen und Römer*, p. 355).

16 See Quintilian, *Institutio oratoria*, 4 vols (London: Heinemann, 1966–69), VIII.vi.6, and Cicero, *De oratore* (London: Heinemann, 1942), III.155.

17 Cicero, *De oratore*, III.155.

18 Ibid., III.161. Cicero's *De oratore* is one of the main reference texts of Nietzsche's *Darstellung der antiken Rhetorik*, and he quotes it in key passages. See *KGW* II/4, pp. 421–22, 435–37, 440, 445, 457, 483 and 488.

19 See Volkmann, *Die Rhetorik der Griechen und Römer*, pp. 355–57, and Wackernagel, *Poetik, Rhetorik und Stilistik*, p. 394. Nietzsche's understanding of metaphor as *Übertragung* is widely noted in the secondary literature, but rarely examined in detail. See Angèle Kremer-Marietti, *Nietzsche et la rhétorique* (Paris: PUF, 1992), pp. 133 and 236–37; Günter Abel, 'Logik und Ästhetik', *Nietzsche-Studien*, 16 (1987), 112–48 (123–24); Lawrence M. Hinman, 'Nietzsche, Metaphor, and Truth', *Philosophy and Phenomenological Research*, 43 (1982), 179–99 (183); Philippe Lacoue-Labarthe, 'Le détour (Nietzsche et la rhétorique)', *Poétique*, 5 (1971), 53–76 (64).

20 See Gerber, *Die Sprache als Kunst*, vol. 1, pp. 225, 226, 259 and 329. On Gerber's notion of *Übertragung*, see especially Andrea Orsucci, 'Unbewußte Schlüsse, Anticipationen, Übetragungen: Über Nietzsches Verhältnis zu Karl Friedrich Zöllner und Gustav Gerber', in Tilman Borsche, Federico Gerratana and Aldo Venturelli (eds), *'Centauren-Geburten': Wissenschaft, Kunst und Philosophie beim jungen Nietzsche* (Berlin: de Gruyter, 1994), pp. 193–207 (pp. 202–204).

21 See Aristotle, *Rhetoric*, 1406bff; Cicero, *De oratore*, III.155ff; Quintilian, *Institutio oratoria*, VIII.vi.4 and 8; Augustine, *De doctrina Christiana* (Oxford: Clarendon Press, 1995), II.23–57 and III.9–56.

22 See Jean-Jacques Rousseau, *Essai sur l'origine des langues*, ed. Angèle Kremer-Marietti (Paris: Aubier Montaigne, 1974), p.97.

23 Nietzsche borrowed Hamann's *Schriften und Briefe*, ed. Moritz Petri (Hanover: Meyer, 1872) from the University Library in Basle.

24 See Johann Georg Hamann, *Aesthetica in nuce*, in *Sämtliche Werke: Historisch-kritische Ausgabe*, ed. Josef Nadler (Vienna: Herder, 1949–57), vol. 2, p. 199.

25 Stephen Houlgate, *Hegel, Nietzsche and the Criticism of Metaphysics* (Cambridge: Cambridge University Press, 1986), p. 48.

26 See Aristotle, *De interpretatione*, 16a3ff, and Augustine, *De doctrina Christiana*, II.3f.

27 An example would be Otto Caspari's *Die Sprache als psychischer Entwicklungsgrund* (Göttingen: Dieterich, 1864). On the development of phrenology in the late eighteenth and early nineteenth centuries, see Michael Hagner, 'The Soul and the Brain Between Anatomy and "Naturphilosophie" in the Early Nineteenth Century', *Medical History*, 36 (1992), 1–33.

28 See Arthur Schopenhauer, *Die Welt als Wille und Vorstellung*, in *Sämtliche Werke*, ed. Wolfgang Freiherr von Löhneysen (Frankfurt am Main: Insel, 1960–65), vol. 2, pp. 34–35. For some examples of Schopenhauer's use of physiology, see ibid., vol. 1, pp. 129 and 153–55 as well as vol. 2, pp. 367–69 and 430–32.

29 See Schopenhauer, *Die Welt als Wille und Vorstellung*, in *Sämtliche Werke*, vol. 2, p. 369. On Schopenhauer's materialist tendencies, see Christopher Janaway, *Self and World in Schopenhauer's Philosophy* (Oxford: Clarendon Press, 1989), pp. 181–83.

30 See Schopenhauer, *Die Welt als Wille und Vorstellung*, in *Sämtliche Werke*, vol. 2, p. 376.

31 See Herbert Schnädelbach, *Philosophy in Germany, 1831–1933* (Cambridge: Cambridge University Press, 1984), pp. 66–68.

32 See Christoph Asendorf, *Batteries of Life: On the History of Things and their Perception in Modernity* (Berkeley, CA: University of California Press, 1993), pp. 153–55, and Michael Hagner, 'Die elektrische Erregbarkeit des Gehirns: Zur Konjunktur eines Experiments', in Hans-Jörg Rheinberger and Michael Hagner (eds), *Die Experimentalisierung des Lebens: Experimentalsysteme in den biologischen Wissenschaften 1850/1950* (Berlin: Akademie Verlag, 1993), pp. 97–115.

33 For detailed discussions of the understanding of electricity in the eighteenth century, see John Heilbron, *Electricity in the Seventeenth and Eighteenth Centuries: A Study of Early Modern Physics* (Berkeley, CA: University of California Press, 1979), pp. 261–63.

34 See Alessandro Volta's 'Sur l'électrophore perpétuel', *Journal de physique*, 8 (1776), 21–24 and his *Lezioni compendiose sulla elettricità*, in *Le Opere di Alessandro Volta: Edizione Nazionale* (Milan: Hoepli, 1918–29), vol. 4, pp. 389–457.

35 On the debate between Volta and Galvani, see especially Marcello Pera, *The Ambiguous Frog: The Galvani–Volta Controversy on Animal Electricity* (Princeton, NJ: Princeton University Press, 1992).

36 See Johann Wilhelm Ritter's *Beweis, dass ein beständiger Galvanismus den Lebensprozess in dem Thierreiche begleite* (Weimar: Verlag d. Industrie-Comptoirs, 1798) and *Das elektrische System der Körper: Ein Versuch* (Leipzig: Reclam, 1805).

37 See Hans Christian Oersted, 'Über die Art, wie sich Electricität fortpflanzt', *Neues allgemeines Journal der Chemie*, 6 (1806), 292–302; idem, 'Experiments on the Effect of a Current of Electricity on the Magnetic Needle', *Annals of Philosophy*, 16 (1820), 273–76. See also Michael Faraday, *Experimental Researches into Electricity* (London: Taylor, 1839–55) and James Clerk Maxwell, *A Treatise on Electricity and Magnetism* (Oxford: Clarendon Press, 1873). On the general background, see P.M. Harman, *Energy, Force, and Matter: The Conceptual Development of Nineteenth-Century Physics* (Cambridge: Cambridge University Press, 1982), pp. 30–32.

38 On electricity in the physical sciences in nineteenth-century Germany, see K.L. Caneva, 'From Galvanism to Electrodynamics: The Transformation of German Physics and Its Social Context', *Historical Studies in the Physical Sciences*, 9 (1978), 63–159, and M.N. Wise, 'German Concepts of Force, Energy and the Electromagnetic Ether: 1845–1880', in G.N. Cantor and M.J.S. Hodge (eds), *Conceptions of Ether* (Cambridge: Cambridge University Press, 1981), pp. 269–307. On the institutional background, see R.S. Turner,

'The Growth of Professional Research in Prussia, 1818 to 1848: Causes and Context', *Historical Studies in the Physical Sciences*, 3 (1971), 137–82.

39 See Emil Du Bois-Reymond, *Untersuchungen über thierische Electricität* (Berlin: Reimer, 1848–60), vol. 1, pp. 10–12 and 25–27.

40 See ibid., vol. 1, pp. 31–33 and 160–62; Eduard Hitzig, *Untersuchungen über das Gehirn: Abhandlungen physiologischen und pathologischen Inhalts* (Berlin: Hirschwald, 1874); and Alexander Bain, *Mind and Body: The Theories of Their Relation*, 2nd edn (London: Henry S. King & Co., 1873), p. 117. Bain's study was translated into German as *Geist und Körper* (Leipzig: Brockhaus, 1874).

41 See Pierre Paul Broca, 'Sur le siège de la faculté du langage articulé', *Tribune Médicale*, 74 and 75 (1869), 254–56 and 265–69. Pierre Flourens, *Recherches expérimentales sur les propiétés et les fonctions du système nerveux dans les animaux vertébrés*, 2nd edn (Paris: Balliere, 1842), pp. 87ff; and David Ferrier, *The Localisation of Cerebral Disease, Being the Gulstonian Lectures of the Royal College of Physicians* (London: Smith, Elder & Co., 1878). On the function of these and other topographical models, see Michael Hagner's studies 'Lokalisation, Funktion, Cytoarchitektonik: Wege zur Modellierung des Gehirns', in Michael Hagner, Hans-Jörg Rheinberger and Bettina Wahrig-Schmidt (eds), *Objekte, Differenzen, Konjunkturen: Experimentalsysteme im historischen Kontext* (Berlin: Akademie Verlag, 1994), pp. 121–50, and idem, 'Hirnbilder: Cerebrale Repräsentationen im 19. und 20. Jahrhundert', in Michael Wetzel and Herta Wolf (eds), *Der Entzug der Bilder: Visuelle Realitäten* (Munich: Fink, 1994), pp. 145–60. See, further, Robert M. Young, *Mind, Brain and Adaptation in the Nineteenth Century: Cerebral Localisation and its Biological Context from Gall to Ferrier*, 2nd edn (Oxford: Oxford University Press, 1990).

42 See William Whewell, *The Philosophy of the Inductive Sciences, founded upon their History* (London: Parker, 1840).

43 See John Stuart Mill, *A System of Logic, Ratiocinative and Inductive* (London: Longmans & Green, 1947), pp. 364–66; Cicero, *De Inventione*, I.49 and 51; Quintilian, *Institutiones oratoriae*, V.xi.2; Aristotle, *Rhetoric*, 1357b.

44 See Mill, *A System of Logic*, pp. 557–59.

45 I quote from Ludwig Büchner, *Kraft und Stoff, oder Grundzüge der natürlichen Weltordnung, nebst einer darauf gebauten Sittenlehre in allgemein verständlicher Darstellung*, 20th edn (Leipzig: Thomas, 1902), pp. 156–58, 20 and 181. On Büchner, see Frederick Gregory, *Scientific Materialism in Nineteenth-Century Germany* (Dordrecht: Reidel, 1977), pp. 105–107.

46 After his parallel studies of philosophy and medicine at the University of Leipzig from 1834 to 1838, Lotze became Johann Friedrich Herbart's successor at the University of Göttingen. On Lotze, see Walter Gebhard, *'Der Zusammenhang der Dinge': Weltgleichnis und Naturverklärung im Totalitätsbewußtsein des 19. Jahrhunderts* (Tübingen: Niemeyer, 1984), pp. 222–24, and Thomas E. Willey, *Back to Kant: The Revival of Kantianism in German Social and Historical Thought, 1860–1914* (Detroit: Wayne State University Press, 1978), pp. 40–42.

47 See Rudolph Hermann Lotze, *Allgemeine Physiologie des körperlichen Lebens* (Leipzig: Weidmann, 1851), pp. 385–87, and idem, *Medicinische Psychologie, oder Physiologie der Seele* (Leipzig: Weidmann, 1852), pp. 174–76 and 197–99.

48 Lotze, *Allgemeine Physiologie*, pp. 399–401. For Lotze's remarks about 'organic electricity', see ibid., pp. 389–90.

49 Lotze, *Medicinische Psychologie*, pp. 197–98, 176–77, 178 and 204.

50 See ibid., pp. 175–77.

51 I quote from Otto Funke, *Lehrbuch der Physiologie für akademische Vorlesungen und zum Selbststudium*, 6th edn (Leipzig: Voss, 1878–79), vol. 1, p. 453. See also vol. 1, pp. 454, 457–59 and 514–16.

52 Ibid., vol. 1, p. 586.
53 See Hermann von Helmholtz, *Ueber das Sehen des Menschen* (Leipzig: Vose, 1855), p. 40; Fechner, *Elemente der Psychophysik*, vol. 1, pp. 15–17; and Eduard von Hartmann, *Philosophie des Unbewussten* (Berlin: Duncker, 1869), pp. 83, 89–91 and 126–28. For valuable background information, see Theo C. Meyering, *Historical Roots of Cognitive Science: The Rise of a Cognitive Theory of Perception from Antiquity to the Nineteenth Century* (Dordrecht: Kluwer, 1989), pp. 125–27 and 181–83.
54 See Lange, *Geschichte des Materialismus*, pp. 411 and 481. In fact, he speaks of 'Nervenleitung' (p. 440), 'Nervenimpulse' (p. 454), of the 'Kreislauf der Nerventhätigkeit' (p. 455) and of 'Spannungszustände' (p. 455). On Lange's physiological background, see George J. Stack, *Lange and Nietzsche* (Berlin: de Gruyter, 1983), pp. 198–200.
55 Lange, *Geschichte des Materialismus*, pp. 426–27 and 438. On the relationship between 'brain' and 'soul' in the nineteenth century, see the contributions in Ernst Florey and Olaf Breidbach (eds), *Das Gehirn – Organ der Seele? Zur Ideengeschichte der Neurobiologie* (Berlin: Akademie Verlag, 1993) as well as Andrew Peacock, 'The Relationship between the Soul and the Brain', in F. Clifford Rose and William F. Bynum (eds), *Historical Aspects of the Neurosciences* (New York: Raven Press, 1982), pp. 83–98.
56 Lange, *Geschichte des Materialismus*, pp. 454, 457 and 442.
57 Gerber, *Die Sprache als Kunst*, vol. 1, p. 240.
58 See also *KSA* 9, 1[66], [115].
59 Gerber, *Die Sprache als Kunst*, vol. 1, p. 258.
60 See Maudemarie Clark, 'On Knowledge, Truth, and Value: Nietzsche's Debt to Schopenhauer and the Development of His Empiricism', in Christopher Janaway (ed.), *Willing and Nothingness: Schopenhauer as Nietzsche's Educator* (Oxford: Clarendon Press, 1998), pp. 37–78 (p. 52).
61 See Gerber, *Die Sprache als Kunst*, vol. 1, p. 244.
62 Ibid., vol. 1, pp. 224 and 260.
63 See René Descartes, *Meditationes de prima philosophia*, in *Œuvres*, ed. Charles Adam and Paul Tannery (Paris: Cerf, 1897–1914), vol. 7, pp. 28 and 34.
64 Christian Thomasius, *Einleitung zu der Vernufft-Lehre* (Halle: Salfeld, 1691), p. 95.
65 See Descartes, *Meditationes*, in *Œuvres*, vol. 7, pp. 27–29.
66 See ibid., vol. 7, pp. 28, 44, 53 and 63–65.
67 Ibid., vol. 7, pp. 85–86. See also pp. 78–80 and 82–83.
68 Ibid., vol. 7, pp. 25 and 51.
69 See also *BGE* 17. Nietzsche's outright rejection of Decartes has not gone unnoticed, even though it is mostly discussed as an example of his criticism of 'metaphysics' and rarely related to his linguistic thought. Alan D. Schrift is one of the few interpreters who see Nietzsche's rejection of Descartes in the context of his reflections on language. See his 'Language, Metaphor, Rhetoric: Nietzsche's Deconstruction of Epistemology', *Journal of the History of Philosophy*, 23 (1985), 371–95 (390–91).
70 See Christopher Janaway, 'Nietzsche, the Self, and Schopenhauer', in Keith Ansell-Pearson (ed.), *Nietzsche and Modern German Thought* (London: Routledge, 1991), pp. 119–42 (p. 124).
71 See Gustav Teichmüller, *Die wirkliche und die scheinbare Welt: Neue Grundlegung der Metaphysik* (Breslau: Koebner, 1882), p. 106.
72 See ibid., pp. 23 and 99.
73 Charles Sanders Peirce, 'The Essentials of Pragmatism', in *Philosophical Writings of Charles Sanders Peirce*, ed. Justus Buchler (New York: Dover, 1955), pp. 251–68 (p. 261).

74 This argument is already prefigured in Lange, *Geschichte des Materialismus*, p. 490, where he describes subjectivity as a 'Fiktion', an 'Ideen-Association' or as a 'Verschmelzung der Sinneswahrnehmungen'.
75 See Stephen Toulmin, 'The Inwardness of Mental Life', *Critical Inquiry*, 6 (1979), 1–16.
76 See Alexander Nehamas, *Nietzsche: Life as Literature* (Cambridge, MA: Harvard University Press, 1985), p. 181.
77 See Bain, *Mind and Body*, pp. 41, 43, 117, 143 and 196.

Chapter 5

Reading Nietzsche through Ernst Mach[1]

Nadeem J.Z. Hussain

Nietzsche has often been taken as presenting us with a radical attack on the pretensions of science and reason. Such readings focus on particular strains within the Nietzschean corpus: his perspectivism, his insistence that scientific theories involve falsification, and his claims that scientific theories are merely interpretations of the world. Those who oppose such readings of Nietzsche point instead to his favourable comments about science, scientific methodology, the results of particular sciences, the role of scientists and the importance of the senses. Various science-friendly labels – materialist, empiricist, positivist, naturalist – fly thick and fast in the hope that some of them, or some version of them, will stick to Nietzsche.[2]

However it is often not quite clear what these labels come to – what, in other words, would have to be true of Nietzsche's views in order for him to deserve one label rather than another. There are at least three obvious dimensions of claims that are relevant here: epistemological claims, metaphysical claims and claims in the philosophy of language. There are also significant ways in which the cluster of claims that falls under a particular label has changed through time. Think for a moment of the range of philosophers that might fall under a heading for 'positivism' in an encyclopaedia – Saint-Simon, Comte, Bentham, J.S. Mill, Avenarius, Popper – but, of course, crucially different claims are being made by these different thinkers despite whatever relative continuity and family resemblances may exist.

Furthermore, if we take into consideration the German context in which Nietzsche is writing, we must add – at the very least – sensualism to our list of science-friendly views. This should remind us that we need in particular to watch for the slide from epistemology to metaphysics. The empiricist claim that the senses are the only source of information about the world is, or at least was and is often taken to be, compatible with a range of views in metaphysics. Indeed an empiricist epistemology does not, or again was not taken to, commit one to an empiricist semantics. Transcendental idealism and the endless neo-Kantian variations on the theme only make the picture more complicated. Most of the views that labelled themselves, or have been labelled, as Kantian and neo-Kantian could, in all sincerity, express admiration for the sciences and the senses.

One strong strain in contemporary Anglo-American Nietzsche scholarship would like to take his positive emphasis on the senses, nature and scientific methodology as grounds for interpreting him as similar in many ways to contemporary naturalists. According to such a reading, Nietzsche has a basically empiricist epistemology and has ontological commitments that are more or less straightforwardly read off of whatever he takes to be the best empirically supported account of the world. This interpretation is assumed to gain support from the pervasive influence of materialism in the mid to late nineteenth century.[3]

However, this view does run into some problems. On the one hand, Nietzsche often suggests that the theories of scientists indeed do not straightforwardly report how the world is. Thus he says:

> It is perhaps just dawning on five or six minds that physics, too, is only an interpretation and exegesis of the world (to suit us, if I may say so!) and *not* a world-explanation; but insofar as it is based on belief in the senses, it is regarded as more, and for a long time to come must be regarded as more – namely, as an explanation. (*BGE* 14)

Or consider the following passage:

> One should not wrongly reify 'cause' and 'effect', as the natural scientists do (and whoever, like them, now 'naturalizes' in his thinking), according to the prevailing mechanical doltishness which makes the cause press and push until it 'effects' its end; one should use 'cause' and 'effect' only as pure concepts, that is to say, as conventional fictions for the purpose of designation and communication – *not* for explanation. In the 'in-itself' there is nothing of 'causal connection', of 'necessity', or of 'psychological non-freedom'; there the effect does *not* follow the cause, there is no rule of 'law'. It is *we* alone who have devised cause, sequence, for-each-other, relativity, constraint, number, law, freedom, motive, and purpose; and when we project and mix this symbol world into things as if it existed 'in itself', we act once more as we have always acted – *mythologically*. (*BGE* 21)

Brian Leiter suggests that this passage and other passages are only a stage in Nietzsche's development and points to Maudemarie Clark's work to suggest that, in the end, one doesn't have to ascribe an anti-naturalist position to Nietzsche.[4] Clark gives us a developmental interpretation of Nietzsche and claims that in the end Nietzsche does not accept the falsificationist claims of the kind suggested by the quoted passages.[5]

I find the claim that Nietzsche gives up on the falsification thesis hard to swallow. My aim here, though, will not be to argue that we should ascribe the falsificationist thesis to the later Nietzsche but rather to suggest an alternative reading of Nietzsche that is both science-friendly and allows for a falsificationist thesis. This alternative reading of Nietzsche uses the fact that there were, as I have already suggested, many different ways in which Nietzsche's contemporaries could be friendly to science and the senses. Building on Thomas Brobjer's remarks on Nietzsche's reading of Mach and Avenarius in Chapter 1 of this volume, I shall argue that it is Ernst Mach's neutral monism – Machian positivism, as I will call it – that provides the basis for the correct interpretation of Nietzsche's published texts on central metaphysical and epistemological issues.

The argument will be perforce partial but I hope that, by deploying a set of different kinds of argument, a convincing initial case can indeed be made. First, I will look at how Nietzsche situated his own epistemology and metaphysics in the section 'How the "True World" Finally Became a Fable' in *Twilight of the Idols*. Here Nietzsche lists a progression of historical positions on the relation between the world of our experiences and some purported real, or more real, world. He correctly sees these positions as linked by natural conceptual developments. These developments that supposedly lead to Nietzsche's own position also lead, I will argue, for exactly the same reasons, to Mach's position. This should give us some reason to suspect that Mach's more systematically laid-out position might help us make sense of Nietzsche's more cryptic remarks. Of course, the first argument will not be effective if there is no actual textual support in Nietzsche for the Machian reading and so I will turn to citing and discussing relevant passages from Mach's and Nietzsche's works. Then I shall move on to a consideration of the evidence available for a causal connection between the two. My claim will be that Mach's views, and some very much like them, were quite widespread and influential. This claim receives support, as I shall try to show, from the similarities many contemporaries noticed between Nietzsche's views and those of contemporary positivists like Mach. Finally, I shall try to show how reading Nietzsche through Mach actually helps us to interpret Nietzsche by considering in detail section 15 of *Beyond Good and Evil*, a passage central to understanding Nietzsche's views on the senses and their epistemic role. Reading this passage through Mach provides a framework for an interpretation that manages to account for the many complexities of the passage that are otherwise hard to deal with.

Nietzsche and Positivism

When it comes to Nietzsche's fundamental views on metaphysics and epistemology, what are the interpretative possibilities suggested by Nietzsche's historical context? Recall my laundry list of labels: empiricism, materialism, naturalism, positivism and sensualism, not to mention all the varieties of neo-Kantianism. Does Nietzsche himself give us a clue as to where he might fit within these contemporary views? He supposedly does precisely this in the passage from the *Twilight of the Idols* entitled 'How the "True World" Finally Became a Fable: The History of an Error'. Let me quote the final few stages of this history:

> 4. The true world – unattainable? At any rate, unattained. And being unattained, also *unknown*. Consequently, not consoling, redeeming, or obligating: how could something unknown obligate us?
> (Grey morning. The first yawn of reason. The cockcrow of positivism.)
> 5. The 'true' world – an idea which is no longer good for anything, not even obligating – an idea which has become useless and superfluous – *consequently*, a refuted idea: let us abolish it!
> (Bright day; breakfast; return of *bon sens* and cheerfulness; Plato's embarrassed blush; pandemonium of all free spirits.)
> 6. The true world – we have abolished it. What world has remained? The apparent one perhaps? But no! *With the true world we have also abolished the apparent one.*
> (Noon; moment of the briefest shadow; end of the longest error; high point of humanity; INCIPIT ZARATHUSTRA.) (*TI* 'World')

There are a few obvious initial reactions one might have: (1) stage 4 certainly is supposed to have something to do with positivism; (2) stages 5 and 6 may involve further developments of positivism; (3) Nietzsche presumably sees himself as representing stage 6.

So, now, if we turn to Nietzsche's historical context what do we find? Who are the positivists? And are there positions, so to speak, beyond positivism? And would any of these positions fit with the rest of Nietzsche's texts – in particular, his claims about the senses, science and falsification?

In Nietzsche's immediate historical context, positivism, both as philosophical and social movement, was associated with Auguste Comte (1798–1857). Comte argues that the human mind goes through 'three different theoretical states: the theological or fictitious state, the metaphysical or abstract state and the scientific or positive state'.[6] It is this last state that Comte wants to endorse:

> [T]he human mind, recognizing the impossibility of obtaining absolute truth, gives up the search after the origin and hidden causes of the universe and a knowledge of the final causes of phenomena. It endeavors now only to discover, by a well-combined use of reasoning and observation, the actual laws of phenomena – that is to say, their invariable relations of succession and likeness.[7]

> Everybody, indeed, knows that in our positive explanations, even when they are most complete, we do not pretend to explain the real causes of phenomena.[8]

Positivism is understood as the rejection of trying to go beyond the phenomenal reality we have access to. John Stuart Mill describes the Comtean positivist position as follows:

> We have no knowledge of anything but Phænomena; and our knowledge of phænomena is relative, not absolute. We know not the essence, nor the real mode of production, of any fact .... The laws of phænomena are all we know respecting them. Their essential nature, and their ultimate causes, either efficient or final, are unknown and inscrutable to us.[9]

Now we can see in this positivist position, some of the themes we see in Nietzsche: the importance of the senses, the emphasis on studying the observable world, and the recommended alliance of philosophy with the natural sciences.[10]

But positivism, in this sense of the term, is a stage that Nietzsche takes himself to be going beyond. The Comtean positivist still accepts the distinction between the true world – the thing-in-itself – and the world of appearances – the phenomenal world. The Comtean positivist, however, simply thinks that there is no point in thinking about the thing-in-itself. Thus the Comtean positivist naturally falls under stage 4.[11]

Now, stages 5 and 6 go beyond stage 4. In stage 5 we abolish the 'true' world and in stage 6 we realize that 'with the true world we have also abolished the apparent one' (*TI* 'World'). Who in Nietzsche's historical context would be a natural successor to Comte? I want to argue that the natural successor most helpful for interpreting Nietzsche's own position is Ernst Mach (1838–1916) and, we should, I suggest, hardly be surprised to learn that Nietzsche owned a copy of Mach's *Contributions to the Analysis of the Sensations* (1886).[12]

Textual Evidence

I will turn in the next section to a more detailed discussion of Nietzsche's historical context and the place of Mach's views in his intellectual milieu. However, it makes sense to consider first the evidence presented by the texts themselves. In this section I will briefly survey the similarities between Mach's and Nietzsche's views on central epistemological and metaphysical issues. We shall return to some of these issues in the very last section of this chapter when I argue that reading Nietzsche through Mach helps us understand what would otherwise be problematic passages in the Nietzsche corpus.

So what does Mach say? An autobiographical footnote in Mach's *Analysis of Sensations* should remind us immediately of Nietzsche's 'How the "True World" Became a Fable'. Mach writes:

> I have always felt it as a stroke of special good fortune, that early in life, at about the age of fifteen, I lighted, in the library of my father, on a copy of Kant's *Prolegomena zu jeder künftigen Metaphysik* [*Prolegomena to Any Future Metaphysics*]. The book made at the time a powerful and ineffaceable impression upon me, the like of which I never afterward experienced in any of my philosophical reading. Some two or three years later the superfluous rôle played by 'the thing in itself' abruptly dawned on me. On a bright summer day under the open heaven, the world with my ego suddenly appeared to me as *one* coherent mass of sensations, only more strongly coherent in the ego.[13]

Mach lays out his basic metaphysical picture in the introductory chapter of *Contributions to the Analysis of the Sensations*. He defends a monism according to which the world consists of sensations.[14] But he prefers calling these sensations 'elements' to emphasize that they are not to be understood as belonging to some particular self – or, in his terms, ego – and because they are the most basic building blocks – elements – of the world: 'The primary fact is not the *I*, the ego, but the elements (sensations). The elements *constitute* the *I*. *I* have the sensation green, signifies that the element green occurs in a given complex of other elements (sensations, memories)'. There is thus a field of sensory elements in which certain relatively stable complexes are given single designations, single names: 'Our greater intimacy with this sum-total of permanency, and its preponderance as contrasted with the changeable, impel us to the partly instinctive, partly voluntary and conscious economy of mental representation and designation, as expressed in ordinary thought and speech.' But the 'useful habit of designating such relatively permanent compounds by *single* names, and of apprehending them by *single* thoughts' leads us to make the mistake of thinking that there is 'a *single* thing with *many* attributes'. Thus also 'arises the monstrous notion of a *thing in itself*, unknowable and different from its "phenomenal" existence'.[15] And indeed we make this mistake about the particular complex that we label as the ego, *das Ich*.

Crucially, everything is, so to speak, on one ontological plane:

> Let those complexes of colors, sounds, and so forth, commonly called bodies, be designated, for the sake of simplicity, by *A B C* ...; the complex, known as our own body, which constitutes a part of the former, may be called *K L M* ...; the complex composed of volitions, memory-images, and the rest, we shall represent by α β γ ....[16]

As soon as we have perceived that the supposed unities 'body' and 'ego' are only makeshifts, designed for provisional survey and for certain practical ends ..., we find ourselves obliged, in many profound scientific investigations, to abandon them .... The antithesis of ego and world, sensation phenomenon and thing, then vanishes, and we have simply to deal with the *connexion* of the *elements* α β γ ... *A B C* ... *K L M* ....[17]

As Mach emphasizes, '*the senses represent things neither wrongly nor correctly*. All that can be truly said of the sense-organs is, that, *under different circumstances they produce different sensations and perceptions*'.[18] Mach's illustration of this point makes things clearer:

A cube of wood when seen close at hand, looks large; when seen at a distance, small; it looks different with the right eye from what it does with the left; sometimes it appears double; with closed eyes it is invisible. The properties of the same body, therefore, appear modified by our own body; they appear conditioned by it. But where, now, is the *same* body, which to the appearance is so *different*? All that can be said is, that with different *K L M* different *A B C* ... are associated.[19]

Now, my aim is not, in the end, to defend the coherence, or the plausibility, of Mach's sensory element monism. However, for those familiar with many of the relevant Nietzschean texts, much here should have rung various bells.

Nietzsche makes similar points about the senses. He attacks most of traditional philosophy for having taken the senses as the basis of all deceptions and confusions in philosophy (*TI* 'Reason' 1). Instead, he pays tribute to Heraclitus:

With the highest respect, I except the name of *Heraclitus*. When the rest of the philosophic folk rejected the testimony of the senses because they showed multiplicity and change, he rejected their testimony because they showed things as if they had permanence and unity. Heraclitus too did the senses an injustice. They lie neither in the way the Eleatics believed, nor as he believed – they do not lie at all. What we *make* of their testimony, that alone introduces lies; for example, the lie of unity, the lie of thinghood, of substance, of permanence. (*TI* 'Reason' 2)

The senses do not lie. It is our language, and reasoning, that can lead to confusion. We can see Nietzsche's position as a somewhat more radical version of the Machian account: Mach talks of natural tendencies to get confused by the usefulness of designating things with single names; Nietzsche will talk about inevitable falsification brought on by language. Mach's language is certainly gentler than Nietzsche's. Compare Mach:

If, to the physicist, bodies appear the real, abiding existences, whilst sensations are regarded merely as their evanescent, transitory show, the physicist forgets, in the assumption of such a view, that all bodies are but thought-symbols for complexes of sensations (complexes of elements).[20]

And Nietzsche:

Formerly, alternation, change, any becoming at all, were taken as proof of mere appearance, as an indication that there must be something which led us astray. Today, conversely,

> precisely insofar as the prejudice of reason forces us to posit unity, identity, permanence, substance, cause, thinghood, being, we see ourselves somehow caught in error, compelled into error .... 'Reason' in language – oh, what an old deceptive female she is! I am afraid we are not rid of God because we still have faith in grammar. (*TI* 'Reason' 5)

But despite the difference in tone, the point is essentially the same. Language, and conscious reasoning that must occur in language, misleads, and thus claims about the world, expressed as they must be in language, tend to mislead (Mach) or necessarily falsify (Nietzsche). As Mach says elsewhere:

> Language, with its helpmate, conceptual thought, by fixing the essential and rejecting the unessential, constructs its rigid pictures of the fluid world on the plan of a mosaic, at a sacrifice of exactness and fidelity but with a saving of tools and labor.[21]

Compare this to one of Nietzsche's notes:

> A concept is an invention that doesn't *completely* correspond; but a lot of it does correspond a little: a sentence such as 'two things that are identical to a third are identical to each other' presupposes 1) things 2) identities: both don't exist. But with this invented rigid concept- and number-world man gains a means to grasp a huge quantity of facts with symbols and imprint them in memory. This symbol-apparatus is his superiority precisely because it distances him as far as possible from the individual facts. The reduction of experiences to *symbols*, and the increasing quantity of things which can therefore be grasped, is his *highest power*. The mental as the ability to be a master through symbols of a huge quantity of facts. *This mental world, this symbol-world, is sheer 'appearance and deception'*, just as every 'thing of appearance' already is. (*KSA* 11, 34[131])

Sometimes, in fact, the language they use is almost exactly the same. Mach quotes with approval a famous aphorism from Lichtenberg emphasizing that one should say 'It thinks' rather than 'I think' – a point that Nietzsche makes without explicit reference to Lichtenberg in *BGE* 17.[22] Similar comparisons can be made between Nietzsche's comments on atomism in the rest of *BGE* 17 and Mach's view of atomism.[23]

Perhaps as important as these clearly shared philosophical positions are the places where Mach's views can provide insight into what are otherwise opaque Nietzschean doctrines. A Machian reading of Nietzsche gives us various possibilities for accounting for talk of perspective: the first is the visual way – namely, to use Mach's language, 'with different *K L M* different *A B C* ... are associated'; second, we can take talk of perspective to be essentially talk of interpretation. Within a Machian reading an interpretation of the world, and thus a perspective on the world, is a theory of the world that sets up names for particular clusters of sensory elements and the relations they stand in. Such interpretations in general will involve falsification, since grammar misleads us, perhaps necessarily, to think that our theory refers to objects and picks out explanatory causal relations. A particular claim can be false in a way that can be distinguished from this general falsification: consider the term – the name, as Mach would say – 'desk' that I use to pick out the cluster of mostly brownish elements in front of me. The claim 'There is a desk in front of me' falsifies in that at least 'desk', 'me' and perhaps even 'in front of' involve commitments that go beyond the facts – that go beyond what is indeed true of the sensory elements. The sentence says too much, but part of what it says gets things right. The claim 'There is a desk on top of me' gets things even more wrong and fails drastically for the purposes of, as Nietzsche says, 'designation and communication' and, of course, life (*BGE* 21). Interpretations can thus certainly vary in the degree to which they get the sensory elements right. As Nietzsche would, of course, remind us, getting it right in any case isn't everything. Standard physics and an account of the world in terms of the will to power would be two ways of interpreting the world – two ways of lumping together complexes and picking out relations between complexes – and perhaps even two ways that get things right about equally. But one could always have other grounds for choosing between them.

Historical Context

The suggestions that there are parallels between the ideas of Mach and Nietzsche and that Mach may indeed have influenced Nietzsche have both been made before. In his 1902 work *Nietzsches Erkenntnistheorie und Metaphysik*, one of the very first sustained studies of Nietzsche's epistemology and metaphysics, the Austrian philosopher Rudolf Eisler repeatedly mentions the resemblance to Mach, as well as other positivistic thinkers such as Richard Avenarius, Wilhelm Ostwald and John Stuart Mill.[24] Indeed, he makes clear in his foreword to the book that it was partly the correspondences between 'the "Positivism" of thinkers like E. Mach, W. Ostwald among others which made a critical investigation of Nietzsche's epistemology and metaphysics particularly interesting.[25]

Hans Vaihinger, too, seems to place Nietzsche in the company of Mach and Avenarius. Vaihinger saw Nietzsche as one of his most important predecessors in the cause of his, Vaihinger's, 'idealistic positivism'.[26] In explaining why *The Philosophy of 'As if'* was not published until 1911, despite his having written much of it by the end of the 1870s, Vaihinger gives four reasons why his work could now be introduced and understood: (1) the dissemination of voluntarism (that is, the primacy of the will over the intellect); (2) Mach's and Avenarius's insistence on analysing mental processes in terms of their usefulness for the biological life of organisms and their reduction of things and events to sensory elements; (3) 'the philosophy of Friedrich Nietzsche'; and (4) the spread of pragmatism.[27] Vaihinger does not explicitly ascribe the Machian position to Nietzsche here, but it is clear that he saw the similarities between the two positions.[28]

But the most extensive discussion of the correspondences between Nietzsche's ideas and the positivism of Mach and Avenarius occurs in the work of Hans Kleinpeter (1869–1916). He was what one might reasonably call a follower of Mach. His book, *Der Phänomenalismus* (1913), popularized Mach's philosophical views and placed them within a philosophical tradition of phenomenalism. He examined the parallels between Nietzsche's and Mach's views in extensive detail both throughout his book and in several articles; before publishing his observations, he had even written to Mach himself to point out the kinship.[29] Indeed, the similarities that Kleinpeter notices are basically the ones I have emphasized above in arguing for the helpfulness of seeing Nietzsche through a Machian perspective. Kleinpeter also suggests that Nietzsche may have been reading Mach; in a letter to

Mach dated 9 November 1912, he writes: 'I received the news from Weimar, that Nietzsche read one of your essays in a scientific journal in 1885 and spoke very favourably about it.'[30] In his book he writes that Elisabeth Förster-Nietzsche had personally informed him that when Nietzsche first read Mach, he immediately recognized in Mach a likeminded thinker (*Gesinnungsgenosse*).[31]

Philipp Frank, the physicist and founding member of the Vienna Circle, also emphasized the affinity between Nietzsche and Mach. Frank finds the 'striking agreement' in their epistemological and metaphysical views even more impressive given the sharp contrast in their ethical views.[32]

Not surprisingly, then, John Blackmore, in his biography of Mach, finds it necessary to mention the fact that writers like Kleinpeter and Frank had noticed the similarities between the thought of Mach and Nietzsche, and agrees that both are 'epistemological phenomenalists'.[33] Blackmore, though, is at pains to stress some differences, and I think it is worth taking a moment to consider these. First, Blackmore claims that 'Mach treated sensations as facts and Nietzsche suspected they were only interpretations'.[34] The Nietzsche text which Blackmore cites is the following:

> Against positivism, which halts at phenomena – 'There are only facts' – I would say: No, facts is precisely what there is not, only interpretations. We cannot establish [*feststellen*] any fact 'in itself': perhaps it is folly to want to do such a thing. 'Everything is subjective,' you say; but even this is *interpretation*. The 'subject' is not something given, it is something added and invented and projected behind what there is. – Finally, is it necessary to posit an interpreter behind the interpretation? Even this is invention, hypothesis.
>
> In so far as the word 'knowledge' has any meaning, the world is knowable; but it is *interpretable* in other ways, it has no meaning behind it, but countless meanings – 'Perspectivism'. (*KSA* 12 7[60])

The use of *feststellen* suggests that Nietzsche is making a claim about *our* ability to establish and state some fact about the world and not about the nature of the world itself. Given the nature of language, and given in particular the nature of concepts, a positivism that took our most fundamental statements about the phenomena to be fully and accurately stating the facts about the phenomena would indeed be mistaken. But, as I've tried to show already, this is not a view that we should ascribe to Mach either. And for both Nietzsche and Mach, the world, even if we cannot simply state the facts but can only interpret the world, is still knowable.

The second difference emphasized by Blackmore is the difference in ethical outlook. Here I think Blackmore is right. In one of his very few published references to Nietzsche, Mach makes a disparaging remark about Nietzsche's 'overweening [*frechen*]' *Übermensch* at one point in the later editions of his *Analysis of Sensations*.[35] (Since it is not in the first edition, Nietzsche would not have seen this comment.) Karl Heller suggests that Mach may well have changed the negative opinion expressed here under the influence of Kleinpeter.[36] This seems implausible and is probably unsupported. It is implausible because Kleinpeter appears to be trying to convince Mach that Nietzsche shares Mach's views on the self, on the knowledge and existence of things in themselves, on sensations as metaphysically and epistemologically primary and so on; but surely Mach could accept Nietzsche's agreement with him on these points without changing his opinion of Nietzsche's

Übermensch. The claim is unsupported because we do not actually have Mach's letters to Kleinpeter, and Heller is trying to reconstruct Mach's views from guesses about what Kleinpeter must be responding to in Mach's letters.[37] Heller may well be drawing on his knowledge of all the letters from Kleinpeter to Mach in the Ernst-Mach-Institut in Freiburg; however, the letters from which he quotes only show that Mach had not read, or had hardly read, Nietzsche before Kleinpeter mentioned him and perhaps that Mach had alluded to Nietzsche's unfavourable reputation. Heller quotes the following lines from Kleinpeter in order to show that Mach must have admitted in a previous letter that he had been unfair to Nietzsche: '*I have had exactly the same experience with Nietzsche. Until recently I hadn't read a single sentence by him*. But I have found him to be much better than his reputation would lead one to think'.[38] Kleinpeter then goes on to repeat his view that Nietzsche, like Mach, wants to base everything on sensations. None of this suggests any acceptance by Mach of Nietzsche's ethical views.

I should emphasize that explicit evidence that Nietzsche specifically had Mach in mind is not easy to come by. We should remember that, though Nietzsche does not often publicly acknowledge his debt to other thinkers, this does not mean that there was not a significant influence; consider, for example, the case of Friedrich Lange. Mach's book wasn't published until 1886 and, even assuming that Nietzsche immediately got his hands on it, it could presumably only directly influence *Beyond Good and Evil* and subsequent books. In Chapter 1 of this volume Thomas Brobjer points out that Nietzsche sent Mach his *Zur Genealogie der Moral* at the end of 1887, suggesting that he read Mach that year. Much of what is explicitly laid out in Mach's *Contributions to the Analysis of the Sensations* was, however, already present in his earlier work and in many of his public lectures. Indeed, in the book he points back to this earlier work.

Consider one example. At the end of the long footnote describing his reactions to Kant's *Prolegomena* quoted above, Mach mentions contemporaries such as Avenarius who share his *Empfindungsmonismus* and then refers back to his paper 'Ueber die ökonomische Natur der physikalischen Forschung'.[39] This was originally an 'address delivered before the anniversary meeting of the imperial Academy of Sciences, at Vienna, 25 May, 1882' and was then published in 1882 in the *Almanach der Wiener Akademie*. There Mach summarizes his theory of elements. In a footnote to this summary, he continues as follows:

> I have represented the point of view here taken for more than thirty years and developed it in various writings (*Erhaltung der Arbeit*, 1872, parts of which are published in the article on *The Conservation of Energy* in this collection; *The Forms of Liquids*, 1872, also published in this collection; and the *Bewegungsempfindungen*, 1875).[40]

Indeed, correctly or incorrectly, Mach takes his monism to be a view shared by many other contemporaries in their 'allied thoughts'.[41]

So I think there is quite a bit of support for the claim that both Mach's view and other similar views were 'in the air' early enough to influence Nietzsche. Of course, part of my claim, and Mach's for that matter, is that his particular brand of empiricism, and the more specific monistic claims it involves, is at least one natural development of certain shared conceptions of the role of sensory evidence and the nature of scientific theories that were widespread at the time. The crucial issue is still, of course, whether reading Nietzsche through Mach allows us to make sense of his views – and hopefully more sense than competing interpretations.

A Machian Reading of *Beyond Good and Evil* §15

So does reading Nietzsche through Mach help us make sense of otherwise opaque passages? In this final section I will consider one particular passage, section 15 of *Beyond Good and Evil*, in detail:

> To study physiology with a clear conscience, one must insist that the sense organs are *not* phenomena in the sense of idealistic philosophy; as such they could not be causes! Sensualism, therefore, at least as a regulative hypothesis, if not as a heuristic principle.
>
> What? And others even say that the external world is the work of our organs? But then our body, as a part of this external world, would be the work of our organs! But then our organs themselves would be – the work of our organs! It seems to me that this is a complete *reductio ad absurdum*, assuming that the concept of a *causa sui* is something fundamentally absurd. Consequently, the external world is *not* the work of our organs – ?

Successfully interpreting this passage is clearly fundamental to understanding Nietzsche's metaphysical and epistemological views, including his assessment of the epistemic role of the senses and thus the status of empirical science. It is, however, just as clear that it is not at all obvious how to interpret this aphorism.

Let me note a couple of features of the rhetorical structure of *Beyond Good and Evil* §15. Consider the very first sentence: 'To study physiology with a clear conscience, one must insist that the sense organs are *not* phenomena in the sense of idealistic philosophy; as such they could not be causes!' There are two interesting features of this very first sentence. First, an assertion of what beliefs are needed for someone to study physiology, or rather to do physiology, with a 'clear conscience' – 'Physiologie mit gutem Gewissen zu treiben' – does not require that the asserter actually think that one should have those beliefs. Or, rather, it doesn't require that the asserter think that the relevant beliefs are true. Second, most idealists, particularly the transcendental idealist, would find the claim being made quite peculiar. After all, the domain concerning which causal claims are the most appropriate – the domain for which we are most confident that we understand what we are up to in making causal claims – is precisely that of the phenomenal world. The natural thing to say, if we are speaking in the 'sense of idealistic philosophy', might well be the following: 'We must insist that the sense organs *are* phenomena otherwise they could not be causes (or at least not causes in any sense that we have a clear grip on)!' Nietzsche must surely be putting this sentence in this paradoxical manner on purpose – we are being put on notice that words are being twisted and played with.

The oddity continues in the next sentence: 'Sensualism, therefore, at least as a regulative hypothesis, if not as a heuristic principle.' This only makes the plot thicken. The suggestion clearly is that this claim about sensualism is supposed to follow as a consequence. Sensualism is epistemic empiricism – that is, the claim that all our information about the world comes through the senses. This claim, as Lange points out, does not, of course, commit one to any particular ontological claims or, for that matter, semantic ones. Thus, even if one accepted sensualism, one could still reject materialism and physiology.

Notice, finally, that the suggestion surely is that there are two arguments in *Beyond Good and Evil* §15, where the second argument is a *reductio* and involves appealing to the absurdity of a *causa sui*. Indeed, the structure of the passage suggests that the first argument *doesn't* involve a *reductio* or *causa sui*. Last but not least, the passage ends with a question rather than an explicit conclusion.

So it would be desirable to have an interpretation that is sensitive to the rhetorical and logical complexity of the passage. I will begin by picking up on a clue in the third sentence: 'What? And others even say that the external world is the work of our organs?' This suggests, I think, that the discussion in the first two sentences – the first argument – concerns a purported 'inner world'. Indeed, I want to read this passage as attempting to break down the distinction between inner and outer world, as another confusion haunting us along with the distinction between phenomena and thing-in-itself.[42]

Note that the first argument, as I've already suggested, cannot really be an argument against Kant, since phenomena, in Kant's sense, are certainly quite suited to being causes. So talk of inner and the outer probably isn't just talk of Kantian phenomena and noumena. But of course there are materialist and, perhaps, neo-Kantian pictures on which an inner world of representations, themselves not causes, are causally generated by the sense organs. That Nietzsche is presumably taking such positions as a target becomes clear when we notice the similarity between Nietzsche's argument here and Lange's discussion of such materialist positions. Although, as I have suggested, there is no *reductio* obviously being run in this argument, there is nonetheless lurking in the background a *reductio* that *Lange* runs – a *reductio* that will now sound surprisingly familiar. Lange suggests that the physiology of the sense organs 'leads us to the very limits of our knowledge, and betrays to us at least so much of the sphere beyond it as to convince us of its existence'.[43] Now, Lange thinks that although such physiological investigation into the sense organs may look favourable for the materialists – in that it promises to give us a materialistic account of our knowledge of the world – it is in fact deadly. Physiology shows us that the sense organs *don't* show us how the world really is and, indeed, that our very concept of matter may have nothing to do with what is really there in the world.[44] And thus materialism, as the belief in 'material, self-existent things', is thoroughly undermined: the 'consistently Materialistic view thus changes around, therefore, into a consistently idealistic view'.[45] Lange draws the following conclusions:

> 1. The sense-world is a product of our organization.
> 2. Our visible (bodily) organs are, like all other parts of the phenomenal world, only pictures of an unknown object.
> 3. The transcendental basis of our organization remains therefore just as unknown to us as the things which act upon it. We have always before us merely the product of both.[46]

He summarizes the chapter by saying:

> The senses give us ... *effects* of things, not true pictures nor things in themselves. But to the mere effects belong also the senses themselves, together with the brain and the molecular movements which we suppose in it. We must therefore recognize the existence of a transcendental order of things.[47]

Now, surely Nietzsche must have had this section of Lange in mind when writing *Beyond Good and Evil* §15.[48] Nietzsche is not, presumably, going to draw the conclusion in favour of a 'transcendental order of things'. What is interesting, and I think crucial, is that Nietzsche doesn't try to block the argument by showing, on physiological grounds, that the senses do in fact show us the way the world is; in other words, he doesn't take on all the physiological arguments for falsification by the sense organs either here or elsewhere. Instead, he argues that on grounds of consistency – my interpretation of 'clear conscience' – the physiologist must accept the sensations as showing us how things are in themselves. Let me summarize what I take the argument to be, inspired obviously by the discussion in Lange:

(1) According to physiology, the sense organs causally generate an internal world of representations.
(2) If the sense organs aren't showing me accurately what the external world is like, then my term 'sense organs' refers only to certain representations in the internal world.
(3) According to physiology, these representations in the internal world are effects not causes.

From (1) and (3),

(4) The sense organs can't just be representations.

And given (2), we get

(5) The sense organs must indeed be showing me accurately what the external world is like.

Now (5) is sensualism, just as we need it to be.

As I've suggested earlier, this does not, of course, require that Nietzsche actually accept such a physiological account! Notice that if he did, then he *would* have to deal with the problems raised by physiological arguments for falsification. Hence the fact that he does not suggest that he does not himself intend to endorse physiological claims about the role of the sense organs. This in turn explains the features of the second sentence, which Clark does not deal with – namely, that sensualism is accepted as a 'regulative hypothesis, if not as a heuristic principle'. Here, I take it, Nietzsche is speaking in his own voice, but in his own voice he only accepts it as a 'regulative hypothesis, if not as a heuristic principle' which will guide a certain interpretation of the world where an interpretation of the world is to be contrasted with a 'world-explanation' (*BGE* 14).[49]

This is the contrast emphasized in the preceding §14 of *Beyond Good and Evil* and, later, in §21. Accepting some set of claims, say those of physics, as an interpretation of the world is not to regard them as literally describing the actual structure of reality. They do not actually pick out explanatory structures. Talk of cause and effect should not be used for explanation – that would require that our talk of causation actually pick out some causal relation in the world – rather, 'one should use “cause” and “effect” only as pure concepts, that is to say, as conventional fictions for the purpose of designation and communication’ (*BGE* 21). A Machian reading of Nietzsche interprets such claims as reminders that language, including the language of science, falsifies. Once we realize this, we can accept a claim that is part of an interpretation of the world without having to accept that the claim is literally true. Perhaps a truly clear conscience would require believing the claim to be literally true, but nonetheless we do have the option now of accepting it as ‘only an interpretation’ – or as a ‘fiction’.

Even a physiological approach to the world, despite its claims about the senses – the claims Nietzsche would learn from Lange – would, on grounds of consistency, have to take the senses as showing us the way the world really is. Both Lange and Nietzsche are insisting that the sense organs can’t consistently both be causes in the materialist’s sense and fail to show us the way the world is. And, for both Lange and Nietzsche, the way out is to do physiology *without*, so to speak, a good conscience. For Lange, we can do physiology without a good conscience because we can take physiology to apply solely to the phenomenal world and thus the sense organs can happily be causes – but not in the materialist physiologist’s sense of ‘cause’, since the materialist is presenting his theory as a comprehensive account of the world.[50]

For Nietzsche, this is not an option since it rests on a distinction between the thing-in-itself and the phenomenal world. So Nietzsche needs another way out and the passage points to this need, presuming the obvious background in Lange. The Machian account would give him precisely that: a monism of sensory elements allows one to do physiology without a good conscience since one can think of the sense organs as causes – but again not in the materialist physiologist’s sense of ‘cause’. Physiology would be part of the causal interpretation – and thus falsifying interpretation – of the world that is physics.

What has been undermined is the consistency of a view that takes the inner world as generated. And, of course, the Machian account is a view that precisely does not assume that the ‘inner world’ is generated and, indeed, is a view that fits, in its own way, with the conclusion that the senses do show us the way the world is – but not in the physiologists’ sense, of course.

Let us now turn to the second argument in *Beyond Good and Evil* §15. Here we deal with a different claim – the opposite claim to that made in the first argument – namely, that the outer world is the work of our sense organs. ‘Work’ here means ‘causation’. And here, I take it, the target is again a certain kind of contemporary physiologist, perhaps one of those committed to some theory of ‘external projection’.[51] Here, Nietzsche suggests, we are left with *reductio ad absurdum*, namely that we are forced to posit a *causa sui*. This is different from the first argument, since there we were not forced to claim something absurd in itself, but rather ended up with a contradiction in the theory that could be resolved by giving up one or the other of the claims – we were not forced to a conclusion that involved a notion contradictory in itself.

The *reductio* is fairly straightforward so I will not go into it in detail. The interesting feature of the second argument – or rather the second part of the aphorism – is the last question. An obvious response to the *reductio* is to reject the claim that the external world is the work of our organs. One way of rejecting this claim is to *accept* the claim that the external world is *not* the work of our organs.

But I take it that the question at the end of the passage is designed to prevent any straightforward version of this way out. The suggestion clearly is that there is a 'third' way – and a way that in some altered sense of the words involved might still be, perhaps misleadingly, expressed by the claim that the external world *is* the work of our organs. Again, the Machian interpretation provides just such an interpretation. The world of sensory elements is the only world there is; as Mach remarked those collections of sensory elements, *A B C* ..., that aren't my body, *K L M* ..., may for some purposes be plausibly treated as 'external' to my body – this is, as he suggests, what we do in physics. But, of course, to quote Mach again:

> Precisely viewed, however, it appears that the group *A B C* ... is *always* codetermined by *K L M*. A cube of wood when seen close at hand, looks large; when seen at a distance, small; it looks different with the right eye from what it does with the left; sometimes it appears double; with closed eyes it is invisible. The properties of the same body, therefore, appear modified by our own body; they appear conditioned by it. But where, now, is the *same* body, which to the appearance is so *different*? All that can be said is, that with different *K L M* different *A B C* ... are associated.[52]

Thus, changes in our sense organs, understood as particular complexes of sensory elements, are connected to changes in the way things in the external world are, where talk of the external world is now understood as the complexes that do not form my body or my thoughts. Or, as he puts it more dramatically later:

> All elements *A B C* ..., *K L M* ... constitute a *single* coherent mass only, in which, when any one element is disturbed, *all* is put in motion; except that a disturbance in *K L M* ... has a more extensive and profound action than in *A B C*. A magnet in our neighbourhood disturbs the particles of iron near it; a falling boulder shakes the earth; but the severing of a nerve sets in motion the *whole* system of elements.[53]

To sum up in Mach's words:

> There is no rift between the psychical and the physical, no inside and outside, no 'sensation' to which an external 'thing,' different from sensation, corresponds. There is but one kind of elements, out of which this supposed inside and outside are formed – elements which are themselves inside or outside, according to the aspect in which, for the time being, they are viewed.[54]

Conclusion

Let me conclude with a few comments about what I hope to have shown. I began by suggesting that, in attempting to save Nietzsche from being set up as the forerunner for various kinds of radical scepticisms, we are often heartened by his emphasis on the senses and science. This, and much else he says, can remind us of the work of a range of Nietzsche's contemporaries and predecessors, many of whom were quite science-friendly. There were, as I said at the beginning, however, a bewildering range of views at that time and there was, and is, a bewildering range of labels which could reasonably be applied singly and jointly to many of these views: empiricism, materialism, positivism, naturalism, sensualism, not to mention all the neo-Kantian schools and personalities. And again, let me emphasize, many of these views were, in their own, but crucially different, ways friendly to the senses and science. So placing Nietzsche must be done with care – both historical care and, given the ambiguity of many of the labels, semantic care.

I then turned to Nietzsche's own very famous history of metaphysics and epistemology – the history, of course, of an error. He positions himself after the 'cockcrow of positivism' at the end of this error in *Twilight of the Idols*. I suggested that it might make sense to look at his contemporaries who placed themselves, or can obviously be placed, in just this manner. I suggested the name of Ernst Mach. I did this for a couple of reasons: first, as I hope to have shown, because much of what Mach says is strikingly similar to what Nietzsche says, in particular in his attempt to get rid of the distinctions between appearance and reality and the internal and external world; second, a Machian reading of Nietzsche provides insight into what are otherwise opaque texts and doctrines; and, finally, because Mach's own views and career show us how one can take science and the senses quite seriously without straightforwardly accepting 'the common sense picture of the world of relatively enduring middle-sized objects or the scientific world-view'.[55] There is thus useful rhetorical force in using someone whose pro-science credentials are hard to criticize to interpret and articulate Nietzsche's similarly sceptical, though more strident, claims about physics, and our common-sense view of the world, as merely interpretations of the world – and interpretations that simplify and mislead and thus falsify.

Notes

1 I would like to thank Thomas Brobjer, Maudemarie Clark, Dagfinn Føllesdal, Kathleen Higgins, Christopher Janaway, Agnieska Jaworska, Krista Lawlor, Brian Leiter, Greg Moore, John Perry, John Richardson, Tamar Schapiro, Larry Sklar, Robert C. Solomon, Susan Sterrett, Michael Strevens, and Allen Wood for useful discussions of the issues raised in this paper. Thanks to an audience at Yale University especially Karsten Harries and Pierre Keller. Thanks to Lanier Anderson for several very helpful conversations. Thanks to Iain Morrison for his comments on a presentation of some of this material at a Nietzsche conference at the University of Texas, Austin. Finally, I would like to thank David Hills for extremely useful comments on an ancestor of this paper and general discussions of related matters over a period of many years.

2 See, for example, Maudemarie Clark's and Brian Leiter's introduction to *Daybreak*, trans. R.J. Hollingdale (Cambridge: Cambridge University Press, 1997), pp. vii–xxxvii. Clark and Leiter emphasize the empiricist, materialist and naturalist influences on, and sympathies of, Nietzsche. See also Ken Gemes, 'Nietzsche's Critique of Truth', *Philosophy and Phenomenological Research*, 52 (1992), 47–65; Brian Leiter, 'Perspectivism in Nietzsche's *Genealogy of Morals*', in Richard Schacht (ed.), *Nietzsche, Genealogy, Morality: Essays on Nietzsche's 'Genealogy of Morals'* (Berkeley, CA: University of California Press, 1994), pp. 334–57; Robert Nola, 'Nietzsche's Theory of Truth and Belief', *Philosophy & Phenomenological Research*, 47 (1987), 525–62; Richard Schacht, *Nietzsche* (London: Routledge, 1983), pp. 52–117; Kenneth Westphal, 'Nietzsche's Sting and the Possibility of Good Philology', *International Studies in Philosophy*, 16 (1984), 71–90; idem, 'Was Nietzsche a Cognitivist?', *Journal of the History of Philosophy*, 22 (1984), 343–63; John T. Wilcox, *Truth and Value in*

Nietzsche: A Study of His Metaethics and Epistemology (Ann Arbor, MI: University of Michigan Press, 1974), pp. 155–70; and Maudemarie Clark, *Nietzsche on Truth and Philosophy* (Cambridge: Cambridge University Press, 1990).

3 For forceful statements of this reading of Nietzsche as a naturalist, see Clark and Leiter, 'Introduction', pp. vii–xxxvii; Brian Leiter, *Routledge Philosophy Guidebook to Nietzsche on Morality* (London: Routledge, 2002), pp. 1–72.

4 Leiter, *Nietzsche on Morality*, pp. 15–25.

5 Clark, *Nietzsche on Truth*. See also Clark's 'On Knowledge, Truth and Value: Nietzsche's Debt to Schopenhauer and the Development of Empiricism', in Christopher Janaway (ed.), *Willing and Nothingness: Schopenhauer as Nietzsche's Educator* (Oxford: Clarendon Press, 1998), pp. 37–78. I am not suggesting that the talk of interpretations in *BGE* 14 entails falsification, but the naturalist is normally construed as taking science to be successfully in the business of generating explanations.

6 Auguste Comte, *Introduction to Positive Philosophy*, trans. Frederick Ferré (Indianapolis/New York: Bobbs-Merrill, 1970), p. 1. Nietzsche owned the German edition of this work: Auguste Comte, *Einleitung in die positive Philosophie* (Leipzig, 1880).

7 Comte, *Positive Philosophy*, p. 2.

8 Ibid., p. 8.

9 John Stuart Mill, *Auguste Comte and Positivism*, in vol. 10 of *Collected Works of John Stuart Mill* (Toronto: University of Toronto Press, 1969), pp. 265–66. This, too, is a source on Comte that Nietzsche may well have read. Nietzsche's library contained a marked-up copy of John Stuart Mill's collected works translated into German including 'August Comte und der Positivismus'. Lange also refers to this book in his discussions of Comte in *The History of Materialism*, trans. Ernest Chester Thomas, 2 edn (London: Trübner, 1879).

10 *PTA* 5–8; *D* 83, 86, 119; *GS* 54, 57–59, 109; *GM* iii 16; *A* 13–14, 59; *BGE* 23, 134, 230; *TI* 'Errors' 6; 'Reason' 1–3, 6; *TI* 'World'; *EH* 'Clever' 2; *KSA* 13, 15[51], [90].

11 Cf. Bernd Magnus, *Nietzsche's Existential Imperative* (Bloomington: Indiana University Press, 1978), pp. 132–36.

12 Mach produced several later editions that included more material. The 1897 English translation of the first edition already includes material not present in the first German edition of 1886. This new material is then incorporated into, and further expanded, in later German editions. The second German edition did not appear till 1900 and is thus too late to influence Nietzsche. I have cited the English 1897 translation of the first edition. However, all texts cited are present in the first German edition of 1886 (Ernst Mach, *Beiträge zur Analyse der Empfindungen* (Jena: Fischer, 1886)).

13 Ernst Mach, *Contributions to the Analysis of the Sensations*, trans. C.M. Williams (Chicago: Open Court, 1897), p. 23 n.1.

14 Ibid., pp. 10, 12. Manfred Sommer, *Husserl und der frühe Positivismus* (Frankfurt am Main: Klostermann, 1985), p. 18, uses the label *Empfindungsmonismus* for Mach's view. For a quick summary of Mach's view, see Walter Del-Negro, *Konvergenzen in der Gegenwartsphilosophie und die moderne Physik* (Berlin: Duncker & Humblot, 1970), pp. 11–12.

15 Mach, *Contributions*, pp. 19, 3, 5–6, 6.

16 Ibid., p. 8.

17 Ibid., p. 11.

18 Ibid., p. 9 n.1.

19 Ibid., p. 9.

20 Ibid., p. 22. Presumably it is not just physicists who make this mistake.

21 Ernst Mach, *Popular Scientific Lectures*, trans. Thomas J. McCormack, 3 edn (Chicago: Open Court, 1898), p. 192.

22 Mach, *Contributions*, p. 22. See also Georg Christoph Lichtenberg, *Schriften und Briefe* (Munich: Hanser, 1967–72), vol. 2, p. 412.
23 Mach, *Contributions,* p. 152.
24 Rudolf Eisler, *Nietzsches Erkenntnistheorie und Metaphysik: Darstellung und Kritik* (Leipzig: Hermann Haacke, 1902), pp. iii, 26, 35, 57, 60, 62, 65–66, 71.
25 Ibid., p. iii. In this foreword he also emphasizes the similarities between Nietzsche's views and the '"Voluntarism" of modern philosophers like Wundt'.
26 Hans Vaihinger, *Die Philosophie des 'Als ob': System der theoretischen, praktischen und religiösen Fiktionen der Menschheit auf Grund eines idealistischen Positivismus*, 5th and 6th edn (Leipzig: Meiner, 1920).
27 Ibid., pp. xiii–xv. This introduction does not seem to be present in the English translations.
28 Cf. Walter Del-Negro, *Die Rolle der Fiktionen in der Erkenntnistheorie Friedrich Nietzsches* (Munich: Rösl, 1923), pp. 192–97.
29 Hans Kleinpeter, *Der Phänomenalismus, eine naturwissenschaftliche Weltanschauung* (Leipzig: Barth, 1913) and, for example, idem, 'Ernst Mach und Friedrich Nietzsche', *Neue Freie Presse*, 23 February 1913. Kleinpeter's letters to Mach are cited in: K.D. Heller, *Ernst Mach: Wegbereiter der modernen Physik* (Vienna and New York: Springer, 1964). Unfortunately, Mach's letters to Kleinpeter are lost; see John T. Blackmore, R. Itagaki, and S. Tanaka (eds), *Ernst Mach's Vienna, 1895-1930, or, Phenomenalism as Philosophy of Science* (Dordrecht/Boston: Kluwer, 2001), p. 232 n.39.
30 Quoted in John T. Blackmore, *Ernst Mach: His Work, Life, and Influence* (Berkeley, CA: University of California Press, 1972), p. 123.
31 Kleinpeter, *Der Phänomenalismus*, p. 143 n.1.
32 Phillipp Frank, *Between Physics and Philosophy* (Cambridge, MA: Harvard University Press, 1941), pp. 51–54. He cites *KSA* 12, 8[2], 9[40], *KSA* 13, 14[103].
33 Blackmore, *Ernst Mach*, p. 123.
34 Ibid.
35 Ernst Mach, *The Analysis of Sensations* (New York: Dover, 1959), p. 25.
36 Heller, *Ernst Mach*, p. 69 n. 1.
37 Blackmore, Itagaki and Tanaka, *Mach's Vienna*, p. 232 n.39.
38 Heller, *Ernst Mach*, p. 70.
39 Mach, *Contributions*, p. 23 n.1.
40 Mach, 'Economical Nature of Physical Inquiry', in *Popular Scientific Lectures*, p. 210 n. The collection referred to is his *Popular Scientific Lectures.*
41 Mach, *Contributions*, p. 23 n.1.
42 As further evidence for my reading, I point to *BGE* 16, where Nietzsche attacks the idea of 'immediate certainties' – an attack on the notion of privileged access to some inner realm.
43 Lange, *Materialism*, vol. 3, p. 202. Lange thus also fails to follow Nietzsche and Mach to stages five and six outlined in *TI* 'World'.
44 Ibid., pp. 205–19.
45 Ibid., pp. 215, 23.
46 Ibid., p. 219. See also pp. 223–24.
47 Ibid., p. 230.
48 Indeed, it is this argument from Lange that Nietzsche approvingly quotes in a letter to his friend Carl von Gersdorff at the end of August 1866 (*KSB* 2, pp. 159–60) reporting his recent discovery of Lange's *History of Materialism* and recommending the book. In the letter Nietzsche quotes, of course, from the 1866 first edition in which the first sentence of the third step reads differently: 'Our real organization remains therefore just as unknown to us as the real external objects' (Lange, *Geschichte des Materialismus und Kritik seiner Bedeutung in der Gegenwart*, 1st edn (Iserlohn: Baedeker, 1866), p. 493).

I do not think this affects the basic point of the argument in the text. Nietzsche does take Lange's talk of 'real' organization and things to be equivalent to talk of transcendental things-in-themselves; after quoting this section from Lange in his letter, he continues to summarize Lange's point as follows: 'Thus the true essence of things, the thing in itself, is not only unknown to us, but also the very notion of it is nothing more nor less than the final product of a contrast dependent on our organization.'

49 Cf. the use of 'regulative Hypothese' in *KSA* 11, 34[247]. See also *GS* 344, *KSA* 11, 26[263] and the discussion, and references, in Hans Vaihinger, *The Philosophy of 'As if'*, tr. C.K. Ogden (London: Routledge & Kegan Paul, 1935), p. 353 n.2.

50 Whether Lange is really right about this is a good question. Surely Nietzsche's point raises a puzzle for Lange. Would a materialist story understood purely in terms of the phenomenal world avoid the contradiction Nietzsche is pointing to? This depends in part on whether or not Lange's translation retains some version of the causally epiphenomenal inner world or not.

51 Lange, *Materialism*, vol. 3, pp. 208–17; Mach, *Analysis of Sensations*, p. 39.

52 Mach, *Analysis of Sensations*, p. 9.

53 Ibid., p. 14.

54 Ibid., p. 310.

55 Clark, *Nietzsche on Truth*, p. 108.

PART II
NIETZSCHE, SCIENCE AND PHILOSOPHY

Chapter 6

Nietzsche's Critique of Scientific Reason and Scientific Culture: On 'Science as a Problem' and Nature as Chaos

Babette E. Babich

William J. Richardson once suggested that 'on the longest day he ever lived Heidegger could never be called a philosopher of science'.[1] It might seem self-evident that Nietzsche, like Heidegger, was anything but a philosopher of science. But this would be far from the truth: Nietzsche was engaged with the problem of science *per se* and the questions he raised are of ongoing philosophical importance.

From his early to his later work, Nietzsche was preoccupied (the word is not too strong) with manifestly epistemological issues and questions concerning science. Thus in the well-known, but in his lifetime unpublished, essay, *On Truth and Lie in an Extra-Moral Sense*, the extra-moral (that is, beyond blame and ethical fault or good and evil) is the logical matter of truth, metaphorical reference and, even more precisely, the relation between truth and lie in an observational, all-too eco-physiological sense.[2] In the context of a glossed engagement with Gustav Gerber's 1871 *Die Sprache als Kunst* (the same Gerber would go on to publish, not at all coincidentally, *Die Sprache und das Erkennen* in 1884), Nietzsche attends to truth and lie in language, which is also to say, as Josef Simon has argued,[3] that his concern is with valid and invalid statements. For Simon, Nietzsche's propositional preoccupation concerns the possibility of knowledge as such. And Nietzsche offers a number of patent knowledge claims: in addition to his notorious teaching of the eternal return of the same or the totalizing nature of the world as will to power, he makes manifestly metaphysical claims about the death of God, as well as developing critically reflective inquiries into the cognitive status of our knowledge of the world and ourselves as knowing subjects. Furthermore, he makes specific claims regarding logical truth and concerning mathematical representations of phenomena and the theoretical and experimental practice of natural science as such, going as far as to single out physics in particular for a series of radically provocative but unmistakably focused observations. Nietzsche is concerned with truth, the physical world of science, and the potential for, or limits of, knowledge.

All this notwithstanding, it remains the case that those who take Nietzsche seriously are inclined not to take his claims on truth and lie or the possibility of knowledge seriously, let alone his cosmological or ecological theories, nor his claims regarding scientific theory (from Darwin to Boscovich and Democritus to the physio-psychology of diet). More typical is an attention to Nietzsche's powers of 'invention', where Nietzsche is thought not only to have invented a conception of life 'as literature' – to use the convention favoured by Alexander Nehamas and Michel Foucault – but entire subject areas such as the 'pre-Socratics', or even the idea of Dionysus in the birth of 'tragedy'.[4] And, so the commentator's reasoning would appear to go, if Nietzsche can 'invent' whole vistas of fictive conventions, he can also be said to have invented a vision of truth, a vision of science and even a vision of what physicists do (described with a word from his own discipline as an 'interpretation' of, rather than a given 'fact' about, the world) (*BGE* 22). Thus when Nietzsche rhetorically remarks that it 'is perhaps dawning on five or six minds that physics too is only an interpretation and arrangement of the world' (*BGE* 14), or identifies a certain scientific methodology (simplicity) as a deliberate articulation of the 'greatest possible stupidity', or criticizes the 'prejudices' of science or the 'crudity and naïveté' of calculative scientific convention (*GS* 123, 373) or debunks the Enlightenment conviction concerning the opposition between *logos* and myth, declaring that the 'biggest myth is that of knowledge' (*KSA* 12, 2[154]), his claims appear as so much excess rhetoric: more hyperbole than philosophically earnest propositions.

In this way, and as a kind of clearing of the conceptual air, recent interpretive trends suggest that *whatever* Nietzsche was doing (be it rhetoric or cultural criticism or 'therapy' or just and *merely* playing parodically with our own all-too-earnest approach to scholarship), it wasn't/isn't philosophy. Hence (analytic) philosophers are not to worry. Even if Nietzsche appears to be talking about philosophy he is only fooling (experimenting) with ideas – not claiming the mantle of philosopher. In particular, this interpretive tendency is characteristic of current studies of Nietzsche's 'rhetoric' and style (although contemporary conceptions of rhetoric and Nietzsche's own formative specialization in the classical tradition of ancient Greek and Roman rhetoric engage importantly different arenas).

This welter of perspectives means (at a minimum) that it is difficult to thematize the nature of Nietzsche's very stylized interest in science. There is a wide difference between an ironic concern with science and a mistaken one, and both ironic and mistaken perspectives differ from the philosophical concern with science as such. That Nietzsche was interested in science is obvious. Indeed, Nietzsche *and* science can manifestly be conjoined and – to echo the logic of Kant as much as that of Suarez or Aquinas – whatever *is* is consummately possible, where determining the conditions of that same possibility is what requires further critical reflection. Here, the question will be: *does* what Nietzsche have to say about science tell us anything about the project of science, or only, and at a minimum, something about Nietzsche? Only in the former case is Nietzsche's thought of value for the philosophy of science.

Certainly, the substance of Nietzsche's claims concerning science are extreme with regard to science as it is usually conceived. Although there are patent exceptions, to read Nietzsche as a philosophic voice concerning science remains a subversive thing to do. In addition to its subversive aspect, the topic of Nietzsche and science invites equivocation. For whatever draws other scholars to find affinities between Nietzsche's thinking and the abstract theories of modern physics (notably quantum mechanics via Boscovich, but also Einstein), Nietzsche's conceptualization of 'science' is not today's modern, technological and information-age science. I will return to this equivocal aspect below, but first it is important to consider the context of Nietzsche's original discussion in *The Birth of Tragedy*. Two points are relevant. The key locus is what Nietzsche identified as the 'problem of science' in his self-critical, second preface to his first book. The second is his discussion of science as such in *The Birth of Tragedy* itself.

'The Problem of Science Itself' and Alexandrine Science

Nietzsche proclaims that '*the problem of science itself*, science considered for the first time as problematic, as questionable' was at the heart of his book on tragedy (*BT* 'Self-Criticism' 2). Nietzsche's effort 'to view science through the lens of the artist and art through that of life' was addressed not to what we think of as (natural) science today but to the science of classical philology, much as Kant had been able to speak of metaphysics as a science of the future (a future metaphysics). This interest in the 'future' of science was Nietzsche's abiding concern.[5] Hence, and from a foundational point of view, the 'problem of science' necessitates establishing a science of the future, analogous to Nietzsche's vision of a philosophy of the future, which resembles the Kantian project of installing a discipline (be it metaphysics or philology, even philosophy itself) on the critical highway of science. Because the problem of science qua science – that is, the problem of the scientificity of science – also corresponds to the logical problem of reflexivity, the problem of science in general likewise calls for this same critical reflection. But, notoriously, this means that science as such, not excluding natural or presumably 'mathematizable' sciences, cannot be critically conceived (or founded) on its own ground.

For Nietzsche, the achievement of Kant's critical philosophy (for Nietzsche this same critical philosophy included Schopenhauer) was its overcoming of the logical optimism of contemporary science, a logical optimism which is still in evidence today as the positivist confidence that knowledge is both possible (in theory) and attainable (in practice), that 'all the riddles of the universe could be known and fathomed' (*BT* 18). Kant's critical philosophy, conceiving 'space, time and causality as entirely unconditioned laws of the most universal validity', demonstrated, in Nietzsche's view, that these same concepts 'really served only to elevate the mere phenomenon ... to the position of the sole and highest reality, as if it were the true essence of things' (*BT* 18). It is Nietzsche's claim here that Kant's critical perspective is itself a continuation of what Nietzsche calls 'Socratic culture'. This is the Alexandrine (or, to use, Nietzsche's later terminology, slavish) culmination of such a logico-scientific vision: 'the delusion of limitless power'. For Nietzsche, it is on the basis of 'the belief in the earthly happiness of all' that one can trace the genesis of the modern, technological, and ultimately consumerist 'demand for such an Alexandrine earthly happiness, into the conjuring up of a Euripidean *deus ex machina*' (*BT* 18).[6] Hence Nietzsche's proposal will be to turn the Kantian critical project against itself, and so to take it further as a move utterly in keeping with the same tradition.

Nietzsche argues that critical philosophy takes the culture of scientific reason or logic to its utmost, but still fully logical, consequences. Using 'the paraphernalia of science itself', the critique of scientific reason would thus effectively outline 'the limits and the relativity of knowledge generally' – ultimately denying 'decisively the claims of science to universal validity and universal aims' (*BT* 18). In this way, Kant's philosophic legacy signalled the *logical* destruction of 'scientific Socratism's complacent delight in existence by establishing its boundaries' (*BT* 19).

The Birth of Tragedy features a focus on science not because of the modern importance of science but because in the person of Socratic optimism – that is, the confidence in the power of reason, in the spirit of 'logical' optimism – 'science' presided over the death of tragedy. It is possible (Nietzsche argues) to propose a rebirth of tragic culture – precisely where 'the man of theory' recognizes the fatal limitations of his own enterprise (including science) in the absence of the foundations that it itself claims as necessary and which it itself has undermined as such (*BT* 19).

On logical grounds alone, we have noted, empirical science cannot be regarded as self-grounding (Nietzsche would challenge the idea that anything be thought to be self-grounded or, indeed, self-caused). Hence Nietzsche regarded the task of raising the problematic question of science as necessary on critical grounds. The alternative foundation for such a study would be art, and Nietzsche accordingly identified this as the methodology behind his own project. Thus he would not merely refer science to art by taking a leaf from the book of artistic culture, but went further to define art itself in terms of the ultimate perspective of life.

For contemporary readers who are not inclined to speak of 'scientific Socratism' and who do not think in terms of the categories of Alexandrine culture or Euripidean machinations (that is, apart from Nietzsche's Socratism or his Alexandrine vision or critical understanding of Euripides), the question of the meaning of science for Nietzsche can be troublesome. How can one take Nietzsche's reflections on a particular scholarly discipline (even if he does name it the 'science of aesthetics') and apply it to a philosophical reflection on science as such?

It is easy to protest that Nietzsche's claim to have raised the question of the problem of science as such is a non-starter. There is no 'problem of science' as Nietzsche puts it, simply because – as received wisdom tells us – science is itself the basic means for 'solving' problems. In this way, the philosophy of science is a straightforward celebration of, and inquiry into, the enabling conditions of science. Science thus constitutes its own self-evident or axiomatic foundation and, anxieties about Euripidean machinations to the contrary, the achievements of technological modernity therefore prove the (perfectly, positively, optimistic) success of science. In this fashion, the debate can be concluded before it begins. Following a similar line of thought, if it may be said – as one scholar has recently claimed – that Nietzsche cannot be said to 'have' a 'philosophy',[7] how is one to argue that Nietzsche has a philosophy of science, a *Wissenschaftstheorie*?

Nietzsche's philosophy of science, as I have argued elsewhere, would entail rethinking the aim of the philosophy of science, not giving it a woollier, but rather a much more rigorous, focus on science as such. Here, as a further preliminary to such a critical philosophy of science, it is helpful to begin by asking what Nietzsche meant by *Wissenschaft*, the term he employed to speak of what we call 'science'. The German word *Wissenschaft* is not univocal, and it is worth reflecting, in particular, on its often stated but rarely discussed difference from the English word 'science'. For the Nietzsche whose first published text begins with a promise of gaining an advantage for what he calls 'die aesthetische Wissenschaft' (*BT* 1), and who went on to write a still insufficiently appreciated book entitled *Die fröhliche Wissenschaft* (*The Joyful Science*), an inquiry into the meaning of *Wissenschaft* would seem invaluable, even beyond the present context.

Nietzsche alludes to the etymological stem of *wissen* in Greek (*PTA* 3) and in several contexts makes much of the limited focus on a single sense (notably ocular) that counts for the tradition of science since the Greeks. In this context, we can see the importance of Nietzsche's emphasis on the difference that might be made by including the full sensible range of the human body. Thus, in addition to the sense of sight, Nietzsche speaks of hearing and the senses of taste and even smell and touch. It is in this same context of the body as such a complex knowing instrument, widely attuned or keyed to all its senses rather than merely reduced to the single privileged primacy of vision, that Nietzsche could call the body the 'great reason, a plurality with one sensibility, a war and a peace' (*Z* i 'Despisers of the Body'), as contrasted with the intellect which he downplays as our 'four-square little human reason' in the materialist context of empirical science (*GS* 373).

The German word *Wissenschaft* dates from the fourteenth century, a conventional term coined in a theological and mystical context as a translation of the same *sciens, scientia* from which the English word *science* is derived. But the English 'science' (rooted in the Latin, *scire* (to know) and related to *scindere* (to cut, divide)) does not render the more complex set of associations implied by *Wissenschaft*. *Wissenschaft* is a German word-form which maintains a powerful array of etymological connections via *wissen* linking it to the Old High German *wizzan* and Old Saxon *wita*, but also the English *wit* and *wot*, as well as to the Sanskrit *veda* and the ancient Greek *oidá*, as well as the Latin *videre*.[8]

By contrast, the English term 'science' is not (as 'wisdom' is) an Anglo-Saxon word. Yet it is instructive that 'science' has had its current 'non-arts' connotation only since the eighteenth century. Certainly, although *Wissenschaft* also and increasingly shares this same limiting focus on the mathematical and natural sciences, it has an inherently broad usage, both in Nietzsche's time and our own. Thus the *Wildhagen–Héraucourt German–English Dictionary* refers first to natural science before listing the extended definition of '*Wissenschaft*' as 'learning, scholarship, erudition, and knowledge'. Although, as first coined, *Wissenschaft* originally referred only to simple knowledge – *Der kleine Pauly* notes, for example, Goethe's '*davon hab' ich kein' Wissenschaft*' – it currently corresponds to the collective pursuit of species of knowledge. In distinguishing between science and *Wissenschaft* in English and German usage, with important consequences for the discipline of the philosophy of science in particular, this complex difference continues to make all the difference. The *Wahrig* dictionary thus defines '*Wissenschaft*' as 'ein geordnetes, folgerichtig aufgebautes, zusammenhängendes Gebiet von Erkenntnissen', while the *Shorter Oxford English Dictionary* defines 'science' as 'the state or fact of knowing; knowledge or cognizance *of* something specified or implied' and features a separate definition of '*Wissenschaft*' as the '(systematic pursuit of) knowledge, science, learning, scholarship'. *Wissenschaft* defined in terms of an ordered, systematic and coherent disciplinary arena of knowledge corresponds only to the last sub-entry in the *OED*: 'The kind of organized knowledge or intellectual activity of which the various branches of learning are examples.' As the noun corresponding to *wissen*, *Wissenschaft* carries the connotations of the 'ways' or conduits of knowing – as heard in English with the archaic *wis* (to show the way, to instruct) or *wist* (to know) – far more than it is associated with simple 'knowledge' (*Erkenntnis*).

More relevant to our current concern than etymology alone, it is important to underscore the breadth of professional *Wissenschaften*, as these are more numerous than those that are gathered under the rubric of 'science'. Hence, although Nietzsche's identification of himself as a scientific practitioner strikes an English speaker as odd, it was not out of place in his day and would still be an accurate description in contemporary Germany, where, along the Diltheyan axis of *Geistes-* and *Naturwissenschaften*, one can speak of *Musikwissenschaft*, *Literaturwissenschaft*, *Museumswissenschaft* and so on. But, as with all equivocations, the subtle problem of 'what things are called' (in Nietzsche's words) remains. Classical philology is not a science – that is, not as the term works in English. Similarly, unless one is a practitioner of a cult religion (such as Christian Science or Scientology), one does not make reference to 'religious science' but to 'religious studies' and not to 'art science' but to 'art history', whereas German scholarship – and this makes a not inconsiderable difference for the discipline – differentiates between *Kunstgeschichte* and *Kunstwissenschaft*. Nevertheless, we can observe that, in the Middle Ages, the study of music was *scientia bene modulandi*, using a definition of science (a usage 'now rare' as we are reminded by the *OED*) – meaning skill or technique in the (still current) musical sense in which jazz is a 'sweet science' and in which not only could Minnesota Fats have his game of pool down to a 'science' but an English football team manage to carry the sobriquet the 'school of science,' and so on.[9]

In the context of his early (and later) reflections, when Nietzsche proposed to examine 'the problem of science' he refers to this very broad sense of science (*Wissenschaft*) because what he wanted to address was the specifically *scientific* character of science. Thus Nietzsche's talk of science with regard to philology (in his book on tragedy) inevitably exceeded philology in its scope and works, as I have argued, as a genuinely philosophical philosophy of science. Such a philosophy of science requires a specifically critical perspective on, or approach to, science.

For Nietzsche, the critical problem of science derives from science's own self-founding limit. Neither Aristotle nor Newton nor Kant nor, most recently, Gödel would quarrel with this limitation. But Nietzsche goes beyond the issue of critical foundations and the concept of method and extends his critique beyond his own discipline to the natural sciences like physics and chemistry precisely in their mathematical articulation as such. But, as if difficulties of translation were not enough, the historical referents to such scientific kinds as Nietzsche criticizes are, of course, not the same as those designated by the same terms today. Thus Nietzsche speaks of psychology (recall his discordant reference to the 'English psychologists' who have such a mechanical interest in the functioning of altruism at the start of his *On the Genealogy of Morals*) and here and there he makes manifestly biological and medical references. To parse references of this kind, seemingly very like his invocation of science as such, requires a great deal of scholarly sensitivity. We increasingly have the benefit of several valuable analyses of this kind. Hence when Nietzsche speaks of human physiology and makes some fairly strong claims about nutrition and climate, these are claims which, if you are Gregory Moore or Marc A.

Weiner, can be usefully shown to be arising from his own epoch and the prejudicial visions of the day.[10]

But, even in addition to tracing the history of ideas as a history of scientific concepts and conceptualizations, problems remain because, even on their own terms, Nietzsche's views are often perplexing – partly, but only partly, due to his rhetorical style. His comments on Darwin, for example, are notoriously confusing to commentators, inspiring some to find him pro- and others anti-Darwinist. Likewise, his discussion of atomism and force seems to mix Lucretius, Democritus and, almost for good measure, Boscovich. If Nietzsche can happily call Boscovich a Pole for the sake of convenient local comparison with Copernicus, some have argued, then his references to science must also be less than precise. Rhetorically excessive and even irresponsible, Nietzsche's views on science are just bad science (and that would only make his 'bad' philosophy worse).

Not so fast – for we are moving very fast when we denounce as factually inaccurate Nietzsche's contentions on account of their temporal difference from our own era without first examining them in terms of that same period. This, we have noted, is what Nietzsche would call a lack of science – that is, philology, with all its historical sensitivity. And our historical scientificity or sophistication has improved: we know better than an earlier Whiggish tradition of historians (of ideas as much as of the sciences) in the wake of recent sociological and anthropological studies of science and its technologies.[11] And we are plainly just beginning with such a re-evaluation of a received view on the 'genesis and development' of science. This work is preliminary to any such evaluation and in part it has been begun with specific regard to Nietzsche, although the task is a long one. Relevant contributions such as those of Robin Small and Thomas Brobjer and, earlier, George Stack and, still earlier, the work of the chemist Alwin Mittasch will go some way towards making such a contextual consideration of Nietzsche's claims possible. But, as the anthropologists of culture, be they of other knowledge discourses (for example, Latour, Pickering) or of other societies (for example, Geertz, Levi-Strauss), have learned to their pain (or, better, are attempting to learn), one can mean to bring a neutrality to bear on one's studies (and one's subjects) and still fail to apply it. There is, as in all things, a gap between what we know ought to be done and what we do. This is not merely a matter of *akrasia*, *knowing* the good, but failing to *act* on that knowledge. It is also a matter of not knowing what we incidentally, inadvertently and so unavoidably smuggle into our own effort to know not only what is other than ourselves but – and this was Nietzsche's great insight – ourselves as well. For Nietzsche took Kant further than Kant himself, which does not mean that he took himself further than he himself could go. That must be our task, though we ought to remember that the same ineliminable residue – what Lacan liked to call the Real – will remain, even for us.

So far, in raising the question of the meaning of 'science' for Nietzsche, my concern has been to ask how Nietzsche's critique of science might be turned to the advantage of the philosophical discipline of the philosophy of science as such. Along with Heidegger, I take Nietzsche's understanding (and his critique) of science and of logic or truth as quintessentially philosophical for the very rigorous reason (in Heidegger's sense of radical rigour) that, rather than assuming the value of science as such, Nietzsche raises the question of science as a question.

Nietzsche's philosophical critique of scientific reason puts the critical *rationality* of science in question. Thus Nietzsche means to pose the question of science as a critical problem rather than as a patent or resolvable problem. This critical project involves the articulation of the problem of science as such, which means that Nietzsche adverts to the prime difficulty of putting science in question as the difficulty of questioning what is ordinarily unquestionable. Indeed, science as authority and as 'method' *is* the means of critique or critical questioning. For this reason, Nietzsche regards raising 'the problem of science itself ... as a problem, as questionable' (*BT* 'Self-Criticism' 2), as a task to be accomplished over time, not merely as a point to be made or a problem to be remedied.

Given such a long-term expression of the 'problem of science', Nietzsche can assert that he is the first to raise the *question* of the questionableness of science. If he liked to suggest that he aspired to a more radical doubt than Descartes (*KSA* 11, 40[24], [25]) and if he was surely more critical than Kant in calling for the key reflex of the critical project to be turned against itself, Nietzsche nonetheless differs from the Enlightenment project of philosophical modernity in general (from Kant if not Descartes), because he *does not* exclude his own deliberately provocative solution as a problem at the limit of critical reflection. This is what Nietzsche calls knowledge's *tragic* terminus. In Nietzsche's words, the logical project of both science and the tragic myth turn out to be 'shipwrecked' on the very same issue (*BT* 15, 18) – namely, the question of foundational transparency – and this could, in some sense, also be extended to what Nietzsche later criticizes as the limit of the subject's self-knowledge in general. Thus the problem of the subject is itself the problem of critique. In this way, any critical project is irrecusably subject to distortion precisely because an organon, even a reflective organon, cannot be turned on itself. Thus the reflexive limit of the critical project entailed that what Nietzsche termed 'the problem of science' could itself not be recognized (conceived or thought) on the ground of science (*BT* 'Self-Criticism' 2).

Nietzsche's Critique of Judgement: Science in the Light of Art

As an unremittingly questioning (that is, *philosophic*) approach to the question of science, Nietzsche's grounding question of theoretical or scientific knowledge takes Kant's project of critique further than Kant had done, for the simple, arch-critical reason that Nietzsche does not assume the actuality of synthetic *a priori* judgements (cf. *BGE* 4). For Nietzsche, strictly speaking, such judgements were not possible for human beings: we have 'no right to them' and hence 'in our mouths', he would write, 'they are plainly false judgements' (*BGE* 4; cf. *BGE* 11). In his first *Critique*, although he asks how (qua actual) the judgements of physical science are possible, Kant himself shrinks from putting science itself in question. And if Kant is deliberately uncritical with regard to science, we are even more restrained today. To criticize science is thought to be anti-science, anti-modern, and ultimately irrational. Hence, when a physicist such as Stephen Weinberg or Richard Feynman or Alan Sokal (who deconstructed the scientific claims of deconstruction for the sake of truth, justice and American science) speaks,[12] the authority he commands as a man of science resembles nothing so much as the authority of the priest's in the era of religious belief (and Nietzsche was among the first to see this critically, whereas Comte would seize upon the same resemblance in just the opposite way). Thus even the philosophic reception of the scientist's claims tends to reserve criticism, taking a position of justifying defence. A relatively uncritical or trusting adulation of science has characterized intellectual culture since the days not only of Newton but also of Lucretius. Indeed, there are those who will argue – with Nietzsche at the forefront – that science itself is the true legacy of Plato's academy and his proto-Cartesian idealization of geometry as the *sine qua non* for wisdom itself.[13] Among philosophers of science, though not an advocate of Nietzsche's critique of scientific reason, Paul Feyerabend is alone in his call for a radical scepticism concerning the claims and the activities of science – a scepticism invoked in the critical name of science itself.[14] This was exactly the problem of science as it preoccupied Nietzsche, who claimed that the hallmark of the nineteenth century itself was the victory not of science but of the scientific method, as a victory even over itself (*KSA* 13, 15[51]).

Nietzsche's critical undertaking challenges the possibility of *any* knowledge of the philosophic (epistemic) foundations of science (as art) conceived in the light of what he speaks of as art. Regarded as an art, science is a kind of *techne* (and Heidegger's critical philosophy of technology employs a similar parsing of modern science and technology). As such an artful practice, science is a means for constructing what is held to be scientific and therefore true. Against physicists who speak of decoding 'nature's conformity to law' – as if 'nature' *followed* 'laws' in the image of the ideal of democratic politics – Nietzsche charges that the very *idea* of such conformity 'exists only thanks to your interpretation and bad philology: it is not a fact, not a "text"' (*BGE* 23). If Nietzsche begins with philological science in his first book – this is what he means by the 'science' of aesthetics – he certainly does not limit himself to the art of classical criticism in his later texts where he explicitly invokes physics and chemistry.

As an art or *techne*/technique, science is a means for winning what will count as truth. Thus Nietzsche argues for a parallel between the ascetic practices and ideals of both religious and scientific projects in the third section (and not only there) of *On the Genealogy of Morals*. Nietzsche describes the ascetic striving for truth as alienating and alien: 'Man does not exist by nature in order to know. Two faculties required for different purposes – *truthfulness* – and *metaphor* – have engendered the inclination to truth' (*KSA* 7 [19[178]). For Nietzsche, rather than an ordinary perceptual correspondence – which is, for him, no more than the idealist fantasy of seeing and saying the truth – the drive to know depends on the humanizing (anthropomorphic) inclination or direction of ego-logocentric engagement with the world: 'Ultimately, every law of nature is a sum of anthropomorphic relations.' Nietzsche emphasizes 'especially number' (*KSA* 7, 19[237]), as if to challenge the modern scientific ideal of quantificational objectivity. This we may regard as Nietzsche's Protagorean qua Procrustean Delphic principle: 'the basic thought of science is that man is the measure of all things' (*KSA* 7, 19[237]). Or, as he puts it elsewhere: 'All natural science is only an attempt to understand the human, the anthropological; or, more precisely, to return always to the human via the longest and most roundabout way' (*KSA* 7, 19[91]).

Nietzsche's reflections on truth and lie return again and again to the problem of logic in the wake of Socrates' transformation of the philosophic enterprise because, with Socrates, 'truthfulness gains possession of logic'. By the time we get to Aristotle, the challenge of 'the infinite difficulty of classification' (*KSA* 7, 19[216]) finds its determinative resolution in the principle of non-contradiction, as Aristotle underscores this first principle as an axiomatic assumption apart from any necessity for (or, indeed, any possibility of) demonstration.[15] This same principle sets the terms for what Nietzsche calls the conflict between art and knowledge.

And the outcome of this conflict is decided at the outset.[16] There is (can be) no articulable or logical basis for conflict between the irrational and the rational on logical grounds. By the same token, there is no contest (no *ratio*, no *analogy* or *comparison*) between art (qua irrational) and knowledge (or science). Thus while Nietzsche claims that both art and science trade in illusions, traditional philosophers of science, together with the vast majority of scholars in other disciplines and above all scientists themselves, hear in Nietzsche's assertions (when they advert to them at all) only the chaotic impressions of what they sometimes think to recognize (inaccurately enough) as the 'Dionysian' (that is, the irrational). This discounting is not a result of argument but of conviction, to use Nietzsche's language. Such philosophers of science simply trust that the scientists do indeed *know* that the object/aim of science is to determine the pure and simple truth. Sociological studies of science, compounded with the new history of science cited earlier, make a difference in this conception, but not enough difference; traditional philosophy of science refuses to raise Nietzsche's question of the problem of science as such.[17]

Nietzschean Physics as the Dynamics of 'Chaos in all Eternity'

If Nietzsche begins his concern with science as a disciplinarily specific one, a project taking him to a critique of logic and to the limitations of reason, he expands his critical concern to address the limitations of natural science as a discipline charged to 'know' the real world. This focus finds its expression in his notorious claim that the nature of the world is nothing but 'will to power' (*BGE* 36) and 'chaos' its 'total character' (*GS* 109).

Nietzsche thus describes the 'total character of the world' as 'chaos in all eternity – in the sense not of a lack of necessity but of a lack of order, arrangement, form, beauty, wisdom, and whatever other names there are for our aesthetic anthropomorphisms' (*GS* 109). But at the same time, Nietzsche also describes chaos as the veritable source of creative potential within culture: 'I tell you: one must have chaos in one, to give birth to a dancing star' (*Z* 'Prologue' 5). In the second half of this chapter, I shall ask what Nietzsche means by coordinating these seemingly conceptually different claims. For, against the assertion of the chaos of the natural world, any pragmatist could counter that the world displays enough regularity for the physicist Alan Sokal to challenge anyone calling the laws of gravity mere 'interpretations' to take one's leave from a New York City skyscraper by way of an upper-story window. More metaphorically, with regard to the relation between chaos and culture, although sometimes the inevitable accompaniment to creative ardour, chaos is not typically regarded as a *precondition* for creativity. Earlier I inquired into Nietzsche's understanding of the logical problem of science; in what follows I consider his conception of the natural world not merely as expressed in dynamic terms as will to power but very patently and prosaically enough as chaos. Nietzsche's understanding of chaos is opposed to the contemporary understanding of chaos as drawn from the Judaeo-Christian religious tradition, as well as to the understanding of chaos resonant in the (popular) science idea of chaos theory.

In the words of Nietzsche's Zarathustra, 'to give birth to a dancing star' one needs to have retained within oneself the same chaos that exemplifies the world 'in all eternity'. Yet we recall that chaos simply is a yawning, a gap, so that even in Hesiod's *Theogony*, after its first mention, this primordial, self-sufficiently generative chaos recedes, only to be mentioned once more in the context of the battle between Zeus and the Titans and in its persistent displacement at the edge of the world, beyond both Olympian gods and Titans.[18] Chaos in its antique provenance, unlike the contemporary physical concept of entropy, does not increase its range. Thus to our modern ears, in a culture Nietzsche so presciently diagnoses with scientific nihilism and with leisurely, thoroughly mediatized decadence, Nietzsche's classical conception of a creative chaos is increasingly at risk, so that the 'ultimate' men of today can only reply to Zarathustra with blinking incomprehension, asking: 'What is love? What is creation? What is longing? What is a star?' (*Z* 'Prologue' 5).

Chaos has primacy in Hesiod's *Theogony*. It simply comes to be without antecedent: 'Ἤτοι μὲν πρώτιστα Χάος γένετ'.[19] After Chaos, there arise the unbounded gods of the beginning, divinities of aorgic nature according to Hesiod's account, including *Gaia* or broad-breasted earth, dim *Tartarus* and winged *Eros*. First of the mothers of being, Chaos gives birth to the deities of darkness: Erebos and the same Night that, in the Orphic tradition recounting these origins, lays the silver egg in the lap of black darkness, from which is born the god of many names, golden Eros or Phanes, who brings everything hidden to birth or to light.[20] The *Theogony* includes this Orphic resonance: it is Night's incestuous union with Erebos that annuls the character of their respective obscurity, yielding brightest Aether and the day. Thus, Chaos, rather than the masculine world-ordering process of cosmological genesis, is here generatively primordial.

It has been suggested that the Hesiodic meaning of chaos (as chasm or yawning gap) can perhaps be traced back to an Orphic account which names the chasm (or night).[21] Yet the names are less important than the consequences. For the pattern of spontaneous (and feminine) creativity is immediately, and from the outset, quashed even in such archaic accounts. This is especially the case, we could say, for the Greeks: take, for example, the story of Eros that tells of insurgent male power ordered – or unhinged – by desire. Thus the creative power of genesis that is originally female becomes – or is made – male Zeus swallows his first wife, Metis, thereby incorporating not only her wisdom (lodged in his belly, her counsel was always his own) but also her feminine nature. Thus the Zeus who gives birth to Athena fully armed, sprung from his forehead, is the same Zeus who can rescue the heart of the dismembered Zagreus from the burnt ashes of his Titanic tormenters, blasted by his thunderbolt. It is the same Zeus who is transferred by the mediating influence of wine into the womb of Semele, his human lover, there engendering the child Dionysus, only to reduce Semele herself to ashes with the same lightning flash of his godly countenance, to sew the not yet fully formed child into his own thigh, finally giving birth to Dionysus of the two gateways, twice reborn in the wake of both titanic and mortal ash. Such a Zeus was a god of the prime male principle made a prime principle precisely by means of its violent incorporation of feminine creativity. Now the father gives birth to the son, now the sky gods, who are always male, form human beings in their own image.

For Nietzsche, what matters is not to pay homage to the old story of the primordial goddess, or the literal mothers of being, but instead for us to attend to the powers of wild nature in creative self-genesis, in becoming ourselves a work of art. To create oneself, giving birth to a dancing star – a wheel rolling out of itself – one needs the 'chaos and labyrinth of existence' (*GS* 322). This imperative holds not because order kills (the letter, the spirit) but because the chaos in creative question is primordial nature itself: that which is older than all other deities, that of which the most darkly aorgic deities are born, as divinities capable of bringing their own opposites out of themselves. The chaos or nature here described is the wild, untamed and uncontrollable force (Hölderlin's divine, or aorgic, *apeiron* nature). And we, so Nietzsche's Zarathustra tells us, still dispose over this creative power. Yet because we are also, in our age, closest to losing this same power of nature, losing the creative force of chaos, through a blindness that is as much a blindness about ourselves as about nature, we need Zarathustra's reminder.

As self-engendering *physis,* chaos thus corresponds to a generative excess or plenum, conceived as the Theogonic Ur-*chaos* or the Anaximandrian *apeiron*. Such a generative or creative conception of chaos is a feminine aspect of unthinkably consummate, self-sufficient creativity and such a feminine first principle is common to more than one cosmology. Yet the primordial conception of feminine creative potencies (from the notion of chaos to the fantasies of the cultures of the goddess or matriarchy) is eclipsed, as we have seen, from the moment of inception: all accounts of the genesis of the dark children of chaos are quickly elided by the more fertile and various Earth who, herself giving birth to her own lover, sets in motion the dominant account of the succession of masculine progenitors. As the story of the birth of the gods, the theogony becomes a story of genesis (and paternity).

Even more, today's chaos is a pell-mell representing the failure of order, an order *reduced* to disorder (reflecting the temporal schematism articulated in both the religious tradition of Genesis and the thermodynamic conception of entropy). Derived from an aboriginally masculinist vision, the Judaeo-Christian tradition regards the lifeless depths recounted in Genesis as the impotent, and featureless, waters of an irreal and feminine abyss prior to the first divine moment of the creation of the world. In this convergence of religious and scientific perspectives, chaos from either point of view is, and remains, a patently non-creative, manifestly negative concept.

Nature as Chaos, Nature as Art

By equating chaos with nature, understood in its original Greek or primordial sense, Nietzsche repudiates the traditional Western opposition between nature and art. As an absolute will to power, without remainder – 'und nichts ausserdem' – the native chaos of the world is a raw, uncountenanceable and untrammelled realm beyond the imposition of order for the same Nietzsche who teaches the rule (and the illusion) of perspective. Recalling Spinoza's *deus sive natura*, Nietzsche's 'chaos sive natura' (*KSA* 9, 11[197]) de-deifies nature while, at the same time, stripping nature of its rational ground or foundation – its ultimate concord and commensurability with human reason. Nietzsche regards the modern confidence in science as needing critical demythologization as much as any other belief.

Outlining the history of the world of illusion or appearances as the very history of the true ('real') world of Platonic fantasy and Aristotelian productivity, Nietzsche raises the key question regarding the rational underpinnings of the ideal real: as the scientific order perceived in nature as a reflection of the phantasms of human sensibility and human conceptual power. There is no *reality* or nature knowable apart from a thoroughly humanized nature. Thus Nietzsche criticizes the realist conviction that 'the world really is the way it appears' as inevitably naive: 'As if reality stood unveiled before you only, and you yourselves were perhaps the best part of it' (*GS* 57).

For Nietzsche, then, rather than a means to reach the inherently unknowable truth of things as they are in themselves (*an sich*), natural science is a continuous and effective (that is, an exactly successful and consummate) process of what he calls a 'humanization *in summa*' (*KSA* 11, 25[445]). Nature is a human invention and yet, at the same time, it is nature that works its artistry through us and upon us, so that our inventiveness and our artifices are not non-natural – however much, in the case of science, we use this inventiveness for the purposes of the mastery and control of nature (*KSA* 11, 26[170]). With the artistic, inventive expression of such scientific 'means', we express our own nature, the same essence reflecting the inherent truth of nature as 'will to power' or chaos (in all eternity).

Yet, we perceive neither nature's chaos – nor could we ever do so given our perceptual and conceptual apparatus – nor can we recollect any sense of the chaos of impressions within us – this last given the coordinate limitations of our human psycho-physiology. This is a favourite theme for Nietzsche, who declares that nature threw away the key to the welter of physiological activity within our bodies (*TL* 1, *KSA* 1, p. 877), and it forms a fundamental component of his pre-Freudian critique of psychological identity or the subject. In the more straightforward instance of a face-to-face encounter with the natural world, Nietzsche writes:

> As I walk about in open country, I am always amazed to think how everything acts on us with such supreme precision: the forest in this way and the mountain in that way; that, referring to the whole of sensation, there reigns within us not the slightest confusion, misapprehension, or stammering. And yet the greatest uncertainty and chaotic aspect must abound .... (*KSA* 9, 11[252])

In the same way as we do not (cannot) attend to our own range of perceptions, neither do we attend to the full complexity of the things themselves as experienceable objects. Using the metaphor of reading a text (where, as Nietzsche says, the text does not merely disappear in the reading or beneath the interpretation, but the reader instead 'picks about five words at random out of twenty and "guesses" at the meaning that probably belongs to these five words'), we tend not to see what is actually in front of our eyes. We perceive what we already 'know' or believe, 'rather than registering what is different and new in an impression' (*BGE* 192).

Nietzsche draws a parallel to our perception of a thing as obvious (and apparently as stolidly static) as a tree. He argues that, even with regard to such a sizeable object of everyday perception, we misapprehend the tree itself, never seeing it 'exactly and completely, with reference to leaves, twigs, colour, and form; it is so very much easier for us to fantasize some approximation of a tree'. Whether confronted by routine impressions or 'in the midst of the strangest experiences we still do the same: we make up the major part of the experience' (*BGE* 192). We overlook, overleap and so invent our experience in general (for Nietzsche, we are '*accustomed to lying*'). And this circumstance is not ameliorated by adding reference to the complexity of things that are too minor to attract our notice – the wild variety of insects and spiders and plasmodiæ streaming on the bark of the same tree or crawling and flying in the jungles at our feet.

As Ryogi Okochi has rightly underlined in his comparison of Nietzsche's conception of nature and Eastern views,[22] nature is *not* a correlevant *object* for human comprehension. Hence, regarding the world as will to power to all eternity – that is, naming nature as chaos – Nietzsche emphasizes both its distance from our capacity to comprehend nature in itself (this is Nietzsche's routine Kantianism) and its inherent creativity (again recalling the archaic Greek conception of chaos). As chaos, nature itself is interpretive and, on this same level, nature itself is invention, replete with subjectivities, wills. Nature as a whole is a constant and thoroughgoing interpretation and 'necessary perspectivism by virtue of which every centre of force – and not only man – construes all the rest of the world from its own viewpoint, i.e., measures, feels, forms, according to its own force' (*KSA* 13, 14[186]). Giving measure and giving form, testing, and reacting, nature is art.

Everything turns on the question of art for Nietzsche. The difference between the art of the human and the art of nature is the difference between the *artless* art of nature (which lacks all purposiveness) and the *artful* art of human invention that is both artistically consummate (or artful *art*) or else poorly executed or aesthetically artificial. Only culture, and especially the art of practical, technical, purposiveness or *techne*, can call itself art (although it does not always do so). Everything else, be it God or be it nature, lacks artistic awareness, as artless or *natural art*. This same artlessness is ultimately the key to the (active) creative process. Hence Nietzsche agrees with Kant's claim that the highest art is the artlessness that is (or appears to be) an unconscious or *natural* art – hence unaware of what it is or does. This one may name innocence. It is Nietzsche's goal to recover this innocence, this divine naturalness, for human creativity. The result would be a renaturalized humanity (itself only possible on the basis of a redeemed nature, liberated to its own chaos or independence from the order of human control).

From beginning to end, Nietzsche affirms the origin of knowledge in error and illusion, and hence, or ultimately, in art. Thus in an early reflection on this problem, Nietzsche describes the ideal beauties of perceived 'nature', and he paints natural sublimity in the following words: 'It was evening: air streaming with the scent of evergreen, one's gaze opened out upon grey mountain ranges, snow shimmering on high, spanning above, becalmed, blue skies' (*KSA* 8, 23[178]). And contrary to the straightforward, albeit poetic, descriptiveness of this vista in terms of the senses themselves, Nietzsche challenges the reality of any such pure or direct perception of nature. That refusal of unfiltered or direct sense access to the world around us is the point (Kantian as well as psychological) of his teasing assault upon the sobriety of 'realist' perception of the 'real' world in *The Gay Science*: 'That mountain there! That cloud there! What is "real" in that?' (*GS* 57). For Nietzsche:

> A thing of this kind we *never* see as it is in itself, but always overlay it with a tender spirit-membrane – that is what we see instead. Inherited sensations, our own feelings are roused by such things of nature. We see something of ourselves – to this extent, the very world itself is our representation. Forest, mountains, more than a concept indeed, but our own experience and our own history: a piece of ourselves. (*KSA* 8, 23[178])

There is no world apart from what he explicates again and again as the insensible chaos of our sensations and the veritable chaos of nature itself. And chaos must always be ordered. For us, this describes ordinary perception but, for Nietzsche, it is interpretation and it is not true because it cannot touch the chaotic truth of a world apart from our imposed interpretations, a world of abundance or excess. This is what Heidegger captures with a seemingly paradoxical formula, asserting that for Nietzsche '*Truth* is the lack of *truth*'.[23] What Heidegger means to underscore here is the exactly aletheic character of Nietzsche's critique of truth and his phenomenological critique of perception and experience. Thus Heidegger is not charging Nietzsche with tacit contradiction – claiming (as true) that there is no truth. Instead, the very question concerning the truth of nature requires an inquiry into the conditions of possible perception as prerequisite conditions of the very possibility of knowledge (be it of nature or ourselves).

Nietzsche invokes the simplified world of everyday and theoretically mediated perceptions as a matter of sensible, aesthetic 'refinement', even in the case of science. Scientific laws and scientific entities are creatively, aesthetically (inventively) possible on the basis of flattened differences, and taking such exclusion precisely (calculably) into account, thus bracketing potential exceptions (Nietzsche's remark that 'the little errors do not enter into account' underlines that 'margins of error', 'standard deviation' and the theoretical finesse of 'curve fitting' are exactly operative techniques useful for excluding or *discounting* these same 'little' mistakes). As science operates upon the levels of both pure theory and empirical measure, the scientist is a theoretical technician or artisan of exactitude.

By these means exactitude is, and can be, imposed upon an empirically variable world of becoming or change. Science is the technique (art) of uncovering the Platonic ideal of truth in the phenomenal world (both theoretically and experimentally). Far from praising the achievements of science as such a *technites*/artisan of chaos/nature, Nietzsche criticizes the flatfootedness of the artifice of science – branding science an 'error', just as he portrays it as dedicated to the pursuit of 'the principle of "the smallest possible effort" and the greatest possible stupidity' (*BGE* 14). Nietzsche takes this claim to its hyperbolic extreme because our era is characterized by its unquestioning scientific faith. Recall his critique of the putative opposition between scientific and religious ideals in the closing sections of *On The Genealogy of Morals*, where faith in the latest, greatest and most efficacious ascetic ideal of all – that is, the faith in science itself – replaces religious faith. For Nietzsche, claims of scientific truth (conventionalizedly objective, natural scientific knowledge claims) are *artlessly* artificial claims asserting an exactly *non-artistic*, non-constructed 'truth'. Nietzsche's exigence derives from his dedication to the truth not of the ideal world of mathematizable science, but the very real or chaotic world.

Because this world is precisely reality itself, in flux away from what has been and towards what is not yet, empirical nature is always all about change. It is, like ourselves, part of that species of being hostage in time to the immutable reality of the dynamic mode of becoming – as both what is *and* what is not. Although the later Nietzsche mocks the philosophers' 'idiosyncrasies', most evident in what he calls their 'Egyptianism', as 'their hatred of even the idea of becoming', he had earlier acknowledged the very human condition that would end in this hostility to 'change, age, as well as procreation and growth' (*TI* 'Reason' 1). At the start of his unpublished essay, *The Pathos of Truth*, he writes: 'We observe every passing away and perishing with dissatisfaction, often with astonishment, as if we had witnessed therein something fundamentally impossible .... Every New Year's Eve enables us to feel the mysterious contradiction of being and becoming' (*KSA* 1, p. 756). But because of the 'stone' fact that no fact is stone, the supreme law of philosophical knowledge lacks any purchase on the empirical world because what is is never beyond change or time.

Given this tragic, but very human, critical reflection on science, when Nietzsche celebrates the sciences of nature in antiquity and in the works of contemporaries such as Robert Mayer and Ernst Mach, he celebrates natural science – but without accepting science's own ideal of itself as neutral instrument of pure knowing. In spite of its logical self-image (eagerly polished and continually renewed by the reigning philosophy of science), science, especially and particularly natural science, is and can only be an empirical, experiential or experimental discipline. But, for Nietzsche, this means that science must ultimately be a discipline of the body – and of the earth.

Thus construed as an ultimately sensual rationality – which bodily or earthly rationality was for Nietzsche not irrational *per se*, but a more complex, more nuanced rationality than that of the intellect alone – (physical) science is not a science of ideal conceptions (paradigmatic or theoretical projections) but a constantly artful or poietic science, a praxical *techne* of recondite reality becoming in time. This practical or real technical object of science is the dynamic 'truth' of what works as *real* (whether artful, inventive, efficacious belief or 'willed' illusion), not the ideal and literal 'truth of truth'. Nietzsche does not stop here (he is not a pragmatist of an Emersonian, Quinean or, indeed, Davidsonian sort) but adds that the so-called (scientific and logical) truth of truth (ideal truth) is purely unreal: not figuratively, but literally insofar as that truth is theoretically ideal. This logical ideality is the truth of tautology: that is, *logical* truth. The becoming and changing nature Nietzsche recalls to us, both as scientists and as philosophers of science, is and can only be an illogical nature, and he summons us to that same natural world for what he ultimately regards as the honest sake of science: 'We possess scientific knowledge today precisely to the extent that we have decided to *accept* the evidence of the senses – to the extent that we have learned to sharpen and arm them and to think them through to their conclusions' (*TI* 'Reason' 3). Taking his own thinking on logic, truth and reality to its 'conclusions' or 'ultimate consequences', Nietzsche describes his anti-atomistic ideal of the non-lawlike working of the will to power as an *alternative* account of 'nature's conformity to law' (*BGE* 22). For Nietzsche, 'knowledge' as a purely logical enterprise strives for what is humanly or naturally impossible (all natural knowledge is inherently anthropomorphic and hence the human and the natural are inevitably convertible). The impossibility of a logical description of the world is thus not an impossibility enjoined by Nietzsche's terms of analysis but, and precisely following upon, the logical, unchanging terms which philosophers set as the (unconditioned) conditions of knowledge.

Nietzsche challenges the tendentious grammatical distinction between subject and object, active and passive (*D* 121), organic and inorganic, life and death (*GS* 109). Thus when he asks 'if the perspectival is an essential property' he can answer: 'this would be possible *if all being were* **essentially** *a perceiving something*' (*KSA* 12, 5[12]). In this light, we can read a subsequent reflection: 'Assuming, however, that we assign certain values to things, after we have forgotten that we were the givers, these same values in turn work *retroactively* upon us' (*KSA* 12, 5[19]). Because 'every centre of force adopts a *perspective* towards the entire *remainder*' (*KSA* 13, 14[184]), there is always a reciprocity between interpretations and perspectives. On the basis of the relational interaction of the world conceived both in its subjective and objective dimensions as will to power, and its ultimate interactive involvement with itself ('the world viewed from the inside ... would be "will to power" and nothing else' (*BGE* 36)), to regard the world 'from the inside' as Nietzsche says – not from the assumption of our own interest but rather as it might be seen according to its collective and varied interests – is to construct the object subjectively. This is Nietzsche's perspectivalism.

Today's vision of demythologized, insensitive object-nature is a systematic expression of the cultural illusion of science in the West and the Enlightenment's reverse mythological vision of a mechanical nature revealed through the sobriety of scientific sophistication. Here we note that the privilege of the subject and its ideal of objectivity are errors (delusions): 'Once you know that there are no purposes, you also know that there is no accident' (*GS* 109). In Nietzsche's vision of the world, necessity and purpose are dissociated. But in the ideal order of nature's conformity to law, what opposes purpose is accident. Without this ideal, without purpose, there is no accident and only chance (with its archaic necessity) remains.

The Tragic Critique of Knowledge: The Truth of Logic (Lie) and Life (Chaos)

We have seen that, as a critique of science, Nietzschean genealogy can be seen in terms of more than one focus. Not only does Nietzsche articulate a genealogy of morals (out of reactivity or *ressentiment*), but he also began his life's work with a genealogy of tragedy (from ancient dithyrambic modes to its contemporary decline as an 'art-form'). Now we see that he likewise offers a genealogy of logic and even reason itself. Nietzsche similarly outlines a genealogy of science – out of the spirit of myth and magic and alchemy but ultimately on the occluded paradigm of religion. In the case of logical thinking, as we have seen, he finds the origins of logic on the side of metaphor, metonymy and the other tropes of language groping towards an apprehension of a reality in transit, an illogical real. Thus logic, for

Nietzsche, begins in illogic; logic develops out of, or in, its opposite (cf. *BGE* 2, *GS* 111, etc.).[24] How?

Concurring with every account of the prehistory of philosophy, Nietzsche reminds us that from myth we advance to *logos*. But a post-lapsarian history of logic following upon a lost moment of golden insight into, or 'experience' of, truth is sheer invention. In sense perception, as in our anthropomorphic linguistic and epistemological constructions, we literally have only what we call the false – that is, perception, representation, and not the thing in itself. Nietzsche describes our practical knowledge of real things in the real world as 'essentially illogical' because it incorrigibly involves 'the *identification of things which are not the same*, of things which are only similar' (*KSA* 7, 19[236]). Nietzsche's reserve here opposes what he calls the 'philosophical' prejudice contra change/becoming as an unflinchingly, exactly epistemological (and not a moral) reservation. Our knowledge accounts of the real or natural world incorrigibly leave out or omit 'what is individual' by subsuming it under a 'concept', and with this subsumption what we call 'knowledge begins: in *categorizing*, in the establishment of *classes*' (*KSA* 7, 19[236]). For Nietzsche, *real* knowledge of individual or real things – that is, the modern knowledge of natural science – must be *essentially illogical* if it is to have any kind of approximative or tentative correspondence with the real world, as a practical or real science of nature.

For the sake of the philosophical question of truth and logical rationality, Nietzsche raises the question of science as the question of the measure of the world of real – not ideal – things. Hence Nietzsche continually adverts to the difference between ideal types – like elemental carbon – and real material instances. Not only does he refuse the association (for example, he points to the irreducible, non-chemical difference between diamond, coal, and graphite – *KSA* 13, 14[187]), but he also challenges the operative application of simple mathematics (or arithmetic) to describe the biological or indeed ecological process. On a variety of occasions, he points out that in the real world of real things, two halves do not add up to or equal a whole – as true for an apple pie as it was in Solomon's judgement of an early child custody case and as it would apply (using one of Nietzsche's favourite metaphors) to the lowest worm (*KSA* 13, 11[259]). In the case of a particularly simple worm cut in half – in the enthusiasm for vivisection that is perpetuated in today's laboratories but above all in today's instructional methods – two halves make two wholes. For Nietzsche, this still non-calculative sum is far from the rule in the case of garden-variety worms. Mostly, the sundered parts add up to nothing at all. In statistically rare cases, the two halves can represent a lucky subtraction, reducing to just one somewhat shorter worm. That is, providing both halves do not die (the most likely scenario), the two halves make not two but one truncated worm. Worm segments are not the same as line segments. To take another similarly simple invertebrate, the starfish suffering the loss of one arm regenerates what only an aesthetically challenged individual (or scientist) could call another arm: like the lizard's tail or hand, the stump is a badly stunted replication of what was originally lost. Nietzsche is fond of pointing to the example of cell division to make the same point in another direction as well as to invoke his general economy of abundance in expression (*KSA* 12, 1[18]; see also *KSA* 12, 9[145], [151]).

For Nietzsche, 'all of life is based on semblance, art, deception, points of view, and the necessity of perspectives and error' (*BT* 'Preface' 5). Nietzsche saw that the critical self-immolation of knowledge ('the truth that one is eternally condemned to untruth' (*KSA* 1, p. 760)) at the limit of the critical philosophic enterprise must be combined with the sober notion that insight into illusion does not abrogate it and, above all, does not mean that illusion lacks efficiency. To the contrary, 'from every point of view', Nietzsche argued, 'the *erroneousness* of the world in which we live is the surest and the firmest thing we can get our eyes on' (*BGE* 34). Nor, as we have traced the etymology of science/*Wissenschaft*, is this visual metaphor an incidental one here.

As he reflects in his early notes on the 'Conflict Between Art and Knowledge', Nietzsche argues that the task of philosophy is the poetic task of a physician of culture. In order to speak in this way, art and knowledge are not to be ordered as disparate categories or separate cultural realms but in continuity with one another. Since science cannot subject itself to critique – that is, as the problem of science cannot be posed on the ground of science – Nietzsche proposes the perspective of the healing power of art. As 'physician of culture', the philosopher is to be an artist of science, a composer of reflective thought, refusing the calculations of science as the sickness deadly to the 'music' of life (*GS* 372, 373). Against such deliberate deafness, letting 'those who have ears hear' (*GS* 234), the 'only' help for science will not be more science (or better scientific understanding) but the therapeutic resources and risks of art (*KSA* 7, 19[36]).

Nietzsche regards his own pursuit of knowledge as thoroughly, intrinsically, if exactly joyously 'scientific'. What Nietzsche means by thinking in the critical service of science (as the 'mastery' of science) can only be expressed in its contextual connections to topics in other kinds of philosophic reflection traditionally regarded as distinct. Thus we have seen that Nietzsche links epistemology and politics in an aesthetic, rather than practical or moral, resonance. This aesthetic connection for Nietzsche means that the problem of science must be seen as the problem of art and life. This is neither a perspective opposed to (anti-)science nor is it traditionally on the side of science. Instead, Nietzsche's interpretive touchstone contrasts what affirms or serves life with what denies or opposes life, which last is what Nietzsche means by nihilism. The problem of Nietzsche's philosophy of science addresses the problem of life where 'knowledge and becoming' mutually and incorrigibly exclude each other (*KSA* 12, 9[89]; cf. 7[54]). Writing that 'our art is the reflection of *desperate knowledge*' (*KSA* 7, 19[181]), Nietzsche sets art and knowledge on the same level. For this reason, both art and knowledge can be used either in the service of life or else against it. But Nietzsche's claim that '*science* can serve either goal' (*GS* 12) does not amount to a positivistic or naive expression of science's celebrated neutrality. As a logical or theoretical project, science is the kind of illusion (or convention) that remains inherently nihilistic. Because science as such is not objectively neutral, it must always be reviewed critically not on its own basis but rather on the ground of what makes science possible, and that is what Nietzsche names the 'light' of art. That optic – or perspective prism, to allude to a Goethean metaphor for Nietzsche's own approach to science – is life in its complexity and not distinct from its tragic sense.

Notes

1 William J. Richardson, 'Heidegger's Critique of Science', *The New Scholasticism*, 42 (1968), 511–36.
2 See my *Nietzsche's Philosophy of Science: Reflecting Science on the Ground of Art and Life* (Albany, NY: SUNY Press, 1994).
3 Josef Simon, 'Grammar and Truth: On Nietzsche's Relationship to the Speculative Sentential Grammar of the Metaphysical Tradition' in Babette E. Babich and Robert S. Cohen (eds), *Nietzsche, Theories of Knowledge, and Critical Theory: Nietzsche and the Sciences, II* (Dordrecht: Kluwer, 1999), pp. 129–51.
4 See Alexander Nehamas, *Nietzsche: Life as Literature* (Cambridge, MA: Harvard University Press, 1985); Tilman Borsche, 'Nietzsches Erfindung der Vorsokratiker', in Josef Simon (ed.), *Nietzsche und die philosophische Tradition* (Würzburg: Königshausen & Neumann, 1985), pp. 62–87; James I. Porter, *The Invention of Dionysus: An Essay on the Birth of Tragedy* (Stanford, CT: Stanford University Press, 2000); idem, *Nietzsche and the Philology of the Future* (Stanford, CT: Stanford University Press, 2000).
5 It was because he wrote as a philological *scientist* that Nietzsche regarded his methodological considerations directly relevant to the 'problem of science'.
6 Translation modified. Kaufmann's translation has 'Alexandrian' rather than 'Alexandrine'.
7 James I. Porter, *The Invention of Dionysus*, p. 4, writes: 'I would say that Nietzsche has no philosophy in the proper senses of "having" or "philosophy"'.
8 As a philologist, Nietzsche was characteristically conscious of this root connection between vision and scientific knowledge – hence his focus on the ocular tendency of science in general – but especially natural science. See *KSA* 11, 25[389]. It might be worth investigating the degree to which this ocular conception inspired both his focus on what he called the 'science of aesthetics' in his first book, his emphasis upon the importance of the haptic sense in the physical sciences (cf. *TI* '"Reason" in Philosophy' 3), and his special attention to taste, a focus he earlier played back to its etymological association with wisdom as such (see *PTA* 3, *KSA* 1, p. 813).
9 I thank Greg Moore, one of the editors, for this helpful addition. Moore remarked that the Everton Football Club (in Liverpool) used to be known as the 'School of Science', such, he explained, 'was the technical brilliance of their soccer'.
10 See Gregory Moore, *Nietzsche, Biology and Metaphor* (Cambridge: Cambridge University Press, 2002) and Marc A. Weiner, *Richard Wagner and the Anti-Semitic Imagination* (Lincoln, NE: University of Nebraska Press, 1997), especially Chapter 5.
11 See Steven Shapin and Simon Schaffer, *Leviathan and the Air Pump: Hobbes, Boyle and the Experimental Life* (Princeton, NJ: Princeton University Press, 1985); Peter Dear, *Discipline and Experience: The Mathematical Way in the Scientific Revolution* (Chicago: University of Chicago Press, 1995); Lawrence Principe, *The Aspiring Adept: Robert Boyle and his Alchemical Quest* (Princeton, NJ: Princeton University Press, 1998). See also Ludwik Fleck, *The Genesis and Development of a Scientific Fact*, trans. F. Bradley and T. Trenn (Chicago: University of Chicago Press, 1979 [1935]).
12 For references and a critical assessment, see my discussion of this 'hoax' in 'The Hermeneutics of a Hoax', *Common Knowledge*, 6 (2) (1997), 23–33.
13 See David Lachterman, *The Ethics of Geometry* (London: Routledge, 1993).
14 Paul Feyerabend, *Farewell to Reason* (London: Verso, 1987).
15 Aristotle, *Metaphysics*, 1005b 15–25 in *The Complete Works of Aristotle*, ed. Jonathan Barnes, 2 vols (Princeton, NJ: Princeton University Press, 1984).
16 The logical field of science and philosophy has traditionally been regarded as a battle (eristic) where exactly the best argument wins, and 'losing' claims are consequently silenced as irrational or irrelevant. Logic determines the standards for such a contest.

Hence the conflict Nietzsche insinuates between art and knowledge must be played out on the terms of a logical contest.

17 For a resumé and prospectus of the relevance of such sociological studies, see Rom Harré, 'Science as the Work of a Community', in Babette E. Babich (ed.), *Hermeneutic Philosophy of Science, Van Gogh's Eyes, and God: Essays in Honor of Patrick A. Heelan, S.J.* (Doredrecht: Kluwer, 2002), pp. 219–29. See also, in the same volume, John Ziman, 'No Man is an Island', pp. 203–17.

18 Hesiod, *Theogony*, ed. M.L. West (Oxford: Clarendon Press, 1966), 116, 814.

19 Ibid., 116.

20 Carl Kerenyi, *The Gods of the Ancient Greeks* (London: Thames and Hudson, 1951), pp. 16–17.

21 Evidence for this is given by no less an authority than Aristotle (*Metaphysics*, 1071b 27), who implies that night is an alternative name for chaos.

22 Ryogi Okochi, 'Nietzsches Naturbegriff aus östlicher Sicht', *Nietzsche-Studien*, 17 (1988), 108–24.

23 Heidegger, *Nietzsche I*, (Neske: Pfullingen, 1961), pp. 620–21.

24 See Simon, 'Grammar and Truth' and, for a related discussion, Holger Schmid, 'The Nietzschean Metacritique of Knowledge', in Babich (ed.), *Nietzsche, Theories of Knowledge, and Critical Theory*, pp. 153–63.

Chapter 7

What Nietzsche Did During the Science Wars

Robin Small

I am going to discuss Nietzsche's relation to some debates of his time which centred upon natural science. 'Science wars' – this phrase is the title of a book about recent debates – tend to be not so much about the content of scientific theories, as about their further interpretations.[1] Today's science wars are about the relation of gender, ethnicity and social class to the status of scientific knowledge. In past centuries they were often about national differences, and I shall take that as a starting-point, even though it will soon be seen that this is a superficial expression of the issues. In particular, I will consider some disputes involving the nations which, by the late nineteenth century, were the two leading representatives of natural science: England and Germany. (Incidentally, I intend to use the words 'England' and 'English' loosely – purely for convenience, because this is how they are used by every German writer, including Nietzsche.)

My opening question is: what did Nietzsche think about these controversies? What is interesting about his position, I think, is that it often uncovers the deeper issues, even if this was not always evident to Nietzsche himself. The rest of my discussion will spell out this point in more detail, and add some further observations along the way. For example, I think that Nietzsche's varying attitudes towards English thought are reflected with special clarity in the course of his relationship with Paul Rée, who came to figure in his mind – quite reasonably, I think – as a typical adherent of the English moral psychology which he wanted to replace with something quite different: his own 'genealogy' of morality.

The background to all this is the rivalry between Germany and England in natural science, and the reflection of that competition on a theoretical and ideological level. It is only fair to say that these 'science wars' were slightly one-sided: German writers were very concerned with the challenge of English scientific empiricism, but English writers on science did not spend much time on German philosophy. In fact, they tended to pass much of it off as a foreign eccentricity: 'sermons from mystical Germans', as W.S. Gilbert (who knew what went down well with his English audience) put it in *The Mikado*.

However, the conflicts were often not between German and English writers, but between German writers and other German writers. There were many in Germany who seized on the cosmopolitan culture of natural science as a force for modernization and liberalism, at a time when German political reform had met with serious and lasting setbacks. Popular science flourished in Germany during this period, in part because it provided a forum for debate which was not available elsewhere. At the same time, there were conservative and even reactionary voices within this discourse, trying to resist the influence in Germany of scientific theories which seemed to be bringing with them some dangerous cultural values. So another question about Nietzsche's position is directed towards this context, involving the debates which he found very close at hand in German academic and cultural life.

In the face of the scientific challenge, German philosophy was on the defensive. By the middle of the century it was evident that the idealist tradition of *Naturphilosophie* had become irrelevant. Nobody studying Faraday's discoveries in electricity and magnetism, for example, could have much patience with Schelling's lengthy *a priori* 'constructions' of the same phenomena. The dogmatic metaphysical approach was not so much criticized as completely set aside. Germany also had its own materialists, such as Ludwig Büchner, whose best-selling book *Force and Matter* went through 21 editions between 1855 and the end of the century. These writers were working outside the universities, as were the followers of Schopenhauer. The fairly strange doctrines of Eduard von Hartmann and the even stranger ones of Philipp Mainländer found a widespread audience, and yet, despite its popularity with readers, Schopenhauer's philosophy did not adapt well to physical science, as any reading of Hartmann's *Philosophy of the Unconscious* will quickly confirm.

One way out of this impasse was a radical but straightforward move: back to Kant. Neo-Kantianism began in 1865 with the youthful Otto Liebmann's polemic *Kant and the Epigones*, in which every chapter concludes with the slogan 'Therefore we must return to Kant' (*also muss auf Kant zurückgegangen werden*).[2] Liebmann proposed a transcendental idealism without the thing-in-itself, which he describes as 'Kant's main error'. The crucial encounter between a revitalized Kantianism and a materialism riding on the prestige of physical science came with F.A. Lange's *History of Materialism*, also published in 1865, and reissued in an enlarged and updated form in 1873–75. The influence of this work on Nietzsche is well established and, in fact, the subject of a whole monograph.[3] Looking at the contemporary philosophical writers Nietzsche read, one finds that nearly all of them chose to identify themselves in some way or other with the Kantian tradition, with varying degrees of plausibility. Many also appealed to Leibniz as a suitable predecessor, because his philosophy seemed compatible with modern science, while adding the 'inner' dimension noticeably absent from conventional materialism. What is more, Leibniz supplied a precedent for German resistance against an English scientific hegemony. Hence, the old rivalry between his supporters and those of Newton was revived by a number of German writers. Their strategy was directly relevant to two scientific controversies arising out of England: one about William Thomson's 1852 paper predicting a 'final state' of thermal equilibrium, and the other about Darwin's theory of evolution by means of natural selection, published in 1859. I intend to say more about each of these in due course.

Returning to Nietzsche, let us consider some relevant factors in his response to the science wars. On the face of it, one might expect him to be a harsh critic of materialism in philosophy, as he is of utilitarianism in ethics. Yet Nietzsche has strong inclinations which lead him in another direction. One of these is his admiration for ancient materialism, to which Thomas Brobjer has already drawn attention in Chapter 1 of this volume. His early interest in Democritus is somewhat hidden, since it never resulted in any finished product, despite an interesting series of notebook sketches. Even the unpublished, although quite polished, essay *Philosophy in the Tragic Age of the Greeks* breaks off before reaching the atomists, leaving us with a very partial picture of Nietzsche's relation to the early Greek thinkers who, he asserts, cover the possible range of basic philosophies. At the same time, he was an admiring reader of Lange's *History of Materialism*, and draws on Lange in interpreting modern atomism as a revival of ancient Greek atomism – although adding his own characteristic spin by suggesting that scientific atomism is inferior in 'spirit' to its ancient counterpart. Another early influence is the natural philosophy of Boscovich, who described his theory as 'midway between that of Leibniz and that of Newton', since it combined Leibnizian points with Newtonian forces.[4] Yet another point to note is Nietzsche's admiration for the sensualistic philosophers of the French Enlightenment, and for later French writers such as Stendhal. Finally, we should mention his friendship with Paul Rée, a thinker very much under the influence of Spencer and Darwin: this factor is particularly important, and needs to be looked at in detail.

At this point one might wonder what would draw Nietzsche to show sympathy with the German side in the science wars. But what about section 252 of *Beyond Good and Evil*? It starts out: 'They are not a philosophical race, these English' and the following section denounces what it calls 'this damnable Anglomania of "modern ideas"' (*BGE* 252–53).

The mixture of conservatism and nationalism shown here is, in fact, a feature of Nietzsche's thinking in various periods, even quite late ones. To some extent, it represents something which his commentators seldom (or rather, never) acknowledge. In *Beyond Good and Evil* Nietzsche describes Kant as 'Königsbergian', but he could himself often be described as Naumburgian. The *Untimely Meditations*, for example, contain pages of quite conventional opinions. Some of his aphorisms (and the same is true of Paul Rée) express a commonplace cynicism, or a bourgeois worldly wisdom. I do not claim that this plays a major role in his approach to the 'science wars', but from time to time his remarks about materialism or Darwinism do follow the standard German line. This trait comes out clearly in his Basle period, partly in his writings on the future of German education, which echo the cultural conservatism of the period, but especially in his public intervention in a major scandal of the day: the furore that broke out in early 1872 over the polemical manifesto of the Leipzig physicist Friedrich Zöllner.[5] This casts a revealing light on Nietzsche's attitudes at the time, as well as being the single occasion on which he took a public stand over an issue involving natural science. Hence, I am taking it as his first engagement in the 'science wars'.

Three Debates in Natural Science

Zöllner had prefaced his book *On the Nature of Comets* with a long and fierce attack on the German scientific establishment, which he thought was far too close to its English counterpart.[6] The crucial *theoretical* issue concerns the importance of epistemology as a part of scientific thinking. Zöllner postulates a fundamental and characteristic difference between the intellectual talents of the English and

Germans: 'One could assert that the English mind is constituted predominantly inductively, the German mind in contrast predominantly deductively.'[7] This is the starting-point for an attack on what Zöllner calls 'the primitive state of epistemology in England'. In part this is due to a 'scientific proletariat' which cares nothing for tradition, but a deeper cause is the decline of the deductive aspect of scientific thinking since Newton's day and its replacement by a strongly inductive tradition, culminating in Michael Faraday's work.

The Germans show their superiority clearly in scientific methodology, according to Zöllner. For instance, whereas Darwin and Wallace postulated the principle of the continuity of organisms on empirical induction, German researchers have adopted a more deductive approach, 'in much greater generality, more from the standpoint of epistemology than from that of immediately observed facts'.[8] Darwin often had this effect on German readers: they accepted his findings, but not his way of arriving at them. Marx and Engels, settled in England yet still writing (and thinking) in German, were impressed by Darwin's work and praised it in print, while complaining to each other about its 'crude English method' and 'clumsy English style of argument'.[9] Even the fairminded Lange refers, in a rather backhanded compliment, to 'that splendid and so often successful one-sidedness which we find with especial frequency amongst Englishmen'.[10] There remained the embarrassing fact that Darwin and Wallace had made the actual discovery, and it is significant that practising German scientists such as Hermann von Helmholtz were not impressed by Zöllner's attempts to derive scientific findings from epistemological principles.

In January 1873 Nietzsche came to Zöllner's defence in public debate, after reading a journal article which praised Zöllner's book, while recommending another book which cast doubt on the mental stability of Richard Wagner. In response, he wrote an angry letter to the editor of the Wagnerian periodical *Musikalisches Wochenblatt*, complaining about this slur on the 'noble name' of Zöllner, and the letter was published – unfortunately, since I doubt that an older Nietzsche would have wanted to see the text included in his collected works as it is.[11] *Nature*, in contrast, published a review of Zöllner's book written in a spirit of good-natured irony – or perhaps with the sort of straight-faced humour which sometimes baffles or infuriates the non-English reader.[12] Zöllner (who kept a close eye on press notices of his work) was displeased with this review, but seems to have overlooked Nietzsche's supportive statement; presumably he was not a reader of the *Musikalisches Wochenblatt.*

Zöllner has been interpreted as a conservative force in German culture.[13] Rather strangely, he later became a follower of a very Anglo-Saxon tendency in nineteenth-century thought: spiritualism. He conducted séances with visiting British and American mediums and defended the reality of their phenomena. From the point of view of scientific methodology, this was a long way from systematic experimentation, let alone the kind of reasoning from epistemological principles that Zöllner had earlier championed, although he did attempt to rationalize the theory by arguing on epistemological grounds for a fourth dimension of space. Nietzsche was appalled by this development, but made no public statement about it, apart from one very cryptic reference to 'a scholarly half-madman' in *Thus Spoke Zarathustra.*

The main scientific issue raised in Zöllner's *On the Nature of Comets* concerns the role of the mathematical apparatus used in a scientific theory. If the equations work – that is, provide accurate correlations between observations – does this mean that they extend our knowledge by revealing the nature of the connections between the phenomena? Two writers who denied this were William Thomson and P.G. Tait, whose *Treatise on Natural Philosophy* (1865) was the bible of Newtonian orthodoxy for the nineteenth century. Thomson and Tait allow that mathematical models can run ahead of established knowledge, often with beneficial results. For example, we do not know how to interpret the mathematical formulae for the wave theory of light, but they certainly seem to work. However, they named the electrical theories of the Leipzig physicist Wilhelm Weber as an example of a mathematical model which does more harm than good. Weber believed that the force exerted by moving electrical particles on others differs from the force exerted by resting particles. Thomson and Tait objected that there was no experimental evidence for this claim and that, worse still, it was inconsistent with the principle of the conservation of energy. Helmholtz had already gone into print in 1870 making the same criticism, so he was hardly likely to object to this point in sponsoring the German edition of their book.[14] Zöllner was concerned to defend his colleague against the English attack, and accused Helmholtz of a 'lack of tact' in helping to translate the offending passage into German. In one sense, he may have won this dispute, since sections 385 and 386 were omitted from the second English edition of Thomson and Tait's *Treatise*, published in 1879. But his own position within German science was permanently damaged by these actions.

This controversy over the role of mathematics in natural science not only continued but heated up considerably in the 1890s – a period some time after the one I am discussing – with Ludwig Boltzmann being accused by English scientists, again led by Thomson and Tait, of substituting mathematics for thinking and engaging in what they called 'mathematical terrorism'.[15] Once again, the real struggles were within the German scientific community, this time with Ernst Mach as champion of the empiricist cause. But this further dramatic (and, in the end, tragic) controversy lies outside our discussion. Nietzsche's view is nevertheless worth noting. Like many philosophers, he preferred the idea of mathematics to actual mathematics. Hence, he was not opposed to 'mathematical terrorism', claiming that the strongest objections to traditional atomism had come from mathematicians – and describing Boscovich in a letter to Heinrich Köselitz as 'not an idealist, but a mathematician' (*KGB* III/1, 20 March 1882). His idea is presumably that traditional atomism is inconsistent with a universal law of continuity which Boscovich takes as the basic premise for his entire theory: 'The mathematical physicists cannot use solid atoms for their science: hence they construct a force-point world that one can calculate with' (*KSA* 11, 40[36]). In this sense, Nietzsche comes out against the empiricism of the English school of physical science in favour of a far more theory-driven approach.

Let us now move forward nearly a decade to the year 1881, when Nietzsche's interest in natural science revived, and look at his relation to the heated dispute surrounding the work of Robert Mayer. This was an issue of priority, concerning the discovery of the conservation of force (or energy, in modern terminology), for which historians of science nowadays assign equal responsibility to James Joule and Robert Mayer. Strictly speaking, Mayer's publications came first but, since they appeared in chemistry rather than physics journals, they failed to attract notice. In

1880 Eugen Dühring's polemical biography, *Robert Mayer, the Galileo of the Nineteenth Century*, was a bestseller.[16] In this book Dühring uses Mayer to attack two targets. One is the English nation and the other is the German scientific establishment – in particular, Hermann von Helmholtz, whose widely read 1847 lecture 'On the Conservation of Force' had credited the discovery to Joule alone.

Dühring had an advantage over other writers on this subject: he had met Mayer a few years previously and interviewed him at length. This encounter occurred just after Dühring's dismissal from the University of Berlin, primarily for attacking the professors there on several scores, including their failure to support Mayer's claims. After an exchange of letters, Mayer came to the spa resort of Wildbad, where Dühring was on holiday with his family. In the course of several meetings, Dühring learned about Mayer's earlier history of mental illness and the harsh treatment he had received. Dühring was surprised to find Mayer a provincial general practitioner with a strong regional accent and conservative opinions about religion and politics, while Mayer seems to have been apprehensive about encountering a notorious Berlin firebrand. However, after Dühring explained that he was not a supporter of the Social Democrats, they got along quite well. Mayer died the following year, so Dühring was able to give his version of their conversation without fear of contradiction.

Nietzsche was informed of the high sales of Dühring's book by his publisher Ernst Schmeitzner, who was also Dühring's publisher at the time. As for the book's content, his friend Heinrich Köselitz supplied a strong recommendation, and followed this by sending Nietzsche a copy of Mayer's collected papers. As far as I know, Nietzsche did not read Dühring's book, but his response to reading Mayer is interesting. His first reaction was very favourable (see his letter to Köselitz, *KGB* III/1, 16 April 1881), but this was succeeded by a quite different judgement a year later, when he described Mayer in correspondence as 'a great specialist – and no more' (*KGB* III/1, 20 March 1882). My guess is that when Nietzsche took a quick glance at Mayer's writings, he noticed a number of succinct formulations and what appear to be axiomatic propositions about the 'equality' of cause and effect, the truth that nothing comes from nothing, and so on. A certain enthusiasm is evident in *The Gay Science*, where a section headed 'Long live physics!' (*Hoch die Physik!*) calls upon all free spirits to 'become physicists' in order to discover what is lawlike and necessary in the world (*GS* 335).[17]

When Nietzsche came to look more closely, however, he would have realized that Mayer is in fact an empiricist of the English kind, who repudiates hypotheses and insists on staying close to observation and experiment. Hence, Mayer resists the theory that heat is just a kind of motion, even though it can be equated with motion.[18] Rather, he says, the transformation of heat into motion 'is stated only as a fact, and in no way is the thing given an explanation'.[19] Similarly, when ice is transformed into water, 'the fact is simply so and remains independent of the unfruitful speculations on how and why'. When Nietzsche writes about current scientific thinking, he often expresses a certain approval of this kind of self-imposed limitation, especially when it takes a quantitative form – that is, he acknowledges that it gets useful results. On the other hand, he still wants to engage in speculation about the underlying origins of observed regularities and correlations.

Nietzsche's comments on Mayer were made only in correspondence, but they throw useful light on his attitude towards current scientific debates. He turns out to be more in tune with Dühring's approach than one might expect, given the consistently negative tone of his comments, which include a detailed commentary on *The Value of Life*, and what seems to be a personal lampoon of Dühring in *Thus Spoke Zarathustra* (see *KSA* 8, 9[1] and Z ii 'Great Events'). Dühring advocates a 'rational physics', presenting himself as a realist but not an empiricist. He believes in 'principles of thinking' which may be taken as binding upon reality – most notably, the so-called 'law of definite number', which amounts to a denial of any actual infinitude, a thesis that Nietzsche generally supports, though not in its application to the case of past time. (Whenever Nietzsche denounces a contemporary author in a particularly hostile and sarcastic manner, one can usually predict that some idea or argument plainly borrowed from that source will turn up in his own writing: Strauss, Hartmann and Dühring all provide cases in point.) Thus, Nietzsche's encounter with Robert Mayer confirms the attitude displayed in his earlier endorsement of Zöllner's scientific programme – that is, an opposition to English empiricism, a belief in the value of epistemology for scientific theorizing, and a fondness for wide-ranging speculation.

The final dispute in physical science I will mention involves the cosmological issue arising out of the second law of thermodynamics. Nietzsche was familiar with this because it was widely discussed at the time, primarily as a result of Hermann von Helmholtz's 1854 lecture, 'On the Interaction of Natural Forces', based in turn on William Thomson's recently published paper 'On a Universal Tendency in Nature to the Dissipation of Mechanical Energy'.[20] Thomson had been cautious in his stated conclusion, and merely observed that it cast doubt on the future of humanity, since at some future time the earth would be unable to sustain human life in its present form. Helmholtz, in contrast, boldly predicted the end not just of humanity but of the entire universe.[21]

Did Nietzsche ever read this lecture? Possibly not, but its main conclusions were quoted at length in a work he certainly did read – Otto Caspari's short book *Thomson's Hypothesis of the Final Thermal Equilibrium in the Universe*.[22] His most concentrated engagement in the literature of popular science, occurring in the autumn of 1881, was set in motion by a different work of Caspari, a collection of critical reviews which put him on to various other authors.[23] The essay on William Thomson is a strange piece, if only because Thomson barely makes an appearance in it. Caspari raises the old question of the eternity of the world, and solves it along Leibnizian lines by postulating an organic model of the cosmos. He rejects mechanistic materialism and comments that '[i]t is highly characteristic for an Englishman to carry the mechanical theory of steam engines over into the universe'.[24] In fact, Thomson made no assertions about the universe as a whole, although it is a logical step from what he says about the earth to Helmholtz's broader conclusion. Caspari responds by postulating processes in which heat differences may re-emerge. His main thesis, however, is that the universe is an enormous organism. One feature of an organism, according to Caspari, is that it is capable of coming to an apparent standstill and then becoming active again without any external cause.[25] Hence, the universe may seem to run down completely but then start up again of its own accord.

Nietzsche's response to this 'macrocosmic organicism' was extremely hostile. In *The Gay Science* he warns, 'Let us beware of thinking that the world is a living being', and goes on to ridicule any such notion as an extreme kind of anthropomorphism. 'This nauseates me,' he comments (*GS* 109). What he especially objects to is the teleology implicit in the Leibnizian solution to the problem of entropy – that is, the need to attribute both foresight and control to the elements with the qualities required to avoid any final state of force. All this amounts to 'a disguised polytheism' (*KSA* 9, 11[201]).[26] Here Nietzsche uses one of his standard arguments, which goes from the possibility of some event to its necessary occurrence in the course of infinite time: 'If the universe were capable of becoming an organism, it would have done so.' Hence, he concludes, we need to think of it as a whole 'as far as possible from the organic'.

Here we see Nietzsche firmly separating himself in print from the German critics of English materialism who drew upon Leibniz's philosophy. His own attempt to answer the challenge of a 'final state' by constructing a model of eternal recurrence makes only those assumptions which are consistent with a non-teleological account of natural processes. What he says in *The Gay Science* implies an interesting thought experiment: what (if anything) is the *opposite* of an organic model of the universe? Not thinking of it as a machine, he argues, because a machine is governed by lawlike regularities.[27] Presumably the answer lies in the concept of chaos, which he seems to think might provide a third option, an alternative to both sides in these disputes – but I think that opens up a new discussion, so will not pursue it here.

Paul Rée and English Moral Psychology

Let us now turn away from disputes in physical science to biological science and the implications of the theory of evolution.

The greatest Anglophile of Nietzsche's acquaintance was Paul Rée, his close friend between 1876 and 1882. Nietzsche's varying attitudes towards English science and philosophy are closely related to the ups and downs of his relationship with Rée. It is true that Rée also admired and emulated the French moralists of the eighteenth century, as did Nietzsche. In his lectures on 'The Future of Our Educational Institutions' he warned Germany against trying to import French culture, but was soon expressing admiration for French thought, especially in *Human, All Too Human*, possibly owing to Rée's influence. On the other hand, the full extent of Rée's Anglophile inclinations may not have been evident when he and Nietzsche first met. Most of his earliest, anonymously published book, *Psychological Observations*, consists of aphorisms along the lines of La Rochefoucauld or Chamfort, usually pointing out some selfish motive behind human behaviour. However, the book ends with a short essay on vanity, which argues that vanity – defined broadly as the desire for the good opinion of others, apart from any immediate benefit this may bring – is an innate human trait which has been encouraged in two ways: first, in the individual by the formation of habits from early childhood; and, second, in the human race by group selection, since pleasure in the approval of others contributes to solidarity which in turn gives any tribe or race an advantage over other groups in the struggle for existence. There is an overlap between these two explanations, since Rée accepts (as Darwin did to some extent) the Lamarckian claim that characteristics acquired by individuals, either by their habits or through the influence of the environment, may be passed on to their descendants as innate qualities.

Apart from this admiration for Darwin's theory of evolution, Rée's preference for English writers does not come through in his early works, because the pessimistic influence of Schopenhauer is still evident. However, in his last, posthumously published book, simply entitled *Philosophie*, Rée repudiates Schopenhauer and states categorically: 'The greatest philosophers are Berkeley and Hume.' This judgement is backed up with endorsements of Berkeley's scepticism about the reality of matter and Hume's scepticism about the necessity of causation.[28] These were not new opinions on Rée's part; many years earlier, he had told Nietzsche about his agreement with Hume's account of causality (*KGB* II/4, 10 October 1877). Rée never changed his ideas, or even added to them. His 1885 book on conscience simply plugs in a great deal of factual material from legal history and anthropology to illustrate the same model of moral thinking.

Rée's second book, *The Origin of the Moral Sensations*, was largely written during the autumn of 1876, when he and Nietzsche were working together in Sorrento. This book presents an argument which has two stages.[29] First, it sets out to explain the occurrence of altruistic (or as Rée calls them, 'non-egoistic') feelings in human beings. Second, it tries to explain the interpretation (or misinterpretation) of these feelings and motives which constitutes a *moral* valuation. The first argument about altruism parallels the argument about vanity in Rée's earlier book. Altruism is a natural and innate human drive, although usually much weaker than egoism. However, over the course of many generations it has been selected and strengthened. The basis for this argument is found in Darwin's *The Descent of Man*, published in 1871, though the psychological ideas owe much to English writers such as John Stuart Mill. Given these points of reference, it is not surprising that Rée was keen to have his book published in English. Nietzsche helped in this aspiration by making the acquaintance of the editor of *Mind*, George Croom Robertson, while both were on holiday at the resort of Rosenlauibad in the summer of 1877.

This encounter with Croom Robertson is interesting in other ways. Nietzsche wrote to his mother reporting 'a very pleasant acquaintance with an English scholar' (*KGB* II/5, 10 August 1877). Professor Croom Robertson, he informed his sister, had 'the *closest* relations with Darwin, Tylor and Spencer' (*KGB* II/5, 4 August 1877). 'He is a professor in University College, London and editor of the *best* philosophical journal, not just in England but anywhere .... He was very taken with Rée's book,' he told Malwida von Meysenbug, (*KGB* II/5, 7 August 1877). While on holiday, Croom Robertson was still at work, Nietzsche reported: he was translating into English a commissioned article by Wilhelm Wundt, surveying recent German philosophy. Nietzsche quickly wrote to Paul Rée, reporting on the contents of the June 1877 issue of *Mind*, which included contributions from Darwin, Tylor and Spencer. (This list is not as impressive as it sounds: Tylor and Spencer's joint contribution to that issue consisted in a bad-tempered exchange over ownership of ideas.) Nietzsche must have looked at the issue, though whether his command of English allowed him to read much of it is another matter – his description of Darwin's article to Rée as 'highly stimulating' may have been only on the basis of conversation in German with the editor.[30]

Nietzsche's name was already known to Croom Robertson from a review of the first three *Untimely Meditations* in the *Westminster Review* two years earlier,[31] as well as from the article by Wundt which he was translating into English, and which appeared in the October 1877 issue of *Mind*. Wundt places Nietzsche under the same heading as Schopenhauer and Hartmann, notes the appearance of his *Untimely Meditations*, and comments: 'In the writings of Nietzsche and others of the same stamp, the pessimistic mood is combined in a very peculiar way with an enthusiastic devotion to certain ideas closely related to religious mysticism.'[32] The same issue also contained a review of Rée's *The Origin of the Moral Sensations*. Its anonymous writer – in fact, the editor himself, as a personal letter to Nietzsche indicates (*KGB* II/4, 20 September 1877) – wrote: 'The essay is marked by great lucidity of expression, and no common boldness of thinking. It deserves attention both for the scientific value of some of its observations, and as a rather striking specimen of the pessimistic vein of thought now prominent in Germany'[33]

Rée's identification with English thinking becomes particularly important for Nietzsche just a few years later. His own account of moral feeling and valuation is developed in opposition to Rée, in contrast with the solidarity expressed in *Human, All Too Human*. In trying to distance himself from Rée, Nietzsche invariably identifies Rée's theories with those of English writers. He typically does not name these, but it seems likely that foremost amongst them is Herbert Spencer, whose *The Data of Ethics* had appeared in 1879. While it covers many of the themes of *The Origin of the Moral Sensations*, Spencer's tone could hardly be more different. Even when using Darwinian arguments, Rée remains a pessimist with a rather cynical view of human nature as irredeemably selfish, along the lines of Schopenhauer and La Rochefoucauld. Spencer, in contrast, is relentlessly optimistic. It is hard to resist quoting his vivid description of the healthy person who, as he explains, displays a high degree of rational egoism, only moderately tempered with altruism.

> Bounding out of bed after an unbroken sleep, singing or whistling as he dresses, coming down with beaming face ready to laugh on the slightest provocation, the healthy man of high powers, conscious of past successes and by his energy, quickness, resource, made confident of the future, enters on the day's business not with repugnance but with gladness; and from hour to hour experiencing satisfactions from work effectually done, comes home with an abundant surplus of energy remaining for hours of relaxation.[34]

Even English readers may have found this image a little startling. As for Nietzsche, his frequent condemnations of Spencer as an English philistine do not quite disguise the fact that his own descriptions of the 'higher man' (or the *Übermensch*) have some noticeable similarities with the picture drawn in this passage.

In the Preface to *On the Genealogy of Morals*, Nietzsche goes to a great deal of trouble to define his own position in relation to Rée's: 'The first impulse to publish something of my hypotheses concerning the origin of morality was given me by a clear, tidy, and shrewd – also precocious – little book in which I encountered distinctly for the first time an upside-down [*umgekehrte*] and perverse species of genealogical hypothesis, the genuinely *English* type, that attracted me – with the power of attraction which everything contrary, everything antipodal possesses' (*GM* 'Preface' 4). The 'English' identification is repeated a few pages further on, where

Nietzsche reports that he offered helpful advice to Rée, although this was apparently not taken: 'My desire, at any rate, was to point out to so sharp and disinterested an eye as his a better direction in which to look, in the direction of an actual *history of morality*, and to warn him in time against gazing around haphazardly in the blue in the English fashion' (*GM* 'Preface' 7).

The best reason for labelling Rée's theory as an *English* approach would be the extent to which it draws on the psychology of Mill and Spencer, which was largely formulated before the appearance of Darwin's work but lent itself very readily to an evolutionary approach. Spencer himself soon modified his ideas to bring them into line with the theory of natural selection. Rée constantly refers to the habitual association of ideas as an explanatory principle in accounting for the moral sentiments, and appeals to Mill's *A System of Logic* as support for his belief in the power of association.

> Once someone has become accustomed to connecting the idea of praiseworthiness with a particular mode of action and the idea of blameworthiness with its contrary, it will easily seem to him as if he had not become accustomed to making these connections, but had been making them from birth. So it seems to us as if we had been connecting the idea of praiseworthiness with non-egoism, and of blameworthiness with egoism, since our birth.[35]

Spencer's theory, cited with approval by Darwin in *The Descent of Man,* is that the accumulated experiences of many human generations have given rise to habits which are expressed in judgements about what is good and bad, and that these judgements are abbreviated in their form, so that things are described simply as 'good' and 'bad', rather than as good or bad for fulfilling our purposes or interests, which would have been the lesson of the original experiences. Rée follows these ideas closely and supplements them with two evolutionary theories: on the one hand, a Lamarckian belief in the inheritability of acquired habits and, on the other, the Darwinian principle that habits which give an advantage in the struggle for existence are favoured by natural selection, and so are more likely to be passed on.

Much more debatable is Nietzsche's claim that Rée is following the English model when he identifies a valuation of altruism with moral valuation 'in itself'. He probably means to refer to Spencer, since in *Twilight of the Idols* he writes: 'Our socialists are decadents, but Mr Herbert Spencer is a decadent too: he considers the triumph of altruism a desideratum!' (*TI* 'Skirmishes' 37). In fact, Spencer's endorsement of altruism is a very limited one. For instance, he notes that someone who shares in the happiness of others will gain more sources of satisfaction from that fact – a sound observation, but not the usual case for altruism.[36] A scientific morality, in Spencer's view, is simply an understanding of adaptation to the necessary conditions of social life: he uses a rather Nietzschean expression in describing this as a 'moral physiology'.[37] A straightforward identification of moral worth with altruism really points to Schopenhauer, not to English moral philosophy. The starting-point for Rée's *The Origin of the Moral Sensations* is an endorsement of the conclusion of Schopenhauer's essay *On the Basis of Morality* – that is, the thesis that the only actions which deserve moral praise are selfless ones. From there, however, Rée's argument goes in a different direction, towards a naturalistic account of selfless feelings and motives, and this is where the English connection becomes evident.

Nietzsche is on firmer ground when he criticizes what he calls the 'English' theory about morality in the 'First Essay'. This is essentially the theory presented by Rée in *The Origin of the Moral Sensations*, and which Nietzsche had endorsed in section 40 of *The Wanderer and His Shadow.* The central part of Nietzsche's argument has a surprising feature: it is arguably more faithful to a Darwinian approach than the theory it is attacking. This needs some explanation. The case in point is punishment, which is already a prominent theme in Rée's account of morality and becomes central in his later work of 1885, *The Evolution of Conscience*. Nietzsche argues that Rée's account of punishment is far too teleological. Rée exposes our present attitudes towards punishment as irrational, but assumes that because he is writing about human behaviour, he can trace even attitudes which express misconceptions back to behaviour which is rational and purposeful. So the explanation is in terms of purpose after all – not purpose in the present day, but at some distant past time (*GM* ii 12).

Nietzsche responds that we must not assume that the present use of anything explains its origin. To regard the eye as made for seeing, or the hand for grasping, is naive. For 'whatever exists, having somehow come into being, is again and again reinterpreted to new ends'. In the case of a pattern of behaviour such as punishment, this implies that a succession of more or less unrelated 'meanings' have been imposed on a certain practice, so that '[t]oday it is impossible to say for certain *why* people are punished' (*GM* ii 12). Now, this is a very authentic Darwinian notion. It recalls what evolutionists call 'pre-adaptation' – that is, the use of an organ adapted to one function for a different function, such as the development of air-breathing lungs from the bladders providing fish with buoyancy, or of defensive quills from insulating hairs. In the same way, the material reality of practices makes them available to new uses and accompanying interpretations not connected with the causes that brought them into being. Hence, evolution displays a high degree of contingency: it is not a direct progress in a certain direction, such as greater complexity, as in the Lamarckian and Spencerian versions of evolution. However, Nietzsche's explanation of each of these changes involves the quite un-Darwinian concept of a spontaneous 'will to power' which seeks to overcome obstacles for the sake of power. With this Nietzsche avoids what he thinks is a mistaken conception of life as 'a mere reactivity'.

As a number of commentators have pointed out, Nietzsche is not at his best on the subject of Darwinism.[38] Possibly he was not well informed, despite reading Lange's account in his *History of Materialism.*[39] According to the French writer Claire Richter, all of Nietzsche's contacts were with supporters of the Lamarckian tradition – but, then, Richter insists that the theory of evolution is a French theory which should be referred to as Lamarckism, not Darwinism.[40] Like other German writers, Nietzsche concentrates not on the idea of natural selection (though interesting uses of the word do appear in his writings) but rather on the struggle for existence. Typical is the section of *Twilight of the Idols* entitled 'Anti-Darwin' (*TI* 'Skirmishes' 14). Nietzsche's comments here are not particularly well directed – for example, his observation that nature exhibits 'riches, profusion, even absurd squandering' is not only consistent with Darwin but, in fact, part of his argument for the universality of the struggle for existence. And Nietzsche goes on to offer the inevitable and quite conventional comment: 'One should not mistake Malthus for nature.'

Yet, despite his repeated claims to be presenting arguments against Darwin, Nietzsche's own genealogy of morals seems to me to have a strongly Darwinian character when one considers the importance it places upon two elements: *materiality* and *contingency*. These come through prominently in critiques of Paul Rée's attempt to derive the moral sensations from habit, environment and heredity. For Nietzsche, Rée's approach is both too intellectual and too utilitarian. It leaves out the materiality of practice, which has its own historical patterns, often determined 'in a purely chance fashion'.[41] Nietzsche calls this 'upside-down' (*verkehrt*), a description which reminds one of the Marxian criticism of Hegelian idealism. He suggests that the use of punishment for deterrence is a later adaptation of an existing behaviour (*GM* 'Preface' 4).[42] Thus, Nietzsche's account acknowledges the importance of materiality in the evolution of morality, and thereby of contingency, in so far as this is located in what escapes the power of reason to extend its formal realm over reality in general.

We can perhaps explain Nietzsche's commitment to an ultra-Darwinism by noting his reading of W.H. Rolph's book *Biologische Probleme*.[43] In large part a response to Spencer's *The Data of Ethics*, this is written for the non-technical reader. Nietzsche thought Rolph's book too popular in its approach, but approved of its criticisms of Spencer.[44] Summarizing Darwin's theory of natural selection, Rolph emphasizes that adaptation to the environment has no future aim, but is determined purely by the current situation. He writes:

> Adaptation has absolutely no tendency, no striving to provide forms, which from the human viewpoint could be described as higher: it has to develop only the result better, fitter. And if this is not possible under the circumstances, it lets the species die out. It rules with great inconsiderateness and with cruelty.[45]

This seems very close indeed to Nietzsche's conception of nature as the embodiment of indifference and lack of purpose, and in turn to his notion of chaos.

Finally, back to the 'damnable Anglomania of "modern ideas"' which I referred to, quoting from *Beyond Good and Evil*. This passage is certainly a stumbling-block for my claim that Nietzsche is often close to English ideas, so I will make some attempt to explain it away. A preliminary sketch for the passage appears as an ill-tempered notebook entry of 1885, apparently set off by Nietzsche's reading of Spencer's *The Data of Ethics*, in which he attacks first the English and then several writers in German (and one in French) whom he accuses of being under English influence. A prominent feature of these attacks, and indicative of how untypical of Nietzsche they are, is an insinuation that the authors are not genuinely German: Paul Rée is 'a German Jew', and W.H. Rolph a 'German half-Englishman' (*KSA* 11, 35[34]). All this shows an unexpected regression to Nietzsche's youthful German patriotism, not to mention a fairly shallow approach to the history of ideas, as, for instance, in the assertion that 'it was *against* Hume that Kant arose'.

The published version repeatedly uses the phrase 'modern ideas' – yet apparently meaning ideas of the previous century – and identifies them as 'of English origin' (*BGE* 253). What the English lack is apparently 'genuine *power* of spirituality, genuine *depth* of spiritual perception; in brief, philosophy' (*BGE* 252). Nietzsche goes on in the next section to suggest that 'the spirit of respectable but mediocre

Englishmen – I name Darwin, John Stuart Mill, and Herbert Spencer – is beginning to predominate in the middle regions of European taste'. The French in particular, he adds, have completely fallen victim to this fashion. Fortunately, Europe still has the Germans to rely upon in standing fast against 'modern ideas' (*BGE* 253).

What do we make of all this? Nietzsche seems to be saying that the English have won the science wars, but we should not feel happy about this, because it is a victory of empirical science over 'spiritual depth'. However, one needs to keep on reading. At the end of Part Eight of *Beyond Good and Evil*, Nietzsche takes a further step, when he predicts a new European cultural synthesis. 'All Europe wants to become one,' he writes (*BGE* 256). (The existence of a world outside Europe is not mentioned.) And he mentions a number of prominent figures who have anticipated such a movement 'into a new light' and 'toward a new sun' – the last example being, rather unexpectedly, Richard Wagner.

This idea of a synthesis which would presumably include scientific culture – already one step ahead of the humanities in being relatively cosmopolitan – picks up some themes I have touched upon. For instance, it may be prefigured in Boscovich's claim to have formulated a physical theory which bridges the gap between Newton and Leibniz. Incidentally, it is significant that Nietzsche (wrongly) credited Boscovich with the same Polish origin that he sometimes claimed for himself when trying to distance himself from Germany. Boscovich did have the advantage (as I think it is) of coming from a small country – in that sense, Nietzsche was at least on the right track – and, in any case, spent most of his life within the structure of the Jesuit order, which gave him an unusually broad international experience. Nietzsche had to find his own way to become a 'good European', using whatever models he could identify. I think this effort extends into his engagements with natural science, and is reflected in many of the responses and observations I have discussed. My conclusion is that, despite the occasional lapse, Nietzsche had a broader vision than most of the participants in the science wars.

Notes

1 Andrew Ross (ed.), *Science Wars* (Durham, NC: Duke University Press, 1996).
2 Otto Liebmann, *Kant und die Epigonen. Eine kritische Abhandlung* (Stuttgart: Carl Schober, 1865), passim. Nietzsche was also familiar with two of Liebmann's later (and more conventional) books: *Zur Analysis der Wirklichkeit. Philosophische Untersuchungen* (Strassburg: Trübner, 1876), and *Gedanken und Thatsachen* (Strassburg: Trübner, 1882).
3 George J. Stack, *Lange and Nietzsche* (Berlin: de Gruyter, 1983).
4 Roger Joseph Boscovich, *A Theory of Natural Philosophy*, trans. J.M. Child (Cambridge, MA: MIT Press, 1966), art. 10.
5 The full story can be found in Robin Small, *Nietzsche in Context* (Aldershot: Ashgate, 2001), pp. 59–79.
6 J.C.F. Zöllner, *Über die Natur der Cometen. Beiträge zur Geschichte und Theorie der Erkenntnis*, 2nd edn (Leipzig: Engelmann, 1872), pp. v–lxxii.
7 Better known, but remarkably similar, is Pierre Duhem's later image: 'English philosophers are almost wholly concerned with applications of philosophy: psychology, ethics, social science. They have little liking for abstract research and do it poorly. They

proceed less by abstract argument than by the accumulation of examples. Instead of connecting deductions, they accumulate facts.' (*Essays in the History and Philosophy of Science*, trans. Roger Ariew and Peter Barker (Indianapolis: Hackett, 1996), p. 52.)

8 Zöllner, *Über die Natur der Cometen*, p. xxiii.

9 Karl Marx and Frederick Engels, *Collected Works* (London: Lawrence and Wishart, 1975–), vol. 40, p. 551 and vol. 41, p. 247.

10 Friedrich Albert Lange, *The History of Materialism and Criticism of its Present Importance*, trans. E.C. Thomas. 3rd edn (London: Routledge and Kegan Paul, 1925), vol. 3, p. 53.

11 'Ein Neujahrswort an den Herausgeber der Wochenschrift "Im neuen Reich"', *KSA* 1, pp. 795–97.

12 Anonymous, 'Sensation and Science', *Nature*, 4 July 1872, 177–78.

13 See Christoph Meinel, *Karl Friedrich Zöllner und die Wissenschaftskultur der Gründerzeit: Eine Fallstudie zur Genese konservativer Zivilisationskritik* (Berlin: SIGMA, 1991).

14 William Thomson and P.G. Tait, *Handbuch der theoretischen Physik*, trans. Hermann von Helmholtz and Georg Wertheim (Braunschweig: Vieweg, 1871).

15 See Enrico Bellone, *A World on Paper: Studies on the Second Scientific Revolution* (Cambridge, MA: MIT Press, 1980), chapter 3.

16 Eugen Dühring, *Robert Mayer, der Galilei des neunzehnten Jahrhunderts* (Chemnitz: Schmeitzner, 1880).

17 In his English edition, Walter Kaufmann expresses puzzlement and comments: 'It would surely be much more plausible if Nietzsche had spoken of psychology instead of physics.' (Nietzsche, *The Gay Science*, trans. W. Kaufmann (New York: Vintage Press, 1974), p. 267.)

18 Robert Bruce Lindsay, *Julius Robert Mayer: Prophet of Energy* (Oxford: Pergamon Press, 1973), pp. 72–73 and 223.

19 Ibid., p. 81.

20 William Thomson, *Mathematical and Physical Papers* (Cambridge: Cambridge University Press, 1882), vol. 1, pp. 511–14.

21 Hermann von Helmholtz, *Popular Scientific Lectures* (New York: Dover Publications, 1962), p. 90.

22 Otto Caspari, *Die Thomson'sche Hypothese von der endlichen Temperaturausgleichung im Weltall beleuchtet vom philosophischen Gesichtspunkte* (Stuttgart: Horster, 1874), pp. 23–24.

23 Otto Caspari, *Der Zusammenhang der Dinge: Gesammelte philosophische Aufsätze* (Breslau: Trewendt, 1881), p. 27.

24 Caspari, *Die Thomson'sche Hypothese*, p. 57.

25 Ibid., p. 48.

26 See also *KSA* 9, 11[213], [292]. The last of these 1881 entries reappears in *KSA* 11, 36[15].

27 For his notes on the universe as a machine see, for example, *KSA* 8, 9[2].

28 Paul Rée, *Philosophie: Nachgelassenes Werk* (Berlin: Duncker, 1903), p. 244.

29 For a more detailed analysis, see my introduction in Paul Rée, *Basic Writings*, ed. and trans. Robin Small (Urbana and Chicago: University of Illinois Press, 2003).

30 On the other hand, if Nietzsche did read through the issue, his eye may have been caught by a critical review of Grant Allen's book *Physiological Aesthetics*, since this is an expression he uses – for example, in proposing to write a work entitled 'Toward a Physiology of Art' (*CW* 7).

31 The anonymous reviewer describes Nietzsche as an 'admiring pupil' of Schopenhauer, deplores the tone of his attack on D.F. Strauss, and expresses the hope that some version of positivism will 'triumph in the end over the barren and bewildering metaphysics of

Germany' ('Contemporary Literature. Theology and Philosophy', *The Westminster Review*, 47, 1 April 1875, 501).

32 Wilhelm Wundt, 'Philosophy in Germany', *Mind*, 2 (1877), 509.

33 Anonymous, 'Review of Paul Rée, *Der Ursprung der moralischen Empfindungen*', *Mind*, 2 (1877), 581.

34 Herbert Spencer, *The Data of Ethics* (London: Williams and Norgate, 1879), pp. 190–91 (sect. 70).

35 Paul Rée, *Der Ursprung der moralischen Empfindungen* (Chemnitz: Schmeitzner, 1870), p. 22.

36 Herbert Spencer, *Social Statics* (London: John Chapman, 1851), p. 69.

37 Ibid., p. 58.

38 See, for example, Jean Gayon, 'Nietzsche and Darwin', in Jane Maienschein and Michael Ruse (eds), *Biology and the Foundation of Ethics* (Cambridge: Cambridge University Press, 1999), pp. 154–97, and Gregory Moore, *Nietzsche, Biology and Metaphor* (Cambridge: Cambridge University Press, 2002).

39 For a broad survey of the German literature, see Alfred Kelly, *The Descent of Darwin: The Popularization of Darwinism in Germany, 1860–1914* (Chapel Hill, NC: University of North Carolina Press, 1981), chapter 1.

40 Claire Richter, *Nietzsche et les théories biologiques contemporains* (Paris: Mercure de France, 1911), p. 9. A similar view is expressed in the inscription below the imposing statue of Lamarck outside the Jardin des Plantes in Paris: 'FONDATEUR DE LA DOCTRINE DE L'ÉVOLUTION'.

41 Darwin condemns any mention of 'chance' or 'accident' in evolutionary thinking as an incorrect expression which corresponds only to our ignorance (*The Origin of Species by Means of Natural Selection*, ed. J.W. Burrow (Harmondsworth: Penguin, 1968), pp. 173, 241 and 358). But not all Darwinians would agree with this opinion.

42 Nietzsche's description of deterrence as 'terroristic' recalls Dühring's *Der Werth des Lebens* (Breslau: Trewendt, 1865), p. 139.

43 W.H. Rolph, *Biologische Probleme, zugleich als Versuch einer rationellen Ethik* (Leipzig: Engelmann, 1882).

44 He remarks that Rolph has succeeded in overturning 'that combination of stupidity (*bêtise*) and Darwinism which Herbert Spencer has put into the world under the title *Data of Ethics*' (*KSA* 11, 35[34]).

45 Rolph, *Biologische Probleme*, p. 14.

Chapter 8

Between Mechanism and Teleology: Will to Power and Nietzsche's Gay 'Science'

Christa Davis Acampora

This chapter broaches the general issue of how Nietzsche imagined the relation between science and art and their tense union in future philosophy. It does so by focusing upon Nietzsche's concern to find an alternative framework for conceiving organic development. Much has been written about Nietzsche's views of the opposition of art and science, and although it was common for some time to read Nietzsche as either reducing science (including the 'sciences' of knowledge and morality) to aesthetics or auguring the eclipse of science by art, increased interest in Nietzsche's familiarity with contemporary scientific developments and his efforts to draw upon and incorporate them in his own work have cast his projects in a somewhat different light.[1] I wish to build upon this recent research in order to advance reconsideration of Nietzsche's practice and anticipation of a possible future *fröhliche Wissenschaft*. As I shall argue below, once we understand how Nietzsche approached his scientific studies and their intercourse with perennial philosophical concerns, we may better appreciate the role Nietzsche envisioned for himself (and those who might follow) as supplying novel concepts and conceptual frameworks that creatively open new possibilities for willing. The latter are vital, since willing, for Nietzsche, is how meaning emerges, and it orients the ways in which we organize our lives. Gay 'science' aims to effect orientations that afford the possibilities for joyfully redeeming – and, I shall argue below, thereby altogether transforming – human existence.[2]

This is an endeavour that can be witnessed virtually throughout Nietzsche's *oeuvre*, occurring not only during his so-called 'positivist' period but also even before his ideas about future philosophy crystallize in his later writings. Nietzsche's consideration of the problem of organic development illustrates this well. His critiques of both teleological and mechanistic accounts of organic development have certainly been discussed in the secondary literature.[3] But considerably less attention has been devoted to Nietzsche's proposed alternative to these views. Though it is generally accepted that Nietzsche's so-called 'theory of will to power' represents his attempt to go beyond existing models of organic development, an account of its relation to teleology and mechanism – how his alternative meets the challenges facing those models it is intended to replace and how it preserves the features he admires – is lacking.[4] This chapter is intended to address this shortcoming. In what follows I shall illustrate how Nietzsche's solution to the problem of organic development, to the need to offer an account that avoids the pitfalls and dangers of teleology and mechanism, draws upon the principles that provide the basis of his conception of 'gay science' and serves as an illustrative example of its practice. Opening up a space between mechanism and teleology requires Nietzsche to take seriously nineteenth-century science and to reflect upon the relation between philosophy and science, in which philosophy is conceived as supplying and refining conceptual frameworks that science might utilize – not as creations *ex nihilo* but rather as concepts that emerge from reflections upon what science reveals about human sensation and physiology and the character of cognition.

Nietzsche's solution to the problem of development reveals precisely these tensions. In what I shall call Nietzsche's 'Heraclitean solution' to the problem of development, Nietzsche not only illuminates his emerging notion of will to power[5] but also exemplifies his philosophical practice of the *Kunst der Auslegung*, the art of interpretation. This I claim is at least one of the unique features of the kind of philosophical activity as 'gay science' that Nietzsche envisions as opening possibilities for the direction of future development. Hence, I shall suggest in my conclusion that Nietzsche conceives his alternative account of development as not only *statically* descriptive but also *actively* transformative. In other words, what is potentially at stake in providing an account of development is not merely describing a state of affairs or the process by which they regularly unfold but rather providing direction for the emergence of other forms of development.

The Problem

One can find ample evidence in Nietzsche's texts, published and unpublished, to prove that he rejects conceptions of nature that render it as striving for some sort of specific end or as operating like a great machine. The same applies for what we designate as individual organisms. Nietzsche's interest in theories of evolution is marked by these concerns. On the one hand, he believes that the idea of what a human being is (particularly in its relation to other animals) proffered by evolutionism marks an advance over the special status accorded to humans in theology and those forms of philosophical thinking that arbitrarily place the human at (or at least very near) the pinnacle of being. The view that humans are thoroughly natural creatures subject to change, growth and decay to the same degree as other organic forms represents, for Nietzsche, progress in the exercise of 'good conscience' in the pursuit of knowledge. It is a view that better accords with empirical observation and is somewhat less influenced by the habit of allowing moral prejudices to direct our perceptions and conceptions of what is true and real. It is on this ground that Nietzsche claims to admire Descartes (see, for example, *Antichrist* 14 and the original preface to *The Gay Science*) and some mechanistic theories that reduce moral and aesthetic questions to those of physiology.

The case against teleology is more complex than this, however. It reaches beyond Nietzsche's rejection of God or his purported relativization of values. His suspicions concerning the concept of the subject, the existence of discrete individuals and the nature of causality are all eventually implicated in Nietzsche's ongoing struggle against teleology. Even early on in his intellectual life, Nietzsche appears to think that the idea of the individual is a fiction, a concept born of habit. This habit, Nietzsche suggests, not only in his notebooks but also in his published writings, issues from grammar, which requires positing a conscious subject, 'a doer behind the deed' on the basis of which we also infer the existence of an intentional will and intelligent designer (*KSA* 12, 2[139]). It is through this same process that Nietzsche thinks we arrive at a conception of causality: things occur because something brings about their happening; there must be an agent behind all change. Our beliefs in the existence of individuals (rather than complexes of forces), the willing human subject and cause and effect are articles of faith, not constituents of knowledge. If we remain unquestioningly committed to these ideas and permit them to guide our scientific inquiries, we will retreat further into the subjective anthropomorphic world of the human rather than acquire the knowledge science seeks.

Similar concerns haunt mechanism, since its various forms often have recourse to the same metaphysical concepts. Although Nietzsche believes it practises somewhat better intellectual hygiene, mechanistic theory, particularly in its reliance upon inferences about causal relations and its postulations of laws, risks falling prey to the very same ills that plague teleological accounts of development (see, for example, *BGE* 21).

Moreover, the mechanistic world-view is potentially detrimental to our understanding of values. Both teleology and mechanism impose univocal accounts of the possible significance of existence: where the former asserts that all values derive from the end towards which everything is supposed to strive, the mechanistic view holds that all existence is essentially blindly (meaninglessly) unfolding in pursuit of nothing other than its mere perpetuation (see, for example, *KSA* 12, 7[54]). Christoph Cox puts it well when he reminds us that Nietzsche thinks that even mechanistic accounts are not free of the very prejudices that he is concerned to purge from metaphysics and moral philosophy.[6] Cox elaborates a convincing argument that Nietzsche's objection to mechanistic physics is grounded in his 'rejection of being'. Having given up all forms of dualistic metaphysics that allow one to distinguish appearance from reality, substance or matter from form and the like, Nietzsche cannot embrace a view of the world that ultimately rests upon seeing nature as consisting of 'things', whose motions are accounted for largely in terms of reactivity in a system whose organization and purpose require the supposition of some sort of master design or ordering principle.[7] Moreover, the notion that natural selection occurs on the basis of 'an increasingly better fit between organisms and the environment' also sneaks in a teleology that assumes some sort of overarching design which the world – shaped by evolutionary forces in the process of adaptation – ever more closely approximates. Hence, as Cox phrases it, mechanistic physics and evolutionary biology 'are still *not naturalistic enough*' for Nietzsche. They still import what Heidegger would later call 'ontotheological' concepts in their explanations of 'motion, change, and becoming'.[8]

Also at stake in Nietzsche's critiques of teleology and evolutionary theory are the possibilities for the meaning of human being. One view makes too much of us – nearly divine and divorced from the rest of creation – and the other makes too little – the blinking last man, descendant of the grinning ape, who by fortune finds the path of least resistance in the 'struggle for existence' and thereby manages to secure nothing more than his sheer 'preservation', mere survival. Both unnecessarily restrict the aims and ends of humankind; neither allows a role for creativity in the development of organisms and their possible futures. The real problem, as Nietzsche sees it, is to broker some sort of compromise between these two positions by developing a conception of human being that: (1) situates it within the world of becoming, but (2) puts it in dialogue with the empirical sciences, and (3) allows for the possibility that we might also be able to raise the bar for indicating the goals for which the human might strive, thereby making room for meaningful art. In such a case the human would not be bound by any particular *telos* but would still be able to find and direct meaningful purposive activity.

Nietzsche formulated lasting views about teleology and mechanism following his reading of F.A. Lange's *History of Materialism* and his subsequent firsthand acquaintance with the positions of various biologists, physicists, psychologists, mathematicians and cosmologists, including Boscovich, Helmholtz, Zöllner, Kopp, Huxley, Pouillet, Roux, Ladenburg, Maedler and others concerned with the philosophical and social implications of such matters, including Spencer, Spir and Goethe. As other commentators, including Thomas H. Brobjer in this volume, have significantly documented the particularities of these influences, I shall not catalogue them here. These earlier insights amplify Nietzsche's understanding of a problem that he thinks he inherits from antiquity and with which even modern science continues to wrestle, namely the problem of purposiveness. Sometime between the summer of 1872 and early 1873, Nietzsche writes: 'The horrible consequence of Darwinism, which, by the way, I hold to be true. All our veneration is based on qualities we take to be eternal: moral, artistic, religious, etc. We do not come a single step closer to explaining purposiveness by appealing to the instincts' (*KSA* 7, 19[132]). The 'horrible consequence' of Darwinism, of course, is that it implies that there are no eternal qualities and hence it undermines all that we value. Nietzsche thought that the problem of purposiveness was significant enough that, for some time in 1868, he intended to write a dissertation on the topic entitled 'On Teleology since Kant'.[9] The notes and drafts for the project provide further insight into Nietzsche's grappling with purposiveness and raise some important questions that I shall endeavour to address below. It is nevertheless worthwhile introducing them here, since they bear on Nietzsche's understanding of his task of mediating teleology and mechanism.

What motivates Nietzsche's prospective study is the question of whether it is necessary, as Kant claims, for us to suppose a teleology in the natural order. He plans to show that 'mechanism bound with casualism yields this possibility' of conceiving nature differently. Most interesting is what Nietzsche considers to be the stakes involved in a refutation of Kant. What would be the *advantage* of meeting Kant's challenge that it is impossible to conceive development without recourse to causality and purposiveness? Nietzsche claims: 'The elimination of teleology has a practical value. Only then is it possible for the concept of a higher reason to be removed: thus we are already satisfied.'[10] Now, it seems fairly straightforward how getting past teleology undermines the need for supposing a higher reason, but what is the practical value of that? Well, it would seem to dissolve some of our theological problems; we would no longer need to furnish proofs for God's existence. But what would it enable us to do, practically speaking? For what does it free us?

Nietzsche does not get around to elaborating the practical advantage he sees in refuting Kant, but there is some suggestion that it opens possibilities for other pursuits; it would free us to pursue other paths. In the context of an established hierarchy of purposes, certain routes of pursuing those aims are open whilst many others are closed. Nietzsche cites Kant's claim in the *Critique of Judgement* §65 that the idea of the effect that teleology seeks to explain stems from the concept of the whole. But, for Nietzsche, the latter is something we invent: 'There are in reality no individuals. Moreover, individuals and organisms are nothing but abstractions. They are unities manufactured by us into which we transfer the idea of purpose.'[11] To conceive of the whole as the purposive cause of the parts is to impede the possibility for understanding and appreciating change (for example, how one thing becomes another) and limits prospects for new developments. The latter, we might say, open different possibilities for being, for 'ways of living'. Thus Nietzsche writes: 'A thing lives – thus its parts are purposive: the life of the thing is the purpose of its parts. But there are countless different ways to live, i.e., countless forms, that is to say, parts.'[12] I take it, then, that the practical value of demonstrating that Kant is wrong when he claims that we cannot help but consider nature in terms of purposes is that, by showing that we may garner different perspectives on the world, we thereby extend the possibility that we might pursue some of them. In doing so, we open up different possibilities for interpretation, different possibilities for the organization of our lives, different ways of living. This, I shall claim below, is precisely what Nietzsche thinks is at stake in providing an alternative account of development. It is not just a matter of 'getting it right' or hitting upon the truth but also of facilitating the kind of growth for which it aims to account.

Although Nietzsche's views on numerous topics change and develop in the two decades that follow, the need to find a way out of the concerns raised in the unrealized plan for the dissertation persist. These concerns culminate in Nietzsche's writings of the 1880s in which one finds a relentless interest in developing a single account of change, growth and development that would be applicable to both the realm of morality – broadly conceived and freed from its religious moorings – and physiology, informed by his scientific studies, particularly in the field of embryology. He wonders about the physical effects of religious beliefs and practices and about the effects of diet, nutrition, climate, disease and health on the mind.[13] Finding both mechanistic and teleological theories wanting, Nietzsche seeks to devise an interpretation of his own, one that might effect a reunion of art and science. His exemplar in this regard was (his own) Heraclitus.[14]

Nietzsche's Heraclitean Solution to the Problem of Development

Both Christoph Cox and Wolfgang Müller-Lauter have recognized that Nietzsche's alternative account of development is indebted to his study of Heraclitus. Both also acknowledge that Nietzsche's proposed 'solution' rests upon his claim that struggle is the most immediate and pervasive phenomenon of existence. Cox even goes so far as to tie such views to Nietzsche's interest in the *agon*.[15] What I wish to explore in this section and those that follow is more precisely how Nietzsche considers the relationship between his studies of development and its applications in the domain of culture and the production of meaning – that is, whether it is possible to arrange creatively the forces of the organism for the production of excellence and health in the same way that the Greeks (allegedly, deliberately) harnessed the productive power of the good Eris in the context of ubiquitous contest (see *HC*, *KSA* 1, pp. 783–92). In this case, Nietzsche's anticipated *gay science* combines both the conscience of the scientist with the exploitation of creative energies manifest in the dynamic process of struggle, resistance, and growth that characterizes all life. It is this prospect that shapes Nietzsche's emerging conception of will to power, which he later imagines would unite the disparate sciences.

In his reading of scientific works – in medicine, chemistry, physiology, and evolutionary theory – Nietzsche comes to recognize prominent conceptual structures that bear a resemblance to the social and individual forms he identifies as having an agonistic structure. Such an organization resembles the model Nietzsche earlier admired as offering insight into the developments of Greek culture and ethics. In his 1872 essay 'Homer's Contest', Nietzsche entertains the hypothesis that the creation of a certain poetic perspective effectively organized Greek culture around the pursuit of excellence. The labours of existence became transfigured as labours for glory once the culture ordered itself in terms of contests in which each person would strive to surpass the other. Homer's contesting heroes, as exemplars of the good (glorious) life, extended the range of possible meanings for the trials of human existence. The meaning of life became transformed in this light, and this revolution shaped all other values accordingly.[16] Nietzsche admires both the contest itself as a mechanism for valuation – as a means for forging, renewing, or transforming standards of excellence that emerge in the course of the meeting of struggling forces – and the specific revaluation Homeric *agon* produces, the poetic reinterpretation it achieves. Although he does not wish to reinstate the ancient *agon* as an agency for renewing the moribund culture of nineteenth-century Europe, he does develop criteria for evaluating different kinds of contest and prospective contestant, and he strives to envisage how future revaluations might transpire.[17]

It is instructive to see how Nietzsche characterizes Heraclitus as a figure who resisted adducing teleological explanations for what is perceived as growth, change and development. In the course of his lectures on the pre-Platonic philosophers, Nietzsche also references contemporary scientific studies, particularly the work of Karl Ernst von Baer, the founder of comparative embryology. He cites Baer's study of perception relative to lifespan and sensory capacity[18] because it articulates a hypothesis that the phenomena of change, development and growth are relative to perception. Creatures perceiving at lower rates of speed have the experience of persistence, but an increase in the speed of perception would result in the phenomenon of constant becoming. Nietzsche reads Heraclitus as articulating the 'higher' perspective that would afford the appearance of *becoming*.

But it is the way in which Heraclitus conceived of strife and applied it comprehensively that most inspires Nietzsche. The notion of 'immanent lawfulness' is especially appealing to him. The logic of the contest, as conceived by Heraclitus, is internal to it. Justice in the agonistic arena is not subject to an externally determined standard of measurement. The game itself has an *internal* aim or purpose, but its aims are not determined according to a larger goal. The contest is a part of a cosmos that has the character of 'a child playing a game, moving counters, in discord and concord' (*PP*, p. 65). The overall character of this play is non-teleological, arbitrary, and 'innocent'. Nietzsche continues:

> ... only in the play of the child (or that of the artists) does there exist a Becoming and Passing Away without any moralistic calculations. He [Heraclitus] conceives of *the play of children* as that of spontaneous human beings: here is innocence and yet coming into being and destruction: not one droplet of injustice should remain in the world. The eternally living fire [Aeon], plays, builds, and knocks down: strife, this opposition of different characteristics, directed by justice, may be grasped only as an aesthetic phenomenon. We find here a purely aesthetic view of the world. We must exclude even more any moralistic tendencies to think teleologically here, for the cosmic child [*Weltkind*] behaves with no regard to purposes but rather only to an immanent justice: it can act only willfully and lawfully, but is does not *will* these ways. (*PP*, p. 70)

Nietzsche contrasts Heraclitus' view with Anaxagoras' 'teleological insight': 'he construes the order of the world as a determinant will with intentions, conceived after the fashion of human beings' (*PP*, p. 72). This insight was picked up by Aristotle, according to Nietzsche's philosophical genealogy, and it then reverberated throughout the history of philosophy in the notion of the opposition of soul and matter: 'a force that knows and sets goals but also wills, moves, and so on and yet is rigid matter. It is strange how long Greek philosophy struggled against this theory' (*PP*, p. 72).

Near the end of the lecture notes, Nietzsche characterizes Heraclitus' non-teleological position thus: 'this playful cosmic child continuously builds and knocks down but from time to time begins his game anew: a moment of contentment followed by new needs. His continuous building and knocking down is a craving as creativity is a need for the artists; his play is a need' (*PP*, p. 72).[19] The play of the child has immanent purposes, directed by the particularities of the play at any given moment, but its shape unfolds without an orchestrating will or design. Within the play, there is a kind of necessity at work, which Nietzsche describes as 'craving as creativity is a need', but that necessity is free from conformity to some law or universal principle. Nietzsche continues:

> From time to time he [the child] has his fill of it [the play] – nothing other than fire exists there; that is, it engulfs all things. Not hybris but rather the newly awakened drive to play [*Spieltrieb*] now wills once more his *setting into order*. Rejection of any teleological view of the world reaches its zenith here: the child throws away its toy, but as soon as it plays again, it proceeds with purpose and order: necessity and play, war and justice. (*PP*, pp. 72–73)

What the vision of Heraclitus allows is a way of conceptualizing the character of the becoming of the world that renders compatible the semblance of order, purpose and regularity, while admitting change, flux and chance. It is possible to think this becoming without imposing a teleological order, precisely what Kant seems to deny. A non-teleological, non-mechanistic metaphysical framework provides Nietzsche with a conceptual model that is also consistent with his reflections on moral psychology and axiology.

Nietzsche's Heraclitean solution allows for the intelligibility of aims, goals and purposes (that is, *necessity*) as immanent to the justice or law that makes a struggle or conflict possible (that serves as the ground for the contest). But it also allows for *chance* as the agonistic process takes on more of the character of play than the execution of design by a supreme being or an omnipotent will. Thus Nietzsche understands phenomena of change, growth and development in the natural world as unfolding in a process analogous to the evolution of morals and values. Heraclitus' playing child is a poeticized interpretation that seeks to capture metaphorically the becoming of *physis*, and, although Nietzsche will also adopt and adapt directly Heraclitus' specific metaphors, it is above all the poetic process applied to the realm of *physis* that seems to be most captivating for him. It serves as a prototype for a science mindful of the fact that it is a perspective – an interpretation, a product of the art of interpretation – a science attentive to its transformative capabilities.

Incorporating Science

Although Nietzsche does not begin to use the term in his published writings until *Thus Spoke Zarathustra*, his earlier account of the Greek *agon* foreshadows the idea of 'will to power'. The urges to strive, struggle, and overcome resistance, the 'terrible drive' responsible for 'fighting and the lust for victory', the irritants of *eris* and envy that plagued Themistocles and Pericles and propelled them to extraordinary accomplishments, the 'monstrous desire' of Xenophanes and Plato to defame their rival Homer, the 'personal struggling impulse' at the root of artistic competitions, the 'base desire for revenge' experienced by Miltiades (*HC*, *KSA* 1, pp. 783, 785, 786, 790 and 791) – each can be read retrospectively as manifestations of will to power (*TI* 'Ancients' 3). Nietzsche later experiments with the conclusions that potentially could follow from the hypothesis that all these drives, which constitute 'our entire instinctive life', ultimately spring from 'the development and ramification of *one* basic form of the will – as will to power' (*BGE* 36). One conclusion that might be warranted according to this hypothesis is that all organic functions are modulations of will to power because it is the source of all efficient force. If so, Nietzsche suggests, it might then be possible to trace the development or evolution of something – 'a thing, a custom, an organ' – as unfolding from 'a succession of ... more or less mutually independent processes of subduing, plus the resistances they encounter, the attempts at transformation for the purpose of defense and reaction, and the results of successful counteractions' (*GM* ii 12). In other words, struggles – or, in some cases, contests – could be interpreted as having occurred in an object of inquiry, thereby facilitating better understanding of what it is and how it came to be than by supposing that its present utility accounts for its development. I take it that this is what underlies Nietzsche's practice of genealogy in *On the Genealogy of Morals* and other later writings – namely, a hermeneutics of the machinations of power, one designed to be as appropriate for values as it is for physics. But at what does this genealogy aim? Does it seek merely to replace one description with another?

In the same section of *Beyond Good and Evil* in which Nietzsche famously suggests that all life might be conceived as expressions of will to power, he calls for an experiment. The experiment consists in asking whether one could account for everything that is supposed to happen in the 'mechanistic world' starting from a 'given' of only our passions and desires. Nietzsche claims that he could, and he then inquires as to whether such an experiment is indeed permissible. It is, he concludes, not only permissible but also *necessary*. 'Conscience of method' dictates the experimentation with multiple kinds of causality (and explanations for the phenomena for which such views seek to account): 'Not to assume several kinds of causality so long as the experiment of getting along with one has not been taken to its ultimate limits … that is a morality of method which one may not repudiate nowadays' (*BGE* 36). Let us further consider Nietzsche's hypothesis of will to power as it relates to causality and how it informs all activities of life.

We find more elaborate details about Nietzsche's reflections on will to power by looking at the notebooks he filled during the time he was preparing *Beyond Good and Evil, On the Genealogy of Morals* and the published writings of the last philosophically productive year of his life. I shall quickly review a handful of the most prominent and relevant ideas, as other critics have treated them in depth elsewhere. The will to power hypothesis does not posit an *Über*-will, or a will that promises one day to overpower all others, but rather conceives the world as comprising forces in conflict. Nietzsche is especially interested in casting these struggles not in terms of political projects but rather as *interpretations*: 'The will to power *interprets*' (*KSA* 12, 2[148]). But what is involved in interpretation? Nietzsche elaborates:

> ... it defines limits, determines degrees, variations of power. Mere variations of power could not feel themselves to be such: there must be present something that wants to grow and interprets the value of whatever else wants to grow. Equal in *that* – In fact, interpretation is itself a means of becoming master of something. (The organic process constantly presupposes interpretations.) (*KSA* 12, 2[148])

At the core of what we call life is this interpretive play: 'the tremendous shaping, form-creating force working from within which *utilizes* and *exploits* "external circumstances"' (*KSA* 12, 7[25]).[20] All that exists is centres of force (*Kraftcentren* or 'dynamic quanta') situated in relation to others; what they are is constituted by these relations. Each such force, human and non-human alike, 'constructs the rest of the world from its own viewpoint, i.e., measures, feels, forms it, according to its own force' (*KSA* 13, 14[186]).

What has hitherto been called 'cause' and 'effect' is nothing more than this perpetual process of interpretation and incorporation:

> It is a question of a struggle between two elements of unequal power: a new arrangement of forces is achieved according to the measure of power of each of them. The second condition is something fundamentally different from the first (not its effect): the essential thing is that the factions in conflict emerge with different quanta of power. (*KSA* 13, 14[95])

Rather than conceiving this interminable process as the action and reaction of two separate entities, Nietzsche suggests that what we have is 'the mutual struggle of that which becomes, often with the absorption of one's opponent; the number of developing elements not constant' (*KSA* 12, 7[54]).[21] Following the hypothesis of will to power, '"Life" might be defined as an enduring form of the *process in which force is established* [*Kraftfestellung*], in which the various struggling parties grow unequally' (*KSA* 11, 36[22]). Accordingly, Nietzsche conceives the individual organism itself as 'a struggle between parts (for food, space, etc.): its development is tied to the victory or predominance of individual parts, to an atrophying, a "becoming an organ" of other parts' (*KSA* 12, 7[25]).

Following Nietzsche's physics of interpretation and incorporation, human beings are conceived as pluralities of affects, which are essentially relational. Each affect has its own perspective in relation to the other affects, and each seeks to have its particular view become the vantage point. All action changes the structure of the relations of affects which constitute us and others as human beings. By living – by taking any action at all – we play a role in creating reality. And the same holds true of our interactions with others. We not only *participate* in shaping social reality; by engaging others, eliciting affects and being involved in relations with them, we are constituted by, and participate in, the constitution of others as well. We are, in sum, the ongoing interpretations of the multiplicity of drives that constitute us and their relations to the struggling drives that constitute others. The questions that Nietzsche appears to entertain at various times in his writings are as follows: Can we formalize at least some of these interactions? Is it possible to direct, organize, or at least facilitate certain kinds of conflicts that would better yield an enhancement of the human type? Rather than relinquishing development to mere reactive response to environmental factors that threaten survival, might it be possible to effect new developments through the imposition of new interpretations? Can we, in any way, direct the ends of the struggles that characterize human existence?

Strange as the latter question may seem, I think this is precisely the idea that Nietzsche entertains. The third essay of the *Genealogy* is organized entirely around a series of case studies of the adoption of ascetic ideals. In asking the question 'what is the meaning of ascetic ideals?', Nietzsche inquires into what we might call the uses and abuses of ascetic ideals for the production of types of humanity or forms of life. He considers the way in which different interpretations effect – or, given the revised description of cause and effect outlined above, *convey* – different ways of living, specifically in the cases of artists, philosophers, women, the 'poorly constituted', priests and saints. He also considers how ascetic ideals tap 'the active and interpreting forces' that characterize existence and bring about the possibility of knowledge, of 'seeing *something*' (*GM* iii 12).[22] What I am endeavouring to identify as 'gay science' is tied up with Nietzsche's anticipation of a kind of selection that adapts Heraclitus' model of playful child, but in which human beings take it upon themselves to create the conditions under which contests over the meaning of humanity can arise.

Is it possible that human beings could actively affect the course of their organic development? And, if so, could they, in effect, bring about a new species?[23] These questions are at once tantalizing, if also bizarre and perhaps somewhat romantic. They might also strike some readers as sitting uneasily with Nietzsche's critiques of teleology and mechanism sketched above, namely his assertion that both viewpoints either overtly or covertly postulate ends towards which human beings strive and in light of which values are fixed. In his *Viroid Life*, Keith Ansell Pearson asks: 'In seeking an alternative conception of "selection" and "value" is Nietzsche not guilty of anthropomorphizing nature and life?'[24] As the standard for the future against which Nietzsche compares previous accounts, Ansell Pearson alludes to the following passage, which is most frequently cited by those scholars who have discussed Nietzsche's naturalism: 'When will we complete our de-deification of nature? When may we begin to "naturalize" humanity in terms of a pure, newly discovered, newly redeemed nature?' (*GS* 109). Ansell Pearson, like most others reading this passage, emphasizes its appeal to 'purity', by which he means uncorrupted by the prejudices and conceptual quagmires that are the legacy of millennia of metaphysics. The account I shall offer below places the emphasis elsewhere: instead of entertaining how the naturalization of humanity becomes *purged* of its deified meanings, I consider the new meaning it acquires through the redemption Nietzsche anticipates.

Ansell Pearson argues at some length that Nietzsche's alternative of will to power 'looks decidedly awkward and hugely problematic. If it is illegitimate to suggest that life and the universe manifest a desire or struggle for self-preservation, on what basis, and with what legitimacy, can Nietzsche claim that the fundamental essence of life is "will-to-power"? Is this also not an anthropomorphism?'[25] Ansell Pearson's question is *apropos*. For if we are to shed the notion of the agent-subject and if we are to move beyond interpretations of the world and humanity's place within it as unfolding according to purposes conceived by an intelligent designer, then how can Nietzsche propose replacing the old frameworks with one that appears to lead us further down these dangerous paths? Insofar as Nietzsche's account seems to infuse the world with intentions modelled on human volition and desires, does he not commit the very same errors he attributed to traditional theological and metaphysical paradigms? Though Ansell Pearson claims that Nietzsche's position appears to be more anthropomorphic than the views he criticizes, he argues that 'a Deleuzian-inspired reading of the will to power would point to its attempt to conceive reality in dynamical and processual terms in which the emphasis is placed on acentered systems of forces, and in which "evolution" is seen to take place in non-linear terms without fidelity to the distinctions of species and genus'.[26] Thus it becomes possible to move beyond 'anthropo-centrism and –morphism … through an improper biology that is faithful to the complex, non-linear, and machinic/pathic character of "evolution"'.[27] Although I am sympathetic to Ansell Pearson's project and his effort to bring about some resolution to a fundamental tension in Nietzsche's work, I wonder whether it is the only way to do so. It leaves unanswered the question of how Nietzsche might have regarded his own anthropomorphism as different from the kind that he criticized. Was he merely inconsistent, holding himself to standards different from those that informed his critiques of others? Was he blind to the degree to which his alternative concepts not only failed to purge God and theo-centric metaphysics from his reflections on nature? Or did he somehow think his own anthropomorphic view differed in kind from those it opposed? It seems to me that there is evidence for the latter position if we follow what Richard Schacht calls Nietzsche's deployment of an 'anthropological optic'.[28]

The 'anthropological optic' facilitates what Schacht describes as 'the program of a de-deification and reinterpretation of ourselves and our world'. It organizes 'a project of comprehending our nature and possibilities' on the basis of a human perspective.[29] This is not to say that it is limited to reducing everything to human terms (as if such terms were fixed) or that it restricts itself to human concerns. It begins, of necessity, by inquiring about the nature of human being and its possibilities for knowing. It advances a perspective on the basis of that conception, and it remains open to the possibility of other interpretations. It seems that Nietzsche might be able to defend himself against the charge of anthropomorphism by claiming that his own view is also informed by a more general conception of the perspectival nature of human knowing; hence it offers a human perspective mindful of the fact that other perspectives from other centres are not only possible but perhaps also justified.

Although I think such a line of defence is possibly open to Nietzsche, I do not think it will carry us the distance. If the same could be said of his adversaries (that they, too, offer an account that stems from a human perspective), then their anthropomorphism alone is not a sufficient ground for mounting a challenge against them. Moreover, it is far from clear that Nietzsche could claim that there is any such thing as a 'human perspective'. If there are no fixed individuals and no subjects, whence comes a species-specific perspective? I think the questions raised by Nietzsche's possible anthropomorphism bear significantly on the matter of how he conceives the task of the philosopher and the relation between science and art, and I shall argue below that Nietzsche's focus on the human stems from his conception of the project of philosophy as redemptive. I shall now turn my attention to his effort to exemplify that relation of science and art alongside the conception of redemption outlined in *Thus Spoke Zarathustra*.

The Zarathustra Paradigm

In *Thus Spoke Zarathustra* Nietzsche aims to articulate an alternative way of conceiving what a human being is, how it develops, and how such conceptions might be relevant for reflections on the aims of humanity as such. In the chapter entitled 'On Self-Overcoming', Nietzsche elaborates for the first time his idea that all existence is characterized by will to power. Zarathustra's speech is addressed to those 'who are wisest' and the 'lover of truth'. One of the aims of the speech is to reveal what lies behind the love of wisdom, to read the pursuit of philosophy as an expression of will to power. The desire to render intelligible what is true, good and real is described as a manifestation of will that ultimately seeks power. Will to power is conceived as the 'unexhausted procreative will of life'. Deploying quintessentially Heraclitean metaphors – such as the river of becoming and the play of the world – Zarathustra makes several points about life and the nature of all organisms, claiming, 'Where I found a living creature, there I found will to power' (*Z* ii 'Self-Overcoming').

Zarathustra observes that life as will to power establishes a dynamic of commanding and obeying: all living beings strive to dominate others or are enslaved. What all life seeks – whether ruler or ruled – is the pleasure of power unfolding: 'The will of the weaker persuades it to serve the stronger; its will wants to be master over those weaker still: this delight alone it is unwilling to forgo' (*Z* ii 'Self-Overcoming'). Even what could be considered the greatest will yield, will risk itself for the sake of power. The dynamic of commanding and obeying that constitutes life as will to power is also one of creation and re-creation. Much like the victor in a contest who aims not only to win according to the standards of judgement that are derived from the results of previous outcomes but rather to serve as the standard bearer of excellence, the 'greatest' must also risk its entitlement to the law. It seeks legislation of the norm and all that is relative to it in order to have even greater opportunities to express that will. The dynamic of life incorporates mutual striving, contextualized valuation and chance – the very elements Nietzsche identifies with the contest. The process does not simply characterize discrete relations. Life itself whispers in Zarathustra's ear that it is '*that which must always overcome itself*'. Everything is connected in the paradigm of self-overcoming.

And it is within this paradigm that Nietzsche's Zarathustra offers an alternative account of redemption, one that I think Nietzsche already has in mind in the passage from *The Gay Science* cited above in the discussion of whether Nietzsche's anthropomorphism is contrary to his project of de-deifying nature (*GS* 109). The *Übermensch*, or the 'overcoming being', that Zarathustra anticipates does not become the measure of all things, but it does entail an activity of esteeming, of willing, that aims at determining value. The project is envisaged as *redemptive* since it seeks to replace or restamp the values that have been held hitherto. A particularly curious feature of Zarathustra's conception of redemption is that it is organized, at least partially, in terms of the past: 'All "it was" is a fragment, a riddle, a dreadful chance – until the creative will says to it: "But I willed it thus!" Until the creative will says to it, "But I will it thus! Thus shall I will it"' (*Z* ii 'Redemption'), Zarathustra's redemption takes the form of a creative *retrospective* willing. One wills the past as if it were one's own responsibility, as if it were the result of one's own willing it to be so. This is not merely reconciliation with the past or passive acceptance of what has transpired; rather, it is at least partially reconstituting the past along the lines of that which one affirms. Thus, the past becomes one's own insofar as the significances and relations that serve as the bases for value become essentially related to one's own perspective. A perspective stemming from a past that has been reconstituted potentially revises or reorients future interpretations.

This is a goal that is quite different from a model of perfection offered as that for which one ought to strive. Zarathustra's conception of the 'comprehensive soul' further illustrates how his alternative conception of development is deployed in his vision of the being engaged in overcoming. Nietzsche describes it as 'the soul that has the longest ladder and reaches down the deepest'; it 'can run and stray and roam farthest within itself' (*Z* iii 'Old and New Tablets'). It enjoys 'the high body, beautiful, triumphant, refreshing, around which everything becomes a mirror' (*Z* iii 'Three Evil Things'). Everything mirrors it not because it has become thoroughly self-absorbed but, rather, because it has become aware of itself as an overcoming-being which is constituted in an ongoing struggle of forces, just as the rest of life is. Thus, 'out of sheer joy' it 'plunges itself into chance'. It challenges itself; it risks itself. It is the soul, 'which, having being, dives into becoming; the soul which *has*, but *wants* to want and will; the soul which flees itself and catches up with itself in the widest circle; the wisest soul, which folly exhorts most sweetly; the soul which loves itself most, in which all things have their sweep and counter-sweep and ebb and flood' (*Z* iii 'Old and New Tablets').

Nietzsche's agonized naturalism conceives of all existence as embroiled in a relation of struggle and becoming, manifesting both 'purpose and opposition to purposes'. Creation neither aims at a single goal nor operates as a complex, well-ordered machine – its paths are crooked, its particular outcomes undecidable in advance, and its growth and decay irregular. It is conflicted, contentious striving, creation and recreation, exuberance and exhaustion.

What is true of creation in the natural world abides in the realm of morals: 'out of this valuation itself speaks – will to power!' (*Z* ii 'Of Self-Overcoming'). What is esteemed as good or reviled as evil is also drawn up in this process of struggle and becoming. As such, good and evil, too, are always overcoming themselves. In this context, the process of valuation is understood as a form of violence, in that it negates, cancels, rejects and expunges previous values and valuations, and yet it produces a kind of strength. Thus 'Life' itself, which seeks strength or power, can imagine this particular sort of destruction as ultimately creative: for it is essentially an engagement of regenerative energy, a kind of *poeisis* – or poetry, perhaps – of existence. It *is* his gay – artistic, graceful, free – science, which takes the form of a *poeisis* both informed by, and itself giving form to, the principles of *physis*. It is what Nietzsche means when he speaks of looking at science from the perspective of the artist (*BT* 'Self-Criticism' 2). But what is involved in adopting this viewpoint, and what are its consequences?

Gay Science

By way of conclusion, I wish to draw on one example that is offered in section 12 of the second essay of *On the Genealogy of Morals*. There Nietzsche entertains the possibility of at least some sort of conscious participation in directing will to power, and it is in such activity that agonal wisdom seems to exercise its greatest potential for him. In this passage, in which he famously argues that development is the history of struggles, Nietzsche explains how his model differs from contemporary mechanistic theories. Entities conceived in this way – in the same way that 'Life' depicts existence in *Zarathustra* – manifest 'purposes and utilities … only as *signs* that a will to power has become master of something less powerful and imposed upon it the character of a function'. Rather than understanding a being's development as unfolding in light of a single or specific goal or in accordance with a system of universal laws, the joints of growth and change exhibit 'a continuous sign-chain' of becoming, a set of orientations that bear out the plurality of possible directions of will to power. Hence, a unified set of causal relations cannot be assumed. Change is more subject to the vicissitudes of chance – the product of risk involved in striving, agonistic interaction of forces – than a progression towards a goal or the unfolding of a logical, systematic order. For any given entity, Nietzsche writes, 'the form is fluid; the "meaning" is even more so' (*GM* iii 12).

If development is thus conceived, Nietzsche thinks, the meaning of something is always up for grabs at any given moment. Every link in the 'sign-chain' is marked by the emergence of a new dominant force that impresses upon, directs and orients a new meaning (that is, a new manifestation of will, its 'perspective'); such is its prerogative. Even within organisms, the parts are similarly engaged in a process of struggle and becoming. Each takes on the character of risk and potential perishing, in exactly the same way that the activity of self-overcoming is described in *Zarathustra*:

> It is not too much to say that even a partial *diminution of utility*, an atrophying and degeneration, a loss of meaning and purposiveness – in short, death – is among the conditions of an actual *progressus*, which always appears in the shape of a will and way to *greater power* and is always carried through at the expense of numerous smaller powers. (*GM* iii 12)

That sacrifice, the magnitude of what is surpassed, Nietzsche suggests, might even serve as a standard of measure for the growth of something greater.[30]

Rather than pursuing a science that investigates what are hypothesized as automatic adaptive responses to changes in the environment governed by laws, Nietzsche questions whether and how one might *effect* an advance in the human species as such. He does not advocate a eugenics that aims at an improvement of the same and directs change along the lines of what is considered as better relative to current prevailing norms and values. Instead, we might describe Nietzsche's project as a kind of pragmatic axiology. How does 'human being' acquire its value? How can that value be transformed and maximized? What would constitute an advance in the meaning of human existence as such? The answers to these questions remain largely undetermined for Nietzsche, but he believes that answering them requires an act of will, a kind of legislation, and the science of *that* has yet to be developed. Nietzsche claims the initiation of that task for himself. His gay science aims to replace the concept of development as adaptation with that of shaping orientations – new goals and ends – by the creation of new interpretations.

Ultimately, what Nietzsche envisages is an integrative mode of philosophizing in which the language of evolutionary science (for example, adaptation, development and so on[31]) is mapped on to, and appropriated for, his project of the poetic transformation of the meaning and making of humanity. Nietzsche's ideas about transformation, incorporation and expulsion are informed by researches into physiological/biological processes.[32] He imagines transformative conceptual unification as wedding the fruits of various scientific disciplines and approaches. Such conceptual development – the creation and offering of 'many kinds of causality', as Nietzsche puts it in *Beyond Good and Evil* – is what philosophy brings to the table. Nietzsche conceives this process as having its hand in poetry or art because it issues not merely from the empirical sciences themselves – although it is, as Nietzsche's own efforts attest, informed by them – but rather from a kind of speculative projection. Nietzsche's future philosophy as gay science advances a kind of conceptual poetry whereby scientific inquiry is *oriented* in such a way as to promote an integrative understanding and the estimation of the value of such investigations. It is animated by an inventive spirit that creates concepts that deliver to scientific pursuits the forms of thinking and conceptual formations that set their researches in motion and that supply the paradigms that make new discoveries possible. A gay science, thus conceived, strives to be even more 'honest' than mechanistic science, although it takes its lead to some extent from that approach as it strives to wriggle out from under the influence of morality. It lies *between*, rather than beyond, mechanism and teleology, as it preserves the problem of the need to offer an account of apparent purposiveness which both perspectives aim to address. And gay 'science' is not quite *science* – or, it is perhaps at least more art than science – since it is pursued mindful of its potency as an art of interpretation (*GM*

'Preface' 8). Gay science is at the same time an art of transfiguration (*GS* 'Second Preface' 3) – it potentially plays a role in *reshaping* (reforming or reconstituting, not merely reinterpreting) both the objects of its inquiry and the inquirers themselves.

Notes

1 See, for example, Jörg Salaquarda, '"Art is More Powerful than Knowledge": Nietzsche on the Relationship Between Art and Science', *New Nietzsche Studies*, 3: 3/4 (1999), 1–11.

2 Like the Provençal troubadours, to whom he likens himself in subtitling his *Die fröhliche Wissenschaft* 'la gaya scienza', Nietzsche strives to introduce new possibilities for language, new means of giving expression to our ideas, which do not merely describe what we already experience but rather make new experiences possible. Gay science accomplishes this through its artistic shaping and reformation of conceptual structures and metaphoric formulations that are drawn from contemporary scientific views and particular phenomena and are then applied to the practice of both science and philosophy.

3 See, for example, Christoph Cox, *Nietzsche: Naturalism and Interpretation* (Berkeley, CA: University of California Press, 1999) and Wolfgang Müller-Lauter, *Nietzsche: His Philosophy of Contradictions and the Contradictions of His Philosophy* (Urbana and Chicago: University of Illinois Press, 1999), which I discuss below.

4 Numerous other treatments of the relation between will to power and its scientific basis and/or relevance to philosophy of science are available. See, for example, R. Lanier Anderson, 'Nietzsche's Will to Power as a Doctrine of the Unity of Science', *Studies in History and Philosophy of Science*, 25 (1994), 729–50; George Stack, *Nietzsche: Man, Knowledge, and Will to Power* (Hollowbrook: Durango, 1994); idem, *Lange and Nietzsche* (Berlin: de Gruyter, 1983); Stephen P. Schwartz, 'The Status of Nietzsche's Theory of Will to Power in Light of Contemporary Philosophy of Science', *International Studies in Philosophy*, 25 (1993), 85–92.

5 I am fully aware of the scholarly controversy surrounding discussion of Nietzsche's so-called concept of 'will to power'. My use of 'will to power' here indicates it as 'emerging', and is not meant to suggest that Nietzsche had any single, definite sense in which he used the phrase, or that he had a full blown *theory* in which will to power would play a central role. Rather, I agree with those who remind readers that Nietzsche's ideas about will to power changed over time, that he never wrote a work in which will to power featured as the central thesis (indeed, that he abandoned at least one plan to execute such a project), and that any account of will to power that rests solely on evidence drawn from Nietzsche's unpublished (and possibly even *discarded*) writings is suspect at best. My goal here is not to offer an account of what Nietzsche meant by will to power, but rather to illustrate how the general idea of will to power as it is characterized in his published writings (and enhanced by drawing on the evidence in his notebooks) grows out of, and is continuous with, longstanding interests that range back to his early publications and plans for others. A well-developed understanding of Nietzsche's conception of the *agon* allows us to do precisely that and situates Nietzsche's developing ideas about will to power is a context that sees such an idea as part of a larger whole.

6 Christoph Cox, *Nietzsche*, especially pp. 213–45.

7 Compare, for example, Nietzsche's claim that the world is not striving towards a final state because it would have reached it already, and this 'fact' (why this should be a 'fact'

is unclear) refutes philosophical and scientific views that postulate such terms of motion. Mechanistic theory is cited as an example (*KSA 13*, 11[72]).

8 Cox, *Nietzsche*, pp. 226, 216 and 220.

9 A recent translation and reconstruction of this project has been prepared by Paul Swift, 'On Teleology since Kant', *Nietzscheana*, 8 (2000), 1–20.

10 Nietzsche, quoted in ibid., p. 8.

11 Ibid., p. 12.

12 Ibid., p. 18.

13 See, for example, *GS* 39. For a discussion of a concept of health that is allied with Nietzsche's conception of the *agon*, see Alfred I. Tauber, 'A Typology of Nietzsche's Biology', *Biology And Philosophy*, 9 (1994), 25–44.

14 I qualify Nietzsche's exemplar as '(his own) Heraclitus' because, like his discussions of Homer and (arguably) Socrates, there is much that is invented by Nietzsche himself in his discussions of Heraclitus. On this matter as it relates to Heraclitus, see Artur Przybyslawski's 'Nietzsche Contra Heraclitus', *Journal of Nietzsche Studies*, 23 (Fall 2002), 88–95.

15 See, for example, Cox, *Nietzsche*, pp. 233–35.

16 I discuss how Nietzsche figures Homer as his exemplary 'revaluator of values' in my 'The Problem of Homer', *Nietzscheforschung*, 5/6 (1999), 553–74, and in 'Nietzsche Contra Homer, Socrates, and Paul', *Journal of Nietzsche Studies*, 24 (2002), 25–53.

17 See my 'Of Dangerous Games and Dastardly Deeds: A Typology of Nietzsche's Contests', *International Studies in Philosophy*, 34 (2002), 135–51.

18 Karl Ernst von Baer, *Festrede zur Eröffnung der russischen entomologischen Gesellschaft im Mai 1860* (Berlin, 1862).

19 In *BT* 24, Nietzsche associates the *Weltkind* with the Dionysian principle.

20 The context of the passage is a longer section that is aimed at criticizing Darwin's views. I designate this interpretive activity as 'play', although will to power, obviously, need not be conceived exclusively as playful. My specific word choice here is based on the fact that Nietzsche's emphasis is on a process that he conceives as essentially *creative*. I reserve the term 'agonistic' for a specific form of playful struggle, one that follows the form of contest and the creative mode of action it facilitates as outlined in 'Homer's Contest'.

21 These views reflect Nietzsche's understanding of the ideas of Wilhelm Roux. See Müller-Lauter, *Nietzsche: His Philosophy of Contradictions*, chapter 9. See also Gregory Moore's *Nietzsche, Biology, and Metaphor* (Cambridge: Cambridge University Press, 2002), especially chapter 1.

22 On the relation between ascetic ideals and science, see David Owen, 'Science, Value, and The Ascetic Ideal' in Babette E. Babich and Robert S. Cohen (eds), *Nietzsche, Epistemology, and Philosophy of Science: Nietzsche and the Sciences, II* (Dordrecht: Kluwer, 1999), pp. 168–78, and Babich, *Nietzsche's Philosophy of Science: Reflecting Science on the Ground of Art and Life* (Albany, NY: SUNY Press, 1994), especially chapter 5.

23 For Nietzsche, the concept 'species' is as much a human invention as the concept 'individual'. See Wolfgang Müller-Lauter, 'On Judging in a World of Becoming: A Reflection on the "Great Change" in Nietzsche's Philosophy' in Babette E. Babich and Robert S. Cohen (eds), *Nietzsche, Theories of Knowledge, and Critical Theory: Nietzsche and the Sciences, I* (Dordrecht: Kluwer, 1999), pp. 168–71. Müller-Lauter argues that, for Nietzsche, what defines and sustains a species is its *habitual* interpretations that have become conditions of existence. Thus, it would seem that, when Nietzsche mentions a particular species, he is also calling attention to the habit that sustains it *as* that species. In this way we might consider being human a habit, one we conceivably might overcome on the way towards becoming something else. It is worth

considering this idea at greater length. Is this anticipated overcoming of the habit of the human intended to introduce a new habit, or, in the course of overcoming the habit of being human, does one strive to overcome habituation itself? If the latter, the *Übermensch* as the form of life on the developmental horizon would not be a new species, but it would not be human either.

24 Keith Ansell Pearson, *Viroid Life: Perspectives on Nietzsche and the Transhuman Condition* (London: Routledge, 1997), p. 105.

25 Ibid., pp. 105–106.

26 Ibid., p. 117.

27 Ibid., p. 120.

28 Richard Schacht, *Making Sense of Nietzsche: Reflections Timely and Untimely* (Urbana and Chicago: University of Illinois Press, 1995), p. 190.

29 Ibid., pp. 190–91.

30 I take it this is what Nietzsche means when he asks about the sacrifice of humanity as a whole, as the passage (*GM* ii 12) continues: 'humanity as a whole sacrificed to the prosperity of a single *stronger* species of man – that *would* be an advance' (Kaufmann's translation emended). Such a reading stands in stark contrast with those who see Nietzsche here embracing the sacrifice of great masses of (presumably less worthy) people for the sake of their noble betters. Whether and how such a sacrifice, in the interest of genuine advancement or evolution, might be effected is considered in *Thus Spoke Zarathustra*: 'All beings so far have created something beyond themselves; and do you want to be the ebb of this great flood …?' (Z 'Prologue' 3).

31 On Nietzsche's appropriation of the language of biology and medicine, see Malcolm Pasley, 'Nietzsche's Use of Medical Terms' in Pasley (ed.), *Nietzsche: Imagery and Thought* (London: Methuen, 1978), pp. 123–58, Scott Podolsky and Alfred I. Tauber, 'Nietzsche's Conception of Health: The Idealization of Struggle', in Babich and Cohen, *Nietzsche, Epistemology, and Philosophy of Science*, pp. 299–311.

32 Robin Small offers an engaging discussion of Nietzsche's use of the terms 'incorporation' and 'expulsion' in 'Disturbing Thoughts and Eternal Perspectives: Some Uses of Symbolism in Nietzsche', *New Nietzsche Studies*, 3: 3/4 (1999), 29–44.

Chapter 9

Nietzsche's Conceptual Chemistry

Duncan Large

In the section of *Ecce Homo* devoted to a retrospective appraisal of *Human, All Too Human*, Nietzsche uses the opening paragraphs for an account of the genesis of the book, which he calls, at the outset, 'the memorial of a crisis' (*EH* 'Human, All Too Human' 1). The crisis to which he is referring is, of course, his break with Wagner, which he dramatizes here as having been 'precipitated' by his disgust at the inaugural Bayreuth Festival of August 1876. In paragraph 3 of this section, though, he admits that, around this time, he was plunged into a more serious existential crisis which led him to question not just his relation to Wagner, but all that he had been doing in his intellectual and professional life over the previous decade. Impatient with himself and with his floundering among philological 'idealities', he was overcome by a longing for 'realities', he tells us, and describes how he roused himself from his ten-year philological slumber when '[a] downright burning thirst seized hold of me: thenceforward I pursued in fact nothing other than physiology, medicine and natural science' (*EH* 'Human, All Too Human' 3).

From the point of view of anyone interested in examining Nietzsche's relation to science, this hyperbolic claim is a key passage in his writings. It betokens one of those periodic occasions in Nietzsche's career when he takes stock and expresses an unease about his own 'calling' (*Beruf*), wondering whether he ought not perhaps to be doing something more meaningful than lecturing on Classics in Basle and writing books that no-one reads. For those who *have* read Nietzsche's books, though, this passage is intended as an explanation of how he came to write a book – *Human, All Too Human*, together with its 'supplements' the *Assorted Opinions and Maxims* and *The Wanderer and His Shadow* – which is so very different from what had gone before (*The Birth of Tragedy* and the *Untimely Meditations*), a difference which is announced programmatically in the very opening paragraph of the 1878 text, entitled 'Chemistry of Concepts and Sensations' (*HH* 1).

It is from this paragraph that I have derived my title, and my aim in this chapter is a very straightforward one, namely to examine the role – or at least a few aspects of the role – that chemistry plays in Nietzsche's thinking, set against the backdrop of his ambivalent relation to the natural sciences in general. For it is surely significant that, from the list of 'physiology, medicine and natural science' which Nietzsche gives us in *Ecce Homo*, he should actually choose chemistry as his emblematic scientific discipline here, as a cure for Wagnerism and for his earlier metaphysical excesses. I want, then, to examine what Nietzsche means by 'chemistry': what it represents to him, and why he should therefore choose it above other scientific disciplines – not only here – as representative of his own method. Analyses of Nietzsche's relation to the natural sciences have tended to focus on other disciplines, with a concentration on physics or medicine, physiology and the life sciences.[1] This is hardly surprising since these are the areas in which Nietzsche read the most and which can be seen perhaps to have the greatest impact on his intellectual development;[2] in fact, the most sustained treatment of my theme – Alwin Mittasch's pamphlet *Friedrich Nietzsches Stellung zur Chemie* – dates back to 1944.[3] My argument will be that Nietzsche's relation to chemistry has been insufficiently addressed so far, and that his apparently sudden enthusiasm for chemistry in *Human, All Too Human* must be appreciated in the context of a preoccupation with the subject which, in fact, spans his intellectual career, for chemistry surfaces at strategic moments as a recurrent theme in Nietzsche's writings.

The passage in *Ecce Homo* in which Nietzsche describes his 1876 crisis actually mirrors very closely his assessment of an earlier crisis, namely the moment of decision he faced in January 1869 on hearing of his impending election to the chair of Classical Philology in Basle – at precisely the point when it seems he was beginning to despair of philology altogether, after years as a star pupil (in Classics, at least) at Schulpforta and under Ritschl's supervision in Bonn and Leipzig. In a letter of January 1869 to his close friend Erwin Rohde, Nietzsche, reflecting wryly on the vicissitudes of fate, breaks the news that has just intervened to dash their common 'dreams of a Parisian future': 'just last week I was going to write to you and suggest that we study chemistry together, that we throw philology where it belongs, among the household effects of our forefathers' (*KGB* I/2, 16 January 1869). Even before Nietzsche's professional philological career gets under way, then, he is conceiving of philology and chemistry as contrasting – if not opposing – modes of existential possibility, of 'Philologie und Chemie im Kampfe',[4] and he is tempted by the latter's seductive lure. Steffen Dietzsch, writing on 'Nietzsches französischer Traum', refers to this letter of 1869 as one of the frequent instances of Nietzsche's beguilement by the possibilities of Paris, a city he never actually managed to visit but which – all the more so for that reason – exercised a persistent and powerful influence on his philosophical imagination,[5] but it seems to me that one can just as well treat this episode as exemplifying 'Nietzsches chemischer Traum', too. Nietzsche's knowledge of chemistry up to this point may have been very slender indeed, and his reading in chemistry always lagged behind his reading in other natural scientific disciplines, yet however hare-brained and abortive this 1869 scheme of studying chemistry in Paris may sound, the fact that he should entertain the notion at all undoubtedly betokens a strong interest in chemistry, at least in principle – on the level of a desire for an alternative mode of existence which Nietzsche subsequently fulfils, I want to argue, by more indirect and metaphorical means.

In commenting on this same episode of 1869, Mittasch muses: 'what if Nietzsche had carried out his plan [perhaps one should say, the aspect of his plan which Mittasch takes as relevant] and begun studying chemistry, for instance with O.L. Erdmann and H. Kolbe in Leipzig: would we be counting him nowadays among our great German chemists?'[6] In his conclusion, Mittasch goes some way towards answering this question when he remarks: 'There can hardly be a similar case in the history of natural philosophy, of a thinker with so little detailed knowledge of chemistry advancing to such far-reaching reflections on the basic questions of chemistry.'[7] This may sound to us now like some rather special pleading for a thinker who, as Mittasch admits, was in matters chemical little more than a talented dilettante; however, my concern is not to argue the merits of Nietzsche's contribution to chemical thought, but rather to analyse the contribution of chemistry to his thought. For Nietzsche did not throw in philology and pursue a career in chemistry, of course; instead, he took up the post in Basle and published *The Birth of Tragedy*, to which I want now to turn.

Nietzsche's Critique of *Wissenschaft*

In the light of Nietzsche's abortive plan to abandon philology in favour of chemistry it is perhaps surprising that only three years later his first major publication should problematize *Wissenschaft* to such a degree. Indeed, when he comes to reassess the book in the preface to the second edition of 1886, 'Attempt at a Self-Criticism', it is this aspect that he chooses to stress, indulging in what we might now call an interesting piece of 'spin doctoring' after the event. For, as he casts his eye back over his first-born of 14 years previously, he deliberately *recasts* his whole endeavour – a book which concerns itself so centrally with problems of aesthetics and the 'eternal aesthetic justification' of the world – as a problematization of *Wissenschaft*: 'What I then got hold of ... – today I should say that it was the *problem of science* [*Wissenschaft*] itself' (*BT* 'Self-Criticism' 2; cf. *GM* iii 25). An important point to bear in mind here, of course – as ever when Nietzsche thematizes *Wissenschaft* – is its broad, more generic semantic field in the German language in contrast to the narrower, more technical sense of the English translation, 'science'. Nietzsche associates *Wissenschaft*, naturally enough, with *Wissen*, knowledge and the process of cognition in general – thus in *The Antichrist* he is able to interpret the drama in the Garden of Eden as hinging on the acquisition of *Wissenschaft* (*A* 48), and back in *The Birth of Tragedy Wissenschaft* is treated as just another compound from *Wissen*, cognate with the *Wissenspyramide der Gegenwart* (*BT* 15: 'pyramid of knowledge in our own time') or such concepts as *Wissensmeer* (*BT* 18: 'ocean of knowledge') and *Wissenskultur* (*BT* 18: 'intellectual culture') – all the products of man's *Wissensgier* (*BT* 15: 'hunger for knowledge') which finds its summation in the *Wissenstrieb* (*BT* 18: 'drive for knowledge') of Goethe's Faust.

In *The Birth of Tragedy* it is Socrates who is the archetypal *Wissenschaftler* and villain of the piece: Socrates is the 'Mystagog der Wissenschaft' (*BT* 15: 'mystagogue of science'), the embodiment of hypertrophied reason. As Nietzsche writes in a contemporary notebook: 'In Socrates the principle of science [*Wissenschaft*] forces its way in' (*KSA* 7, 1[27]). The problem that sets in with Socrates, Nietzsche argues here – 'The Problem of Socrates', as he will put it in *Twilight of the Idols* – is that thought begins to get ideas above its station: it aims at universality and believes it can 'fathom the nature of things' (*BT* 15), which leads to what Nietzsche considers an insidiously optimistic belief in progress that he condemns as a form of escapism, an evasion of the fundamentally Schopenhauerian, pessimistic truth of the gravity and senselessness of existence. As he puts it in the 'Attempt at a Self-Criticism', 'Is scientificity [*Wissenschaftlichkeit*] perhaps merely a fear and evasion of pessmism?' (*BT* 'Self-Criticism' 1).

The pernicious influence of Socrates, the 'theoretical man' *par excellence*, still persists in our contemporary age in the person of the academic scholar (*Gelehrter*), Nietzsche argues in section 18 of *The Birth of Tragedy*, and the scholar is a type whom he will continue to hound relentlessly in later texts. In the second *Untimely Meditation* the target is academic historians; in *Beyond Good and Evil* he devotes an entire division, 'We Scholars', to the problem of 'the scholar, the average man of science' (*BGE* 206), at the same time explicitly declaring his own interest by his use of the first person plural. The polemic continues in *On the Genealogy of Morals* (*GM* iii 23–25), and right up to *Ecce Homo*, where he compares the scholar to a match which has to be rubbed against something in order to catch light: 'The scholar ... *replies* to a stimulus (– a thought he has read) when he thinks – finally he does nothing but react' (*EH* 'Clever' 8). For Nietzsche, then, scholarly 'activity' is but 'reactivity', and the scholar's ideal of 'objectivity' is merely an 'indifference behaving objectively' (*HL* 6). This sense of indifference is for Nietzsche but typical of his *wissenschaftlich* age which is indifferent to values in general and lacks the stamp of any personal style. It is an age which he condemns as nihilistic, as unable to believe in the values it had previously held (*KSA* 12, 5[71]) – an age which, to use one of Nietzsche's favourite criteria, lacks *taste* and finds all previous values merely insipid.

'Chemistry of Concepts and Sensations': Sublimation

Nietzsche's recipe for bringing excitement and, above all, discrimination back to the jaded European palate turns out, at first sight paradoxically, to have the natural sciences as its main ingredient. In his writings after *The Birth of Tragedy* and the *Untimely Meditations* there comes a distinct caesura and a reorientation in his philosophy which has a rehabilitation of science as one of its main features – in an 1877 notebook he even quotes Socrates repeatedly with approval: 'People who have no *scientific* culture are *talking through their hats* when they discourse on serious, weighty matters, and are presumptuous. Socrates is still right' (*KSA* 8, 23[17]; cf. 23[121]).[8] *Ecce Homo* dramatizes the shift, as we have seen, as an aspect of Nietzsche's disaffection with Wagner and with the 'erudition' (*Gelehrsamkeit*) of his profession. The picture is rather more complicated than this account of a simple, catastrophic conversion would suggest, though, on the one hand, even in the early 1870s – and largely in connection with the course of lectures he gave in 1873 on the 'pre-Platonic' philosophers and their science,[9] as Schlechta and Anders showed in 1962 through an examination of the books he borrowed from the University Library in Basle, ordered from his bookseller and so on[10] – Nietzsche was keeping himself informed about contemporary developments in the natural sciences. On the other hand, he refused to embrace the natural sciences uncritically even after this supposed 'turn', for even in the 1880s there are still what he refers to in a note as 'the nihilistic consequences of present-day natural science' (*KSA* 12, 2[127]) to be reckoned with, and as Gilles Deleuze highlights, what Nietzsche objects to is contemporary science's reduction of differences: 'the scientific mania for seeking balances, the *utilitarianism and egalitarianism* proper to science'.[11] This 'mania' shows itself in the shape of three closely allied scientific phenomena which Nietzsche continues to isolate and attack – logical identity, mathematical equation and physical equilibrium – all of which smack, to him, of nihilism (the nihilistic reduction of difference). In the third essay of the *Genealogy*, science is found to be in the service of the ascetic ideal, and here post-Copernican astronomy is in the dock as a paradigm case of man's '*will* to self-belittlement', as having a 'humiliating and degrading effect' (*GM* iii 25).

In response to the nihilism of logic, mathematics, physics and astronomy, then, Nietzsche holds up chemistry as the great restorer of differences. The first evidence of this shift is revealed to the public from the very beginning of *Human, All Too Human*, where 'Chemistry of Concepts and Sensations' is positioned as the first paragraph and thus given programmatic significance. In place of 'metaphysical philosophy', Nietzsche argues here, what is required is a new kind of 'historical philosophy ... which can no longer be separated from natural science', a 'chemistry of concepts and sensations' to teach us that there is no such thing as 'unegoistic action', for example, just as the Kantian notion of 'completely disinterested contemplation' is also an illusion: 'both are only sublimations [*Sublimierungen*], in which the basic element seems almost to have dispersed and reveals itself only under the most painstaking observation'. Nietzsche continues:

> All we require, and what can be given us only now the individual sciences have attained their present level, is a *chemistry* of the moral, religious and aesthetic conceptions and sensations, likewise of all the agitations we experience within ourseves in cultural and social intercourse, and indeed even when we are alone. (*HH* 1)

At the outset of *Human, All Too Human*, then, Nietzsche revalues the natural sciences positively, at the same time differentiating between them and privileging chemistry, the science of selection, as the select science. Philosophical chemistry, for Nietzsche, is an answer to grand scientific abstractions, for it is a critical, analytical and, above all, differentiating science which establishes and explores minute transmutations – Peter Heller describes Nietzsche's conception of chemistry here as a 'doctrine of transformations'.[12] Nietzsche associates chemistry with the kind of genealogical 'questions about descent and beginnings' which, he argues, humanity is all too keen to repress, and with his new method he is seeking in effect to practise a kind of inverted conceptual alchemy, revealing the base origins of what otherwise passes for gold: 'what if this chemistry gave the result that even in this area [of culture and society] the most splendid colours are obtained from lowly, indeed despised materials?'[13]

The message of this (al)chemical manifesto is that contradictions are to be seen as sublimations rather than opposites – Nietzsche's chemical 'principle of the excluded middle' is far preferable to the logical one, for it privileges becoming over being and emphasizes the contiguity of the two 'opposite' poles, as well as the reversibility of the operation which mediates between them. The supposed opposition which Nietzsche is most keen to deconstruct is, of course, that which determines moral judgements, the opposition between 'good' and 'evil'. The next division of *Human, All Too Human*, 'On the History of the Moral Sensations', shows him already beginning to put his new principles into practice with a discussion of the 'Twofold Prehistory of Good and Evil' (*HH* 45), and it finishes with a passage stressing that 'between good and evil actions there is no difference in kind, but at the most one of degree. Good actions are sublimated [*sublimierte*] evil ones; evil actions are coarsened, brutalized good ones' (*HH* 107). It is in the first essay of the *Genealogy* that this insight is given its fullest development, with the twin oppositions of 'good and evil', 'good and bad' being analysed out as sublimations of power relationships; in the second essay Nietzsche then produces his most sustained analysis of the phenomenon of sublimation, treating the manifestations of 'higher culture' as so many 'sublimation[s] and subtilization[s]' of cruelty (*GM* ii 7).

'Gay Science'

Almost 30 years before Freud, then, Nietzsche is using a notion of 'sublimation' which systematically reduces sociocultural, political and psychic events to steps in a chemical process.[14] (At the same time, of course, it still resonates with echoes of the aesthetic sublime, which it subtly subverts.) The term 'sublimation' is, moreover, just one salient example of the chemical idiom which Nietzsche adopts, and which is also in evidence, I would argue, when he talks of the 'reactivity' of the scholar, when he analyses the process of contemporary cultural 'dissolution' (*Auflösung*: *BGE* 200; *KSA* 12, 5[71]), or proposes the 'formula [*Formel*] for my happiness' (*TI* 'Maxims' 44; *A* 1). Nietzsche's use of chemical formulations (or formulae) in his prescriptions itself undergoes a transformation, although, I want to suggest that whereas in the first section of *Human, All Too Human* (1878) he sets out his new chemical principle programmatically in a manifesto for a new method of philosophical *analysis*, by the time of *The Gay Science* (1882) he has worked out a powerful corollary which involves switching from the metaphor of chemical analysis to that of chemical synthesis as a metaphor for a lived philosophy of existential responsiveness and artistic creativity in the laboratory of the world (*KSA* 13, 15[8]).

Nietzsche's 'gay science' is an implicit 'existential imperative', to borrow Bernd Magnus's phrase:[15] in *The Gay Science* itself we are urged, as free spirits, to take risks and to 'live dangerously' (*GS* 335); in the preface to the second edition of *Human, All too Human*, written in 1886, we find ourselves being enjoined to live life resolutely, '*experimentally*' (*HH* 'Preface' 4; cf. *D* 432). As an 'experimental philosophy' (*KSA* 13, 16[32]), then, Nietzsche's affirmative philosophy of heroic *amor fati* in the 1880s still derives its methodology from the scientific laboratory, and the superiority of the natural sciences over other *Wissenschaften* in respect of this experimentalism is particularly evident from a juxtaposition in the third essay of the *Genealogy*. Here the legitimation crisis of *Wissenschaft* – which is denounced as a 'means of self-narcosis' because it still pursues truth rather than pursuing the more adventurous 'truth' of the Assassins, the true free spirits, that 'nothing is true, everything is permitted' (*GM* iii 24) – is itself presented by Nietzsche in experimental, 'free spirit' mode: 'the value of truth must for once be experimentally *called into question*' (*GM* iii 24).

The continued importance of 'philosophical chemistry' to Nietzsche even in 1888 is attested to by his returning to rework the *Human, All Too Human* passage on 'Chemistry of Concepts and Sensations' in the January of that year (*KSA* 14, p. 119), grafting on some of the new positions typical of his later philosophy. He is more extreme now about how his 'historical philosophy' is 'an inverted [*umgekehrte*] philosophy', how it has no traces of a (Kantian) 'An-sich' but is explicitly defined as a 'true *philosophy of becoming*' ('eigentliche *Philosophie des Werdens*'), as a 'chemistry of basic concepts [*Chemie der Grundbegriffe*] ..., these latter presupposed as having become and still becoming'. In general, Nietzsche now grafts on his mature perspectivistic position and supplements the chemical vocabulary with metaphors drawn from optics, psychology and medicine in order to produce a more composite 'research methodology' (*Methodik der Forschung*).

Conclusion

It is nevertheless clear that the premises incorporated in the 'chemical manifesto' of ten years previously remain fundamental to his thinking, and although other scientific vocabularies may have come to the fore in the intervening years – Mazzino Montinari writes of the 'waft of hospital air' that emerges from *Twilight of the Idols*, for example,[16] and Michel Serres, in similar vein, writes of the antiseptic archness of *The Antichrist*[17] – the persistence of a strain of what I am calling 'conceptual *chemistry*' in Nietzsche's writings is not to be overlooked. Nietzsche may not have read as much on the subject as he did on many others, but that does not prevent him drawing on the metaphorical resources of a 'science of transformations' at various key points, just as Friedrich Schlegel (in the *Athenäums-Fragmente*) or, more pertinently in a Nietzschean context, Goethe (in *Elective Affinities*) had before him. In the end, then (and *pace* Mittasch), 'Nietzsche's conceptual chemistry' is evidence less of a preoccupation with cutting-edge natural scientific debate than with the – by now long-established – philosophico-*poetic* potential of transmuting the concepts of chemistry themselves into metaphors.

Notes

1 See, for example: Daniel A. Ahern, *Nietzsche as Cultural Physician* (University Park: Pennsylvania State University Press, 1995); Keith Ansell Pearson, 'Nietzsche contra Darwin', in *Viroid Life: Perspectives on Nietzsche and the Transhuman Condition* (London: Routledge, 1997), pp. 85–122; Alain Juranville, *Physique de Nietzsche* (Paris: Denoël/Gonthier, 1973); Gregory Moore, *Nietzsche, Biology and Metaphor* (Cambridge: Cambridge University Press, 2002); Malcolm Pasley, 'Nietzsche's Use of Medical Terms', in Malcolm Pasley (ed.), *Nietzsche: Imagery and Thought* (London: Methuen, 1978), pp. 123–58; Paul J.M. van Tongeren, 'Physician and Sculptor', in *Reinterpreting Modern Culture: An Introduction to Friedrich Nietzsche's Philosophy* (West Lafayette, IN: Purdue University Press, 2000), pp. 1–13. On Nietzsche and science more generally, see: Babette E. Babich, *Nietzsche's Philosophy of Science: Reflecting Science on the Ground of Art and Life* (Albany, NY: SUNY Press, 1994); Babette Babich and Robert S. Cohen (eds), *Nietzsche and the Sciences*, 2 vols (Dordrecht: Kluwer, 1999); Niels Helsloot, *Vrolijke wetenschap: Nietzsche als vriend* (Baarn: Agora, 1999); Robin Small, *Nietzsche in Context* (Aldershot: Ashgate, 2001); 'Symposium on Nietzsche's "Time-Atom Theory" Fragment', *Journal of Nietzsche Studies*, 20 (2000), 1–81.

2 See Thomas Brobjer's essay 'Nietzsche Reading and Knowledge of Natural Science: An Overview' in Chapter 1 of this volume.
3 Alwin Mittasch, *Friedrich Nietzsches Stellung zur Chemie* (Berlin: Verlag Chemie, 1944). See also idem, *Nietzsche als Naturphilosoph* (Stuttgart: Kröner, 1952).
4 The allusion is to Nietzsche's 1875 notebook on 'Wissenschaft und Weisheit im Kampfe' *(KSA* 8, pp. 97–120), published in English translation as 'The Struggle Between Science and Wisdom', in *Philosophy and Truth: Selections from Nietzsche's Notebooks of the Early 1870s*, ed. and trans. Daniel Breazeale (Atlantic Highlands, NJ and London: Humanities Press, 1979), pp. 127–46.
5 Steffen Dietzsch, 'Nietzsches französischer Traum', unpublished paper presented at the conference 'Nietzsche in Cosmopolis. Der französische Nietzsche', Stiftung Weimarer Klassik, Weimar, December 1994.
6 Mittasch, *Friedrich Nietzsches Stellung zur Chemie*, p. 8.
7 Ibid., p. 34.
8 On Nietzsche's changing approach to Socrates, see Sarah Kofman, *Socrates: Fictions of a Philosopher*, trans. Catherine Porter (London: Athlone Press, 1998), pp. 219–41.
9 See Nietzsche, *The Pre-Platonic Philosophers*, ed. and trans. Greg Whitlock (Urbana and Chicago: University of Illinois Press, 2001).
10 Karl Schlechta and Anni Anders, *Friedrich Nietzsche: Von den verborgenen Anfängen seines Philosophierens* (Stuttgart: Frommann, 1962).
11 Gilles Deleuze, *Nietzsche and Philosophy*, trans. Hugh Tomlinson (London: Athlone Press, 1983), p. 45; cf. Babich, *Nietzsche's Philosophy of Science*, pp. 136–40 ('Science and Nihilism').
12 Peter Heller, '"Chemie der Begriffe und Empfindungen." Studie zum 1. Aphorismus von "Menschliches, Allzumenschliches I"', *Nietzsche-Studien*, 1 (1972), 210–33 (211).
13 On the specifically alchemical associations of Nietzsche's 'revaluation of all values', see Richard Perkins, 'Nietzsche's *opus alchymicum*', *Seminar*, 23 (3) (1987), 216–26.
14 For a comparison of Nietzsche and Freud in this respect, see Paul-Laurent Assoun, 'L'homologie des "psycho-analyses": la métaphore chimique', in *Freud et Nietzsche*, 2nd edn (Paris: Presses Universitaires de France, 1998), pp. 148–52.
15 See Bernd Magnus, *Nietzsche's Existential Imperative* (Bloomington: Indiana University Press, 1978).
16 Mazzino Montinari, 'Nietzsche lesen: Die Götzen-Dämmerung', *Nietzsche-Studien*, 13 (1984), 69–79 (76).
17 Michel Serres, 'Corruption – *The Antichrist*: A Chemistry of Sensations and Ideas', trans. Chris Bongie, in Thomas Harrison (ed.), *Nietzsche in Italy* (Saratoga, CA: ANMA Libri, 1988), pp. 31–52 (p. 32).

Chapter 10

Wonder, Science and the Voice of Philosophy

Tracy B. Strong

Viele versuchen umsonst das Freudigste freudig zu sagen
Hier spricht endlich es mir, hier in der Trauer sich aus.

Many endeavoured in vain joyfully to speak profoundest joy;
Here at last, in the tragic, I see it expressed. (Hölderlin, *Sophocles)*[1]

The *Birth of Tragedy* and the source of philosophy

In the *Poetics*, Aristotle had famously argued that the high point of tragedy was the moment of *anagnorisis*, the moment at which the protagonist recognizes himself for what he is. Thus, in the *Oedipus Tyrannos*, this moment of insight is the catalyst that leads Oedipus to blind himself. We can be known to ourselves, Aristotle seems to argue, and the purpose of tragedy is to produce self-knowledge. Nietzsche, also famously, had noted in the preface to the *Genealogy* that 'we are unknown to ourselves, we seekers of knowledge' (*GM* 'Preface' 1). This is a claim, I believe and have argued, not that we should come to 'know ourselves,' but that claims to self-knowledge are in the end self-defeating and that the purpose of philosophy should not be self-knowledge if by self-knowledge we mean the self knowing the self. If knowledge is perspectival, as Nietzsche avers, then this means at least that final and definitive knowledge of oneself – of one's identity, as we call it – is not only not possible but that the attempt to pursue it will be harmful, even nihilistic. Philosophy, as we shall see, requires and produces acknowledgement rather than knowledge – that is, it requires others and neither a tacit nor an explicit positing of a privileged position. All is maculate.

In the end – or at the beginning – the main focus of *The Birth of Tragedy* is not Wagner but Aristotle. In *The Gay Science*, Nietzsche notes that Aristotle certainly 'did not hit the nail on the head when he discussed the ultimate end of Greek tragedy' (*GS* 80). Arguing in *The Birth of Tragedy* against the centrality of the idea of *katharsis* for the tragic, he claims rather that tragedy produces *Verwandlung* (transformation) or *Verklärung* (transfiguration) and *not* (self-)recognition:[2] the self is not found but achieved; the picture is not that of turning around but of a path. For Nietzsche, successful tragedy constitutes the sealing of a change not so much in what one is, but in the naturalness by which one is able to deal in one's life and history with the historically evolving conditions that affect a culture (see, for example, *PTA* 1).

Ours is not, however, a time readily available to tragedy. If tragedy made the growth of Greek culture possible, we live in an age 'with no culture' (*KGW* III/4, 23[14]). Hence the modern problem, and the problem for philosophy, is first to make culture possible. How may this be achieved? We can learn from what happened in Greece. Such an ability will first require, paradoxically, attaining an involved distance or objectivity from one's own world. Nietzsche writes of an audience that, helpless in its seats, is, like the chorus on stage, unable to affect the course of the dramatic action and will thus not 'run up and free the god from his torments'. As spectators, the audience is in the same inactive Dionysian state as is the chorus on stage. He writes: 'The process of the tragic chorus is the *dramatic* proto-phenomenon: to see oneself [as embodied in the chorus] transformed before one's very eyes [as member of the audience] and to begin to act as if one had actually entered into another body, another character' (*BT* 8).[3] It is thus, he argues, that tragedy effects a cultural transformation in the citizenry-spectators. A potential subtitle to *The Birth of Tragedy* proposed in autumn 1870 reads 'Considerations on the ethical-political significance of musical drama' (*KGW* III/3, 5[42]). As we shall see later, the audience relationship described in *The Birth of Tragedy* is key to Nietzsche's understanding of the political transformation of both culture and philosophy.

Education and Philosophy

The lectures *On the Future of our Educational Institutions*[4] continue Nietzsche's tacit critical exploration of Aristotle. Now, he asks explicitly 'what is it, however, that you call your philosophizing?' (*FEI* 1, *KGW* III/2, p. 154). In the introduction to his Basle auditors – whom he might have thought insulted by a new immigrant's presumption – he assures them that his position in these lectures will be that of a fighter, not of a 'loner' [*Einsamer*], nor that of one who finds that nothing can be done, who sees everything as obvious [*selbstverständlich*]. The belief, then, is that something can be done, and indeed he finds hope in that his project seems to him inscribed in 'nature' (*FEI* 'Introduction', *KGW* III/2, pp. 137–38).

Nietzsche framed these lectures around a supposed return trip to a clearing in the woods near the Rhine that he, or the narrator, made with a student friend. While there, practising with pistols, the friends are interrupted by an old man, said to be a philosopher, and his younger companion, who has turned his back on the life of teaching to which the old man introduced him in favour of a life of solitary non-social contemplation. All of the characters are, in different ways, Nietzsche – as he was, as he is, as he aspires to be, as he is afraid of becoming.[5] The old man is distressed at what he takes to be a duel in progress, and much of the lectures consist of a dialogue between him and his companion about the nature of public high-school education in present-day Germany. Nietzsche argues that far from serving science (*Wissenschaft*),[6] academic specialization now in fact serves the interests of the state. Given the present state of culture – or rather, the lack of it – Nietzsche (as the character who takes one of his voices) argues: 'Philosophy must start not from wonder [*Erstaunen*] but from fright [*Erschrecken*]' (*FEI* 2, *KGW* III/2, pp. 165–66).[7] This is a claim that the practice of philosophy must, in the present day at least, start from new origins. Following on Socrates' claim in the *Theaetetus*, Aristotle had famously announced in the *Metaphysics* that 'it is owing to their wonder that men now begin and at first began to philosophize'.[8] Much as Nietzsche had tacitly argued in *The Birth of Tragedy* that some other understanding of tragedy than that of Aristotle was correct, so he is claiming here that some other basis for philosophy is necessary. These lectures, like the book on philosophy in the tragic age, and despite their public success ('audiences exceeding 300'; see *KGB* II/1, 22 March 1872) and the urging of his publisher, were never published. Indeed, Nietzsche never delivered the sixth lecture and was in fact quite unhappy with some of the earlier ones.

The question of the lectures – what it means to do philosophy at all – is still up in the air. Why must philosophy start from fright? In *Schopenhauer as Educator*, Nietzsche had quoted a passage from Emerson's essay 'Circles' to the effect that when a thinker is let loose on the planet the effects will be destabilizing, bringing about a 'new degree of culture' that will 'instantly revolutionize the entire system of human pursuits'. He goes on to say that the problem with academic so-called thinkers is that they do not frighten us. *It is thus the capacity to be frightened of what thought makes available to us that is lacking* (*SE* 8). (And from this does it follow that any understanding of Nietzsche that is not frightening is miscast? I think it does.) Wonder is apparently either no longer sufficient or no longer appropriate to the cultural situation which Nietzsche finds around him, if only again because, as Nietzsche proclaims, we have 'no culture' (*KGW* III/4, 28[2]). More than that, fright seems to be the antidote to the problem of things being *selbstverständlich* – of it going without saying. Our understanding of the present circumstances has to frighten us before philosophy and culture can be possible.

How, then, are we to be released from the way in which our world conceives of the practice of philosophy? We need to know what it actually means to do philosophy, which means that we need to know what a philosopher is. At about this time Nietzsche announces to his friend Erwin Rohde his intention to write a study of 'the pre-Platonic philosophers' (*KGB* II/3, 11 June 1872). He applies for the chair in philosophy at Basle (and is turned down). In a notebook from the summer of 1872 we find: 'What is the philosopher? To be answered by the old Greeks?' (*KGW* III/4, 19[89]). And some pages later he speaks of the 'desperate inappropriateness [*Ungelegenheit*] to be born a philosopher among Germans' (*KGW* III/4, 19[222]). His notebooks from the period 1869 to 1875 are filled with reflections on the status of philosophy and of doing philosophy. He projects – although never finishes – a volume entitled *The Book of the Philosopher*.[9] Note, however, that the ancient Greeks are not to be studied for themselves, out of a misplaced antiquarianism. Indeed, Nietzsche notes in the introductory lecture to his study of the *Choephori* that he wishes 'to sit in the theatre not as an ancient but as a modern' (*KGW* II/2, p. 34).

Again: how to be released from wonder? And what is the role of science in this release? In a note from winter 1872–73 he indicates that the philosopher occupies a place both beyond the sciences (and thus philosophy 'dematerializes' – that is, takes us beyond the apparent or empirical) and this side of (*diesseits* – 'only of this world') religion (hence it de-deifies and disenchants). He goes on to ask what philosophy should now be or do. His answers are numbered and the first three are as follows: '1) Impossibility of metaphysics; 2) Possibility of a thing-in-itself.

Beyond the sciences. 3) Science as the rescue [*Rettung*] from miracle [*Wunder*]' (*KGW* III/4, 23[7]).

Science can rescue us from concluding that what is has a miraculous nature, that it has its origins in something that is beyond our knowledge and unavailable to it. In other words, going beyond science should make it impossible for one to assume a world beyond that which is – a world of 'meta-physics'. Indeed, science can never show us such a world. Importantly, for Nietzsche this does not eliminate the possibility that one may experience something in itself, as a *Ding an sich*, even if this cannot be attained through science. Philosophy is concerned with that which is, with the ordinariness of the everyday.

Here Nietzsche takes up a theme he followed throughout his life, that of the status of the thing-in-itself. It is part of his continuing engagement with, and radicalization of, Kant. There has always been a question of the status of the thing-in-itself in Kant's thought. There is an intimation here that science will help us take from the problematic that the thing-in-itself has in Kantian thought. For Kant the thing-in-itself is not an object of knowledge, yet it has all the features of things that we do know. Kant's answer to this dilemma is to claim that our relation to the noumenal world is not one of knowledge and that this is not a failure of knowledge. God and the thing-in-itself designate our relation to that world. Kant's concern is thus centred around the relationship of what he sees as two worlds. Nietzsche's intention in *The Birth of Tragedy* is to provide what one might call a transcendental deduction of the thing-in-itself – that is, to experience directly the relation between 'reality' and appearance. Or, as he argues in *The Birth of Tragedy*, the importance of the chorus is to make manifest the relation between the thing-in-itself (the realm not subject to knowledge) and appearance (the realm subject to knowledge) (*BT* 8, 16; cf. *KGW* III/4, 19[146]). It is as a step on this path, along with the work preparatory for what will become *Philosophy in the Tragic Age of the Greeks*, that he begins to read widely – if erratically – in the scientific literature of the time. In a note from 1872 or 1873, he writes: 'From the *natural sciences* one must come to a *thing-in-itself*' (*KGW* III/4, 21[16]). In one of his many outlines for a future project, he cross-references the ancient Greeks with contemporary natural science, especially chemistry (*KGW* III/4, 26[1]). A listing of his library shows that, by 1874, he owned 22 works of a scientific nature. Moreover, about half his borrowings from the Basle library in the summer of 1873 were science books, and, as Thomas Brobjer has discussed elsewhere in Chapter 1 of this volume, Nietzsche would retain a degree of interest in what we call natural science for the rest of his life.[10] In fact, he writes to Gersdorff, he holds his studies in science to be essential to the completion of his study on Greek philosophy (*KGB* II/3, 5 April 1873). Just as the ancient Greeks were men of science, so also to be a philosopher in modern times means, for Nietzsche, knowing modern science.

Why should this be so? Why would a person who, at about the same time, famously proclaims truth to be a 'movable host of metaphors, metonyms, anthropomorphisms: in short, a sum of human relations which have been poetically embellished ... illusions which we have forgotten are illusions' (*TL* 1, *KGW* III/2, pp. 374–75), be interested in science, especially the science of the late nineteenth century, at all?[11]

The first part of the answer comes from the fact that Nietzsche did not think, nor did most of those of his day, that the realm of philosophy was radically separate from that of *Wissenschaft*, nor that science gave humans access to a realm that was not perspectival. Where he did differ from his contemporaries, however, is in the attention he pays to the status of science. Babette Babich has carefully analysed the fact that Nietzsche's understanding of science is, in all aspects, parallel to his understanding of morality and of knowledge in general. In all realms what may not be tolerated is the claim that there is a last word – an absolute answer to the question of truth. Thus in *The Gay Science* section entitled 'To What Extent We Are Still Pious' he writes: 'Science also rests on a faith; there is simply no science "without presuppositions".' He goes on to say that this implies that, in science, one claims that one will not deceive, not even oneself, and 'with that we stand on moral grounds'. Nietzsche's claim about science, as with his claim about knowledge in general, is not that science does not pursue the 'truth' but that it moralizes the object of its pursuit. And in doing so, it 'affirms another world than the world of life, nature and history' (*GS* 344).

The second part of the answer comes from the fact that Nietzsche expects that science can give us 'a *history of the genesis of thought*': it can detach us 'from this ideational world [*Welt der Vorstellung*] ... to a limited extent ... and, for brief periods at any rate, lift us up out of the entire proceeding' (*HH* 16). Science thus plays a role in bringing us to focus on this world and in eradicating the idea of a beyond. He will go on to say that it is the world as *Vorstellung* that is 'full of significance, profound, marvellous and bearing in its womb all happiness and unhappiness' – that is, it has the qualities traditionally attributed to things-in-themselves (*HH* 29).

The Achievement of Philosophical Voice as the Ability to Respond to Another: *Philosophy in the Tragic Age of the Greeks*

In a notebook from winter 1872–73 in which Nietzsche discusses the pre-Platonic Greek philosophers, he writes that the philosopher 'is beyond the sciences: dematerializing' (*KGW* III/4, 23[7]). Earlier he had written that the first effect of science is 'an intensification of world-affirmation', even if its ultimate aim is *Weltvernichtung* – the 'destruction of the world' (*KGW* III/3, 3[11]). By *Weltvernichtung*, Nietzsche means that science destroys the apparent natural conceptuality of the world. But for philosophy to go beyond science gives rise to 'the possibility of the thing in itself' (*KGW* III/4, 23[7]) – that is, we find ourselves in the presence of a realm for which we have no concepts.[12] Elsewhere, he gives an example of this in his discussion of Thales. Nietzsche holds Thales' proposition that 'the principle that all [of all that is] is water' (as reported by Aristotle in *Metaphysics*) to be 'beyond scientific consideration' and a position to which he is drawn by 'a conviction that has its origins in a mystic intuition ... but is meant non-mythically and non-allegorically' (*PTA* 3, *KGW* III/2, pp. 307, 309). In Thales, 'for the first time the man of science triumphs over the man of myth, and then the man of wisdom triumphs in turn over the man of science' (*KGW* IV/1, 6[48]). Science makes possible, although it is not itself, the thinking of being, i.e. of what is the case such that some thing is (for Thales, not crazily, water). Here is the reason that the contemporary would-be philosopher must pursue science, as did the ancient Greek.

In the note cited above from winter 1872–73, now under the caption 'What should the philosopher do at present?', Nietzsche writes: '*Science as the rescue from miracle*' – that is, as denying us any positing of a final world. While science rests on philosophy (and is thus not properly understood as independent in its own right), it continues and makes real the philosophical enterprise, or could do so as long as it is true to its foundations.

Nietzsche's reflections on science are thus integral to his understanding of what a philosophical voice would be, were we actually to encounter one. He opens his essay on the early Greek philosophers by reminding his readers (as he had done in his lecture course) that although there is clearly a great deal of influence on the early Greeks from foreign sources (from Egypt to China), they assimilated and made those influences wholly their own (*PTA* 1, *KGW* III/2, p. 300). His account is not to be a myth of origins, but an account of how the Greeks came 'to fulfil, to intensify, to exalt and to purify these borrowed elements so that only now did they become, in a higher sense and in a purer sphere, inventors'. What they invented were 'the types of those who have a *head for philosophy* [*die typischen Philosophenköpfe*]' (*PTA* 1, *KGW* III/2, p. 301). The premise of *Philosophy in the Tragic Age of the Greeks*, then, is that the pre-Socratics were *philosophers*. By understanding them we will determine whether, in the nineteenth century, it is possible to be a philosopher in German and in Germany, as well as, most importantly, what it means to be a philosopher at all.

It is important to realize here that, no matter how admiring his posture, Nietzsche never advocates returning to the Greeks, nor making modern society over in their image. What they can show us is what philosophy actually is, especially in an age like our own when science tends to overwhelm the possibility of thought.[13] He underlines an entire note written just after he completed *Philosophy in the Tragic Age*: '*It might be a* **duty**: *to recognize that it is impossible to bring back Greece and thus also Christianity and the earlier foundations of our society and politics*' (*KGW* IV/1, 5[156]). What interests Nietzsche about the Greeks is not what they did so much as how they did it; or even more precisely, that it was possible to do what they did. (I am here in opposition to those like Habermas who think that Nietzsche is trying to go back to something pre-rational, or primitive, or originary.)

Whatever it is that Nietzsche finds in these men whom he calls philosophers it is not what he has found in the world around him, at least not in those whom the world today labels philosophers. It is important to recognize that Nietzsche does not think that philosophy is, as it were, a matter of individual choice. In the lecture course he writes: 'The Romans, as long as they developed from themselves alone, are completely unphilosophical. Whether or not one philosophizes has to do with the deepest intertwining of a person and a people .... The Romans are not philosophers for the same reasons that they are not artists' (*KGW* II/4, p. 215). By extension the question is: is philosophy possible in modern Germany? For whatever reason, those who might in nineteenth-century Germany lay claim to the title of 'physician of culture' – men such as Goethe and Wagner – are precisely not those regarded by the world at large as philosophers (*PTA* 1, *KGW* III/2, p. 298). Could a (real) philosopher ever be a cultural physician?

So the question addressed in *Philosophy in the Tragic Age of the Greeks* is one that concerned Nietzsche personally: is it possible today to be a philosopher in

Germany, and if so, how? We might extend it to our own times and countries. (I might note here that one can raise, as with Stanley Cavell I urge that one do, the same question about America or the United Kingdom, and also ask if a professor of philosophy is a philosopher.)

What would that entail – to be a philosopher? Philosophy is not simply the concern with 'philosophical issues'. Rather, for Nietzsche it is a particular stance towards the world: 'The word philosophy, as applied to German intellectuals and writers, of late causes me difficulty: it seems quite unsuitable. I would that one avoided it from now on and spoke, plainly and forcefully, only of thought-industry' (*KGW* III/4, 30[19]). He continues in the next entry by remarking that five individuals who had recently been publicly named as showing that Germany was still a 'nation of thinkers'[14] in fact only 'phosphoresce' – that is, presumably, glow dimly without setting anything alight (*KGW* III/4, 30[20]). The only hints of philosophy in German (Schopenhauer, Hartmann) are generally ignored. He goes on to ask how this situation has arisen (*KGW* IV/4, 30[20]). The question of the possibility of philosophy thus poses itself for Nietzsche in relation to a culture (or the lack of it). He hopes to find the answer in an exploration of how philosophy was possible by an encounter with the 'tragic age of the Greeks'.

To know what it would be to do philosophy is, first, to meet a philosopher and come to know him – or her – as a *person*. In the *Introduction to the Study of Platonic Dialogues*, Nietzsche notes that to understand a philosopher we must have a 'comprehensive psychological portrait [of him] as a touchstone' (*KGW* II/4, p. 148). In *On the Future of our Educational Institutions*, as well as in his inaugural lecture *Homer and Classical Philology* and in *Philosophy in the Tragic Age of the Greeks*, Nietzsche places great emphasis on what he calls the 'personality' of the philosopher. *Philosophy must first and foremost be a kind of encounter between human beings*: it is thus not, in its grounding, about the pursuit of truth. At most it seeks truthfulness or honesty (the 'youngest of the virtues,' Nietzsche notes in *Daybreak* (*D* 456)). In the lectures *On the Future of Our Educational Institutions*, one of the young pistol practitioners asks: 'What is philosophy ... if it prevents a man from being for himself, or from enjoying the select company of a friend – in fact, if it prevents us from becoming philosophers ourselves?' (*FEI* 1, *KGW* III/2, p. 152).[15] Nietzsche will write: '*philosophia academica delenda est*' (*KGW* III/4, 34[45]).

There are two matters to note here. First, it is significant that the pistol practice (something that Nietzsche takes pains to point out was to be of use later in his and his friend's military careers) is conceived by them as integral to a philosophical activity. Presumably it trains one to find the right target and hit it. Second, philosophy will probably at least start as conversation in the select company of a friend. (One is reminded here of what J.L. Austin claims, albeit with an Oxbridge common-room accent, the purpose of philosophy to be. The topic of excuses affords him, he says, 'what philosophy is so often thought, and made, barren of – the fun of discovery, the pleasures of co-operation, and the satisfaction of reaching agreement'.[16]) For them, the commemorative festival is when and where they will form projects and serious plans for their future philosophizing. Their reunion is to be 'all past, all future, with the present only a suspended dash between' (*FEI* 1, *KGW* III/2, p. 152). One might note here that the absence of the present (all past, all future) is probably what keeps them from being able to do philosophy. Indeed, the students in *On the Future of our Educational Institutions* complain that the presence of a philosopher at this place makes philosophy impossible for them: 'The old man laughed. "What? You are afraid that a philosopher will keep you from philosophizing? This kind of thing may indeed happen: you have not experienced it? Have you not had this experience in your university? Surely you follow courses in philosophy?"' (*FEI* 1, *KGW* III/2, p. 154).

They allow that they have not yet taken courses in philosophy but that they certainly shall. What happens, however, is that they remain silent and listen to an exchange between the philosopher and his companion: this is now their first course in philosophy. That they now remain silent, as an audience must, is important and a step forwards in their ability to do philosophy. In the Fourth Lecture the philosopher's companion draws an explicit parallel between the position of the two students (who have been watching and listening to the exchange between him the philosopher) and the chorus for a tragedy (*FEI* 4, *KGW* III/2, pp. 212–13).[17] They allow the philosopher and his companion to act as if they are alone while being in the presence of others. Philosophy has something to do with the possibility of being receptive or open, of letting oneself be. It is as passive as it is active.

To come to philosophy, then, requires first being in the presence of a philosopher. So, the second question is: what is the nature of that presence? What happens when we are in the presence of a philosopher, a real one? Nietzsche begins *Philosophy in the Tragic Age of the Greeks* with the following remark: 'With those who are strangers to us [*fern stehende Menschen*], it is enough to know their goals, so as to value or to reject them in their entirety.' Such people are opposed to those to whom we 'feel proximity, whose activity is more important to us than are their goals' (*PTA* 'Preface', *KGW* III/2, p. 295). Only by taking the second approach, Nietzsche continues, will we be able to find that which is incontrovertible about a philosopher, that which we will experience as the actuality of philosophy: his personal 'pitch, his personal colour [*persönliche Stimmung, Farbe*]'. This is what is given to us as the 'portrait of the philosopher'. What is it, he concludes, about these men that so strikes us that we must always respond to it, no matter what knowledge they or we might have? What voice is the voice of a philosopher – that is, the voice of someone to whom one cannot but respond?[18] In this the achievement of philosophy is aesthetic: a work of art is art because one cannot but respond to it.

Four sets of reflections thus surround Nietzsche's discussion of the philosophical voice. First, the task and activity of philosophy is not constructive – it is not to give answers, not to determine truth – but *destructive*: to make possible, at least in part, the recovering, the transformation of a life, of a culture, of a life together. He notes: 'The *culture* of a people shows itself in the *unified control of the drives of this people*: philosophy controls the drive to knowledge, art the drive to form and ecstasy, agape controls eros and so forth' (*KGW* III/4, 19[41]). Nietzsche will in fact argue that, at present, the philosopher can create no cultures, but that he is the 'most useful when *there is much to be destroyed*, in times of chaos or degeneration .... Now – as there is no culture, what has he to prepare (to destroy)?' (*KGW* III/4, 28[2]). He will elsewhere refer to the philosopher as the *Giftmischer der Kultur*, the poisoner and/but also the apothecary, thus as a *pharmakos* (*KGW* III/4, 23[16]).[19] Nietzsche looks at the Greeks to see how a philosopher might behave towards a culture that is not his enemy. We learn, he avers, that philosophy must work towards '*the destruction of blind secularization*' and 'subjugation of the barbarizing effects of the drive to knowledge'. Note that, for Nietzsche, the notion of the secular is part and parcel of the notion of the religious: both take truth outside of the world. We are to be neither secular nor religious. Likewise, science and its concomitant scepticism tends, when taken to an extreme (as with Socrates), to become rigid and dogmatic (*KGW* III/4, 23[14]).

Second, the activity of philosophy can only occur in, and as the response to, the voice of another. Philosophy is not about solving problems. It is about what happens between persons (as well as between a person and him- or herself), and whatever those relations are about is what human beings are about in the everydayness of their life – their ordinariness. The concept of a response to another is worked out by Nietzsche during these years in *Schopenhauer as Educator* and *Richard Wagner in Bayreuth*. There and in *Philosophy in the Tragic Age of the Greeks*, he is concerned to bring out what he calls the 'personality' of each philosopher – by which he means that which cannot be avoided as long as we do not hold our self as a stranger to the words they lived. We would have then no choice but to respond to them, much as one cannot, as a human being, help but respond to a person as a person. (One can of course respond *not* as a human being: Rousseau saw this and it is the import of what Max Weber saw in the rationalized world of his and our time.)

Third, and in consequence, Nietzsche writes that the beginning of the philosophical mind comes in the amazement that what is actual for us in the everyday or the ordinary is becoming. 'The intellect must not wish only to enjoy this furtively but must become completely free and celebrate saturnalia. The liberated intellect looks clearly at things: and now, for the first time, the everyday appears to it as noteworthy, a problem' (*KGW* II/4, p. 215). At the end of *Richard Wagner in Bayreuth*, he indicates that it is the glory of Wagner's music to:

> ... enlighten the poor and lowly and melt the arrogance of the learned ... [N]ow that it has come about, it must transform the very notion of education and culture in the spirit of everyone who experiences it; it will seem to him that a curtain has been raised on a future in which there are no longer any great and good things except those which all hearts share in common. The abuse which has hitherto clung to the word 'common' will have then been removed from it. (*RWB* 10)[20]

A philosophy that is born of fright of the world as it presently appears to us will be able to move us to delight at that which is. This is, of course, the same dynamic as tragedy. In *The Birth of Tragedy* he writes: 'The cry of horror or the longing moan over an irretrievable loss are intoned from the highest joy' (*BT* 2). What is central to Nietzsche's understanding of tragedy is that this realization of the world presents itself as a source of passion and joy, not of despair.

Finally, philosophy can only be done in a voice that is one's own. This is the significance of Nietzsche's insistence on the particularity of pitch and colour. What, however, does it mean to say that a voice is one's own? The purpose of philosophy is to make culture possible. In *Schopenhauer as Educator*, Nietzsche writes that only the person who 'has attached his heart to some great man' – what he calls here an exemplar – 'receives thereby the first blessings of culture' (*SE* 6).[21] The purpose of culture is the production of what he calls 'genius'. One should tread slowly here. 'Genius' is a term that Nietzsche here almost certainly takes from Emerson, in whose style the *Untimely Meditations* are written. Emerson writes: 'To believe in your own thought, to believe that what is true for you in your private heart is true for all – that is genius.'[22] And already in 'The American Scholar' Emerson had noted that genius was 'the sound estate of everyman'.[23] Genius, we find in Nietzsche, is used interchangeably with what he calls (still in *Schopenhauer as Educator*) 'the furthering of the emergence of true human beings' (*SE* 6).[24] One's *own* voice is the voice of genius, genius as Emerson and Nietzsche understand it. I am reminded of a phrase somewhere in Mark Twain's notebooks: 'All of us have music and truth inside but most have a hard time getting it out.'

So *Philosophy in the Tragic Age of the Greeks* is about philosophy, or about the possibility of philosophy, about the role of philosophy in making culture possible, about what that might mean in terms of a valuation of the everyday, of *Dasein*;[25] it has as a requirement the production of (one's own) genius. Why, then, is it set in the tragic age? Or more accurately, what is philosophy such that it makes the tragic age possible? Part of the answer is to be found in Nietzsche's claim that only in Greece during the 'immense' period between Thales and Socrates, has the philosopher been *at home*, and not a 'chance random wanderer', 'conspiring against his fatherland'. For a philosopher not to be a terrifying 'comet', a culture is needed. Only among the Greeks, who had a culture, did the philosopher 'shine like a star in the solar system of culture' (*PTA* 1, 2, *KGW* III/2, pp. 303, 304). Nietzsche's task, as he sees it, is to '(re-)describe [*umschreiben*] the world in which the philosopher and the artist are at home' (*KGW* III/4, 19[3]). Thus Nietzsche writes that he wants 'to know how philosophy behaves towards an existing or developing culture which is not the enemy' (*KGW* III/4, 23[16]). In order to know this, 'one must know what we call his "age"' (*KGW* III/4, 28[2]).

The culture in which these philosophers were at home was the 'tragic age'. In other words, each of these philosophers embodied an element of what it meant to do philosophy, something that was done in Greece as both the exponent and proponent of a culture. Philosophy and politics and tragedy are close to coterminous. The focus on the tragic age has to do with whether or not the Greeks will successfully incorporate these elements into the world that issues from the Peloponnesian War. In a collection of fragments to which Nietzsche gave the general name *Science and Wisdom in Conflict*, we find this remark:

> One can describe these older philosophers as those who felt the Greek air and customs as a *constraint* and *barrier*: thus as self-liberators (war of Heraclitus against Homer and Hesiod, Pythagoras against secularization, all against myth, especially Democritus) .... I conceive of them as the **precursors of a reformation** of the Greeks: but not that of Socrates .... One set of phenomena carried all of the reforming spirits along: *the development of Tragedy*. (*KGW* IV/1, 6[18])

Tragedy here is a 'means' to carry out a reformation and is to be seen as *made possible by* and as *a continuation of* the individual achievements of the philosophers. Thus *Philosophy in the Tragic Age of the Greeks* is an investigation of what lies behind and leads up to the developments discussed in *The Birth of*

Tragedy. In the latter book, the elements of that which was 'Greek' had remained unexamined. More importantly, *the role of philosophy in making tragedy possible and of tragedy in making the* polis *possible* had disappeared under the destructive Socratic enterprise. Nietzsche in fact writes to Rohde in 1873, informing him that he is working on a book about Greek philosophy that he may call 'the philosopher as the physician of culture' (*KGB* II/3, 22 March 1873). This is why philosophy leads to the tragic age of the Greeks.

What are the elements of a full philosophical voice? Nietzsche's understanding here is formulated in conjunction with that of the early Greek philosophers. These men, it is vital to remember, were not philosophers as we generally understand that word today – that is, philosophers as opposed to something else. They were *physiologoi*, inquirers into what is. And they were also thus what we should call, but no longer do (because we think of them as 'philosophers'), men of science. Indeed, Thales was a navigator, a geometer, a very successful speculator in olive futures, a military engineer, an astronomer, and a ... thinker.[26] The important point here is that the distinction between science as a form of knowledge and philosophy is not one that Nietzsche draws.

Nietzsche expresses here a need for mastery of a twofold tension. On the one hand, these *physiologoi* bring about a 'mastery of the mythical' by 'strengthening the sense of truth against unrestrained poetry [*freie Dichtung*]'; on the other, they bring about a 'mastery of the drive for knowledge' by strengthening the 'mythic-mystical'. Nietzsche associates Thales, Democritus and Parmenides with the first, and Heraclitus, Empedocles and Anaximander with the second. Both of these masterings are associated with the impossibility of 'grounding a popular culture in philosophy'. What philosophy and philosophers can do instead is destroy what Nietzsche calls 'rigid dogmatism'. For, Nietzsche asserts, 'any power (religion, myth, drive to knowledge), when excessive as a rigid domination (Socrates), has barbarous, unethical [*unsittliche*] and stultifying effects' (*KGW* III/4, 23[14]). It can also destroy a kind of blind secularization, which Nietzsche holds to be the inverse of religion. (The idea of the secular requires the idea of religion and the elimination of one entails the elimination of the other.)

As Nietzsche develops the elements of a fully human, this-worldly voice in his diadochical analysis of the philosophers/scientists who make the tragic age possible, we find the following.[27] If the practice of science is not to be antithetical to philosophy, what does it require? This is the point of *Philosophy in the Tragic Age of the Greeks*.

First, the practice of science requires a non-religious and non-miraculous understanding of the world. The point, however, is not to ground the polity on science, but to make possible, as Nietzsche argues in *The Birth of Tragedy*, a regrounding on a mythic understanding. (I take a myth to be an illusion of which one accepts the authority while knowing that it is truthful rather than true.) This is a re-foundation for the *polis*. *Thales* is Nietzsche's exemplar of this; he makes possible, Nietzsche notes, a pan-Hellenism that would have been the only way of preventing the Persian Wars.[28]

Second, it requires an account of the centrality of change and contingency. Nietzsche finds this in *Anaximander* – that which is has no qualities and is the equivalent of the Kantian *Ding an sich*. Without such an account we will be kept from

Dionysian knowledge – that is, the knowledge of the transitoriness of the worlds in which we live. So Hamlet is the prototype of the Dionysian man for Nietzsche.

What we need, then, is an account of the definite. The third part of the philosophical voice is the realization that not only is everything impermanent, but that since all that comes into being must pass away, conflict is endemic to human existence. Conflict – the *agon* that Nietzsche names as central to all aspects of Greek life, and which Christa Davis Acampora discusses in more detail in Chapter 8 of this volume – is also or in fact understood as immanent in existence. Definiteness is thus a denial of existence. (In *Experience*, Emerson says: 'Our love of the real draws us to permanence, but health of the body consists in circulation, and sanity of mind in variety or facility of association. We need change of objects.'[29]) In Greece this voice is that of *Heraclitus*, for whom the world was a game 'of the child Zeus' and of 'fire with itself' (*PTA* 6, *KGW* III/2, p. 322).

Fourth, we need to know that that which is comes not from that which is not, but only from itself (a point that Heidegger was to drive into the *Grund)*. Truth is therefore not a predicate of phenomena, of that which appears. Here the voice was that of *Parmenides*.

Fifth, with this we must acknowledge the centrality of plurality, especially of value. The world is thus non-rational, but is nevertheless one and its unity lies in its appearance, in its beauty. This is the voice of *Anaxagoras*.

The text of *Philosophy in the Tragic Age of the Greeks* breaks off here. We know from the notes that, had Nietzsche finished it, he would have dealt with three other elements that speak in a philosophical voice. Sixth, *Democritus* holds to the necessity of an aesthetic understanding – that is, that what appears to us as a given world is in fact constructed, an artefact of our being. To Anaxagoras, Democritus adds the claim that the world is necessary and comprehensible and focused in this world, in the *polis*, that therefore other beings are around each of us and are part of the world which is our own world – a point not really grasped by Heraclitus.

Seventh, the existence of others as central to science is key to *Empedocles*, who holds that value, beauty and reality are of and in the world only, and that they result from, as well as lead to, different perspectival stances. Nietzsche calls attention to his political reforms and relates him to the Pythagoreans. He is a 'democrat, who has social reform up his sleeve' and is identified with 'love, democracy and communal property' (*KGW* IV/1, 6[38, 50]).

Finally, it is not hard to find each of these elements in the analysis of *The Birth of Tragedy*. All of this could have supported a tragic age. Instead, the last of the philosophers puts an end to it. He is the 'virtuoso of life,' *Socrates* (*KGW* IV/1, 6[17]).[30] 'With Empedocles and Democritus the Greeks were well on their way towards *assessing correctly* the irrationality and suffering of human existence; but thanks to Socrates, *they never reached that goal*' (*KGW* IV/1, 6[25]). Socrates, Nietzsche goes on to note,

> destroyed the naturalness [*Unbefangenheit*] of ethical judgement;
> annihilated science [NB];
> had no feeling [*Sinn*] for art;
> wrenched the individual from his historical bonds;
> promoted dialectical verbiage and chatter. (*KGW* IV 1, 6[26])

In fact, 'after Socrates it is no longer possible to preserve the common weal [*das allgemeine Wohl*]' (*KGW* III/4, 19[20]).

Each of these thinkers *embodies* science and thus philosophy (although Socrates will be a limiting case for Nietzsche). What interests Nietzsche as much as their doctrine – or the metonymic fragment of their doctrine in which they emerge most fully as persons – is that they were able to hold together both life and knowledge, each in their highest intensity:

> This struggle between life and knowledge is all the greater, and their remaining under a common yoke all the stranger, the more powerful both drives are; that is, the fuller and more flourishing life is, and in turn the more insatiable knowledge is and the more greedily it pushes its way towards all adventures. (*KGW* IV/1, 6[48])

In *The Birth of Tragedy* these two drives had been the Apollonian (life) and the Dionysian (wisdom). Tragedy had held them together as society and culture – and as a polity.[31] Here they are held together in the lives of each philosopher such that they are for us, were we to encounter them, '*beautiful possibilities of life*' (*KGW* IV/1, 6[48]). And were we to encounter them, we would say of them 'how beautiful they are', much as someone encountering Greece will also exclaim, as Nietzsche writes at the very end of *The Birth of Tragedy*, 'how much did this people have to suffer in order to become so beautiful' (*BT* 25). The ancient Greek philosophers held these two drives together. Nietzsche presents them as stations on a philosophical path, as the way in which philosophy pulls us along.

However, it is precisely from Nietzsche's appreciation of the polymorphic quality of the philosophers of the tragic that an unexpected conclusion comes. If it is the case that philosophy, art and science are not conceptually different enterprises – that is, if the same strictures apply to each – then the question of doing philosophy is simply one of degree. If we live in illusions that we ourselves have produced, then we are all, to some degree, artists – that is, producers of form. There is, says Nietzsche, 'no separate philosophy, distinct from science: there as here one thinks in the same manner'. For both, 'there is in any case something of the artist in this production of forms with which something thus enters into memory: it discerns a given form, and by this reinforces it'. As a result, he concludes, 'it is only a question of degrees and quantities: *all humans are artists, philosophers, scientists, and so forth*'.[32]

This aristocratic/democratic claim – that all are capable of art, philosophy and science, merely because of the fact that they are (and insofar as they are) human – is Nietzsche's hope. To make it actual is the task which he sets himself from the very beginning of his career as a thinker. As he writes to Heinrich von Stein, in 1882, 'I want to take away from human existence some of its heartbreaking and awful quality'. This, he says, is thc 'task of my life' (*KGB* III/1, beginning of December, 1882).[33] What I am calling a democracy of excellence or perfectibility – that all of us have the ability to be what we are, and that, for several reasons, most of us are not that – is central to Nietzsche throughout his life, though rarely discussed in the secondary literature.

How might one recall humans to the fact, the actuality that they can be what each of them are – 'artists, philosophers, scientists'? The point of the analysis is to show that, to some degree, each of these categories is available and open to each of us. There is an order of rank, but we are all ranked according to the same criteria. One might say that Nietzsche tried to demonstrate precisely this to his contemporaries in writing *The Birth of Tragedy*. That work opens with the claim that:

> We will have conquered much for aesthetic science [*Wissenschaft*] when we do not come only to the logical insight but to the unmediated certainty of vision [*Anschauung*] that the ongoing development of art is bound to the duplexity [*Duplizität*] of the Apollonian and the Dionysian. (*BT* 1)

'Not only ... but': Nietzsche clearly had hopes that his book would be received on both of these levels at once. For him, it was not just an exalted rhapsody about the experience of Greek drama. Thus he will write to Rohde as his friend prepares a response to Wilamowitz with various scholarly citations in support of his undertaking (*KGB* II/, 3 July 16, 1872; see also his letter to Ritschl, 12 August 1872). Upon publication, Nietzsche enthusiastically sent volumes of *The Birth of Tragedy* to his friends and teachers. Responses were slow in coming, for the book was in no way what one might have expected as the first product of the man expected to become the star of the next generation of German philologists. Especially painful was the absence of any response from Ritschl, his teacher. Finally, at the end of January, Nietzsche swallowed his pride and wrote to Ritschl expressing his surprise that he had received no acknowledgement, a silence all the more difficult, he indicated, as his book had 'something of the manifesto about it' and thus required a response (*KGB* II/1, 30 January 1872). Just as he had sought to present the Greek philosophers as persons whom one could not possibly hold at a distance, and to attack German educational institutions as encouraging such *Fernstehen*, so also was *The Birth of Tragedy* not the sort of book, Nietzsche felt, to which an indifferent reaction was possible.

The two modes that Nietzsche strove to hold in tension, one with the other, appeared, to his teacher, hopelessly discordant. In early 1873 Ritschl wrote to Freiherr von Vischer:

> But our Nietzsche: that is really a distressing chapter – as from your letters you have grasped, despite all goodwill for a first-rate person. It is extraordinary how two souls can dwell together in that man. On the one side the most rigorous methodology of scholarly scientific research ... on the other this fantastic-exaggerated, overly clever reaching into the incomprehensible, this Wagnerian-Schopenhauerian, artistic-mysterious-religious rhapsody [*Kunstmysterienreligionsschwärmerei*].[34]

Ritschl wanted Nietzsche to be academically professional, period. He had grasped the twofold quality of Nietzsche's approach, but not the fact that Nietzsche thought that, if the two could not be reconciled, then at least a delicate balance might be maintained between them. Indeed, he thought that his entire project of the early 1870s depended on pursuing the two approaches to knowledge simultaneously, just as he hoped that *The Birth of Tragedy* would reach two audiences at the same time. The first was the German scholarly community. Nietzsche proposed in his work a new understanding of the origins of tragedy and, by implication, a new understanding of Greek history and identity. The second intended audience was the

German public and German culture in general, a public that, as he made clear in the lectures on education, required a more direct certainty in matters of culture. The combination of both would recover what he calls 'the noble conception of a *peuple*, a *publicum*' (*KGW* III/4, 19[26]).

The architectonic for this was laid in *The Birth of Tragedy* itself. The aim of the book is to provide a test by which contemporaries may determine if they are able to become truly cultured – that is, hold life and knowledge in a lasting and beautiful tension. Nietzsche writes:

> Whoever wishes to test himself completely accurately as to how closely related he is to the true aesthetic member of an audience or rather belongs to the community of Socratic-critical persons, has only to examine honestly the feeling with which he receives the wondrous spectacle present to him on stage: does he feel offended in his historical sense, relying as it does on strict psychological causality, does he benevolently concede that it is intelligible to the childish, but alien to him, or does he feel something else. (*BT* 23)[35]

Nietzsche continues this passage with a quiet justification of the mixture of modes with which he has shaped *The Birth of Tragedy*. He thinks that he has found a way around the nihilism of contemporary culture by using knowledge and science against itself, as it were. If we have become historical-critical beings, then Nietzsche will show us that it is precisely this quality that will permit us to move to a new level of aesthetic understanding. Nietzsche thus did hope for a role for traditional scholarship in his project. The reception of the book – not only by Wilamowitz, but by Ritschl and others – as merely romantic inspiration and the prophetic utterance of a scholar lost to Wagnerism was thus disturbing not just to his wounded pride and tarnished reputation, but to the possibility of his whole project.

When Ritschl broke his silence on *The Birth of Tragedy* (with a somewhat distant and strained letter two weeks later), it was revelatory to Nietzsche. The overwhelmingly negative reception of the book brought this course to a halt. Nietzsche began to have doubts about the viability of the project of cultural regeneration that he saw himself undertaking with Wagner. The scene at the beginning of *Thus Spoke Zarathustra* in which the inhabitants of the town embrace precisely that which Zarathustra wishes to warn them against re-creates this realization: it is a distress with the possibility of the success of simply teaching what is needed for cultural health.

Philosophy in the Tragic Age of the Greeks and the lectures on educational institutions are texts that, by their very incompletion, signal the grave doubts that Nietzsche came to have about the viability of his project as he then conceived of it. His doubts were intensified by the incomprehension that his first published book caused in the very people whom he had thought would grasp it. It required him to pursue another direction: a resounding and radical critique of all the institutions and forms that kept his contemporaries from acknowledging philosophy, tragedy and the possibility of culture. In a note from 1886, written as he was preparing to write the 'Attempt at a Self-Criticism' that would become the preface to the new edition of *The Birth of Tragedy*, Nietzsche remarks that he 'overvalued German nature; ... [and] ... did not understand the source of the modern desolation' (*KGW* VIII/1, 2[111]). Bismarck, it turns out, had not undervalued the resources in the German people. And if Bismarck were right, what would prepare the cultural revolution he had envisaged?

Remedying this shortcoming becomes his mission from the mid-1870s until the end of his life, and all of his work from that period must be understood in these terms. The work of *Human, All Too Human, Daybreak*, and the first four books of *The Gay Science* are all investigations into his own blindnesses and obscurities, what it was in him that led him to be surprised by the reception of the project that centred around *The Birth of Tragedy*. Following that, the work of the 1880s may be read as systematic investigations into various human realms. *Thus Spoke Zarathustra* is (among other things, to be sure) an exploration of various human institutions. *Beyond Good and Evil* is a study of how it is possible to make a claim to knowledge in various areas. *On the Genealogy of Morals* is a study of what it, in fact, means when one claims a moral stance. In a like manner, *Twilight of the Idols* is a study of the nature of claims to authority and *Ecce Homo* a critique of what it means to make a claim, that is to write at all as an author.[36] There is thus not a 'young' and a 'mature' Nietzsche, still less a 'middle-period' Nietzsche. A reading of *Philosophy in the Tragic Age of the Greeks* in conjunction with the lectures on *The Future of our Educational Institutions* and *The Birth of Tragedy* shows us the continuity of his philosophical project, its particular relation to *Wissenschaft*, and why he came to see how much more radical his undertaking was than when he had first conceived of it as a precocious academic success.

Notes

1 Michael Hamburger, *Hölderlin*, (London: Harvill, 1952), pp. 102–103 – thanks to Babette Babich for this reference.

2 The opposition to Aristotle has been little commented on but was already noted by Wilamowitz in *Zukunftsphilologie*. See Ulrich von Wilamowitz-Moellendorf, 'Future Philology! A Reply to *The Birth of Tragedy* by Friedrich Nietzsche', *New Nietzsche Studies*, 4: 1/2 (2000), 1–32 (32 n.52).

3 For a full discussion see my *Friedrich Nietzsche and the Politics of Transfiguration* (Champaign: University of Illinois Press, 1999), pp. 161–82.

4 For an account of the circumstances of these lectures see Curt Paul Janz, *Friedrich Nietzsche. Biographie*, vol. 1 (Munich: Hanser, 1978), pp. 444–46.

5 See the discussion in Jacques Derrida, *The Ear of the Other* (Lincoln, NE: University of Nebraska Press, 1988), pp. 22–24.

6 The German word *Wissenschaft* does not translate well as 'science'. See Chapter 6 of this volume by Babette Babich.

7 The word in Aristotle is *jaumazein*. Liddell and Scott give 'wonder,' 'astonishment.' Heidegger has argued that the sense of awe-fullness is also present here. Nietzsche insists on emphasizing what one might call one side of *thaumazein*.

8 Aristotle, *Metaphysics*, trans. W.D. Ross, in *The Complete Works of Aristotle* (Princeton, NJ: Princeton University Press, 1984), vol. 2, 982b, 12–13. The passage from the *Theaetetus* is as follows: 'This sense of wonder is the mark of the philosopher. Philosophy indeed has no other origin, and he was a good genealogist who made Iris the daughter of Thaumas' (Plato, *Theaetetus*, trans. F.M. Cornford, in *The Collected Dialogues of Plato*, ed. Edith Hamilton and Huntington Cairns (Princeton, NJ: Princeton University Press, 1961), 155d). Iris was a messenger the gods employed to send messages to humans.

9 The fragments of this work, together with related material, have been published in French as *Le Livre du philosophe*, ed. A. Kremer-Marietti (Paris: Flammarion, 1969) and in English as *Philosophy and Truth*, ed. David Breazeale (Atlantic Highland, NJ: Humanities Press, 1979). The material appears in its original notebook form in *KGW* III/4, pp. 5–128.

10 Max Oehler (ed.), *Nietzsches Bibliothek* (Weimar: Wagner, 1942) pp. 23–26 and 51–52.

11 I am indebted to the incisive account of this passage in Babette Babich, *Nietzsche's Philosophy of Science: Reflecting Science on the Ground of Art and Life* (Albany, NY: SUNY Press, 1994), pp. 62–64. See also: Günter Abel, *Die Dynamik der Willen zur Macht und die ewige Wiederkehr* (Berlin: de Gruyter, 1984) and Abel Rey, *Le retour éternel de la philosophie de la physique* (Paris: Flammarion, 1927).

12 In this, Nietzsche is a radical Kantian. Though there is no direct proof that I know of, his discussion here could be drawing on Kant's position in *Critique of Judgement*, §§ 45–49.

13 Cf. Heidegger's argument that 'science does not think'.

14 Nietzsche takes up the idea of a 'nation of thinkers' as applied to Germany again in *AOM* 319.

15 Much of the language here is drawn from Nietzsche's reading of Emerson, especially 'The American Scholar'. See *Essays and Lectures* (New York: Library of America, 1983).

16 J.L. Austin, 'A Plea for Excuses', in *Philosophical Papers* (Oxford: Clarendon, 1961), pp. 123–52 (p. 123).

17 This occurs just after the narrator's friend has cited a passage from the 'Walpurgisnacht' section of *Faust I*, where Will o' the Wisp says that he hopes that his awe will steady his path.

18 This has been the concern of the recent work of Stanley Cavell, *A Pitch of Philosophy* (Cambridge, MA: Harvard University Press, 1994).

19 A citation of Derrida should not prove necessary here.

20 See my 'Nietzsche and the Song in the Self', *New Nietzsche Studies*, 1: 1/2 (1996), 1–15.

21 The phrase in the original is 'empfängt damit die erste Weihe der Kultur'. I have slightly modified Hollingdale's translation.

22 R.W. Emerson, 'Self-Reliance', in *Essays and Lectures*, pp. 259–82 (p. 259).

23 Emerson, 'The American Scholar,' in *Essays and Lectures*, pp. 53–71 (p. 57).

24 I give 'emergence' for *Entstehung* in preference to Hollingdale's 'production'. Delicacy in translation is essential to this essay. See Stanley Cavell, *Conditions Handsome and Unhandsome and Emerson's Perfectionism* (Chicago: University of Chicago Press, 1990).

25 See Joseph Lima and Tracy B. Strong, 'Telling the Dancer from the Dance: On the Relevance of Ordinary Language Philosophy for Political Theory', in Andrew Norris (ed.), *Stanley Cavell and Political Theory*, forthcoming.

26 See Mott Greene, *Natural Knowledge in Preclassical Antiquity* (Baltimore: Johns Hopkins Press, 1992), chapter 5.

27 For a more detailed analysis see my *Friedrich Nietzsche and the Politics of Transfiguration*, pp. 152–61 and chapter 6. See also *KGW* III/4, 19[143], 19 [245], 19[328], 21[11], 21[16–23], 23[17–23], 26[1] and IV/1, 6[48–50].

28 So also will Nietzsche later proclaim himself a 'European'.

29 Emerson, 'Experience', in *Essays and Lectures*, pp. 471–92 (p. 476).

30 See the discussion in my *Friedrich Nietzsche and the Politics of Transfiguration*, pp. 112–23 and pp. 168–85.

31 See Nietzsche's essay 'The Greek State' (*KGW* III/2, pp. 258–71).

32 Nietzsche, *Le Livre du philosophe*, pp. 61–63 (my italics). This passage is not in the *KGW*.

33 I would not have remembered this letter had I not read David Allison's fine book, *Reading the New Nietzsche* (Lanham, MD: Rowman and Littlefield, 2001).
34 Quoted in Janz, *Friedrich Nietzsche: Biographie*, vol. 1, p. 511.
35 For a fuller discussion, see my 'Aesthetic Authority and Tradition: Nietzsche and the Greeks,' *History of European Ideas*, 11 (1989), 989–1007.
36 For an elaboration of this, see my 'Nietzsche's Political Aesthetics' in M. Gillespie and T. Strong (eds), *Nietzsche's New Seas* (Chicago: University of Chicago Press, 1990), pp. 153–74.

Bibliography

Works by Nietzsche

Briefwechsel: Kritische Gesamtausgabe, ed. Giorgio Colli and Mazzino Montinari, Berlin: de Gruyter, 1975–1984.
Werke: Kritische Gesamtausgabe, ed. Giorgio Colli and Mazzino Montinari, Berlin: de Gruyter, 1967– [in progress].
Werke: Kritische Studienausgabe, ed. Georgio Colli and Mazzino Montinari, Munich/Berlin: DTV/de Gruyter, 1988.
Werke und Briefe: Historisch-kritische Gesamtausgabe, ed. Hans Joachim Mette, Karl Schlechta and Wilhelm Hoppe, 8 vols, Munich: Beck, 1933–42 [incomplete].

Beyond Good and Evil, trans. Walter Kaufmann, New York: Vintage, 1966.
Beyond Good and Evil, trans. R.J. Hollingdale, Harmondsworth: Penguin, 1973.
The Birth of Tragedy and The Case of Wagner, trans. Walter Kaufmann, New York: Vintage, 1967.
Daybreak, trans. R.J. Hollingdale, Cambridge University Press, 1997.
Ecce Homo, trans. R.J. Hollingdale, Harmondsworth: Penguin, 1979.
The Gay Science, trans. Walter Kaufmann, New York: Vintage Press, 1974.
Human, All Too Human, trans. R.J. Hollingdale, Cambridge University Press, 1986. (Includes *Assorted Opinions and Maxims* and *The Wanderer and his Shadow*.)
Le livre du philosophe, ed. A. Kremer-Marietti, Paris: Flammarion, 1969.
On the Genealogy of Morals, trans. Walter Kaufmann, New York: Vintage, 1969.
Philosophy and Truth: Selections from Nietzsche's Notebooks of the Early 1870s, ed. David Breazeale, Atlantic Highlands, NJ: Humanities Press, 1979. (Includes *On Truth and Lie in an Extra-Moral Sense*.)
The Portable Nietzsche, trans. and ed. Walter Kaufmann, New York: Viking Press, 1971. (Includes *Nietzsche contra Wagner*.)
The Pre-Platonic Philosophers, trans. Greg Whitlock, Urbana and Chicago: University of Illinois Press, 2001.
Twilight of the Idols and The Anti-Christ, trans. R.J. Hollingdale, Harmondsworth: Penguin, 1968.
Twilight of the Idols, trans. Duncan Large, Oxford: Oxford University Press, 1998.
Thus Spoke Zarathustra, trans. R.J. Hollingdale, Harmondsworth: Penguin, 1969.
Untimely Meditations, trans. R.J. Hollingdale, Cambridge: Cambridge University Press, 1997. (Includes *Schopenhauer as Educator*, *Richard Wagner in Bayreuth*, *David Strauss, the Writer and Confessor* and *On the Use and Disadvantage of History for Life*.)

Secondary Literature

Abbey, Ruth (2000), *Nietzsche's Middle Period*, Oxford: Oxford University Press.
Abel, Günter (1984), *Die Dynamik der Willen zur Macht und die ewige Wiederkehr*, Berlin: de Gruyter.
Abel, Günter, (1987), 'Logik und Ästhetik', *Nietzsche-Studien*, **16**, 112–48.
Acampora, Christa Davis (1999), 'The Problem of Homer', *Nietzscheforschung*, **5/6**, 553–74.
Acampora, Christa Davis, (2002), 'Nietzsche Contra Homer, Socrates, and Paul', *Journal of Nietzsche Studies*, **24**, 25–53.
Acampora, Christa Davis (2002), 'Of Dangerous Games and Dastardly Deeds: A Typology of Nietzsche's Contests', *International Studies in Philosophy*, **34**, 135–51.
Ahern, Daniel A. (1995), *Nietzsche as Cultural Physician*, University Park: Pennsylvania State University Press.
Allison, David (2001), *Reading the New Nietzsche*, Lanham, MD: Rowman and Littlefield.
Anderson, R. Lanier (1994), 'Nietzsche's Will to Power as a Doctrine of the Unity of Science', *Studies in History and Philosophy of Science*, **25**, 729–50.
Andler, Charles (1920-31), *Nietzsche, sa vie et sa pensée*, 6 vols, Paris: Bossard.
Andreas-Salomé, Lou (2000), *Nietzsche in seinen Werken*, Frankfurt am Main: Insel.
Anonymous (1872), 'Sensation and Science', *Nature*, 4 July, 177–78.
Anonymous (1875), 'Contemporary Literature. Theology and Philosophy', *Westminster Review*, **47**, 1 April, 501.
Anonymous (1877), 'Review of Paul Rée, *Der Ursprung der moralischen Empfindungen*', *Mind*, **2**, 581.
Ansell Pearson, Keith (1997), *Viroid Life: Perspectives on Nietzsche and the Transhuman Condition*, London: Routledge.
Aristotle (1984), *The Complete Works of Aristotle*, ed. Jonathan Barnes, 2 vols, Princeton, NJ: Princeton University Press.
Asendorf, Christoph (1993), *Batteries of Life: On the History of Things and their Perception in Modernity*, Berkeley, CA: University of California Press.
Assoun, Paul-Laurent (1998), *Freud et Nietzsche*, 2nd edn, Paris: Presses Universitaires de France.
Assoun, Paul-Laurent (2000), *Freud & Nietzsche*, London: Athlone Press.
Augustine, St (1995), *De doctrina Christiana*, Oxford: Clarendon Press.
Austin, J.L. (1961), *Philosophical Papers*, Oxford: Clarendon.
Babich, Babette E. (1994), *Nietzsche's Philosophy of Science: Reflecting Science on the Ground of Art and Life*, Albany, NY: SUNY Press.
Babich, Babette E. (1997), 'The Hermeneutics of a Hoax', *Common Knowledge*, **6** (2), 23–33.
Babich, Babette E. (ed.) (2002), *Hermeneutic Philosophy of Science, Van Gogh's Eyes, and God: Essays in Honor of Patrick A. Heelan, S.J.*, Dordrecht: Kluwer.
Babich, Babette E. and Robert S. Cohen (eds) (1999), *Nietzsche, Theories of Knowledge, and Critical Theory: Nietzsche and the Sciences, I*, Dordrecht: Kluwer.

Babich, Babette E. and Robert S. Cohen (eds) (1999), *Nietzsche, Epistemology, and Philosophy of Science: Nietzsche and the Sciences, II*, Dordrecht: Kluwer.
Baer, Karl Ernst von (1862), *Festrede zur Eröffnung der russischen entomologischen Gesellschaft im Mai 1860*, Berlin.
Bain, Alexander (1873), *Mind and Body: The Theories of Their Relation*, 2nd edn, London: Henry S. King & Co.
Bain, Alexander (1874), *Geist und Körper*, Leipzig: Brockhaus.
Beale, Lionel (1851), *The Laws of Health, in Relation to Mind and Body*, London: Churchill.
Bellone, Enrico (1980), *A World on Paper: Studies on the Second Scientific Revolution*, Cambridge, MA: MIT Press.
Berti, Maria Christina (1997), 'Beiträge zur Quellenforschung', *Nietzsche-Studien*, **26**, 580–81.
Bishop, Paul (1995), *The Dionysian Self: C.G. Jung's Reception of Friedrich Nietzsche*, Berlin: de Gruyter.
Blackmore, John T. (1972), *Ernst Mach: His Work, Life, and Influence*, Berkeley, CA: University of California Press.
Blackmore, John T., R. Itagaki and S. Tanaka (eds) (2001), *Ernst Mach's Vienna, 1895–1930, or, Phenomenalism as Philosophy of Science*, Dordrecht: Kluwer.
Blondel, Eric (1986), *Nietzsche, le corps et la culture: La philosophie comme généaologie philologique*, Paris: PUF.
Bock, Carl Ernst (1866), *Das Buch vom gesunden und kranken Menschen*, 7th edn, Leipzig: Keil.
Borsche, Tilman (1985), 'Nietzsches Erfindung der Vorsokratiker', in Josef Simon (ed.), *Nietzsche und die philosophische Tradition*, Würzburg: Königshausen & Neumann, pp. 62–87.
Boscovich, Roger Joseph (1966), *A Theory of Natural Philosophy*, Cambridge, MA: MIT Press.
Brobjer, Thomas H. (1997), 'Nietzsche's Reading and Private Library, 1885–1889', *Journal of the History of Ideas*, **58**, 663–93.
Brobjer, Thomas H. (2001), 'Beiträge zur Quellenforschung', *Nietzsche-Studien*, **30**, 418–21.
Brobjer, Thomas H. (2003), 'Women as Predatory Animals or Why Nietzsche Philosophized with a Whip', in Christa Davis Acampora and Ralph A. Acampora (eds), *A Nietzschean Bestiary: Animality Beyond Docile and Brutal*, Lanham, MD: Rowman & Littlefields.
Broca, Pierre Paul (1869), 'Sur le siège de la faculté du langage articulé', *Tribune Médicale*, **74** and **75**, 254–56 and 265–69.
Brusotti, Marco (1992), 'Beiträge zur Quellenforschung', *Nietzsche-Studien*, **21**, 390–97.
Büchner, Ludwig (1902), *Kraft und Stoff, oder Grundzüge der natürlichen Weltordnung, nebst einer darauf gebauten Sittenlehre in allgemein verständlicher Darstellung*, 20th edn, Leipzig: Thomas.
Campioni, Giuliano, 'Beiträge zur Quellenforschung', *Nietzsche-Studien*, **21**, 401–407.
Caneva, K. L. (1978), 'From Galvanism to Electrodynamics: The Transformation of German Physics and Its Social Context', *Historical Studies in the Physical Sciences*, **9**, 63–159.

Carruthers, Peter (1996), *Language, Thought and Consciousness: An Essay in Philosophical Psychology*, Cambridge: Cambridge University Press.

Caspari, Otto (1864), *Die Sprache als psychischer Entwicklungsgrund*, Göttingen: Dieterich.

Caspari, Otto (1874), *Die Thomson'sche Hypothese von der endlichen Temperaturausgleichung im Weltall beleuchtet vom philosophischen Gesichtspunkte*, Stuttgart: Horster.

Caspari, Otto (1881), *Der Zusammenhang der Dinge: Gesammelte philosophische Aufsätze*, Breslau: Trewendt.

Cavell, Stanley (1990), *Conditions Handsome and Unhandsome and Emerson's Perfectionism*, Chicago: University of Chicago Press.

Cavell, Stanley (1994), *A Pitch of Philosophy*, Cambridge, MA: Harvard University Press.

Cicero (1942), *De Oratore*, London: Heinemann.

Clark, Maudemarie (1990), *Nietzsche on Truth and Philosophy*, Cambridge: Cambridge University Press.

Clark, Maudemarie (1998), 'On Knowledge, Truth, and Value: Nietzsche's Debt to Schopenhauer and the Development of His Empiricism', in Christopher Janaway (ed.), *Willing and Nothingness: Schopenhauer as Nietzsche's Educator*, Oxford: Clarendon Press, pp. 37–78.

Comte, Auguste (1970), *Introduction to Positive Philosophy*, trans. Frederick Ferré, Indianapolis/New York: Bobbs-Merrill.

Cox, Christoph (1999), *Nietzsche: Naturalism and Interpretation*, Berkeley, CA: University of California Press.

Crawford, Claudia (1990), 'Nietzsche's Physiology of Ideological Criticism', in Clayton Kotb (ed.), *Nietzsche as Postmodernist: Essays Pro and Con*, Albany, NY: SUNY Press.

Darwin, Charles (1968), *The Origin of Species by Means of Natural Selection*, ed. J.W. Burrow, Harmondsworth: Penguin.

Dear, Peter (1995), *Discipline and Experience: The Mathematical Way in the Scientific Revolution*, Chicago: University of Chicago Press.

Deleuze, Gilles (1983), *Nietzsche and Philosophy*, trans. Hugh Tomlinson, London: Athlone Press.

Del-Negro, Walter (1923), *Die Rolle der Fiktionen in der Erkenntnistheorie Friedrich Nietzsches*, Munich: Rösl.

Del-Negro, Walter (1970), *Konvergenzen in der Gegenwartsphilosophie und die moderne Physik*, Berlin: Duncker & Humblot.

Derrida, Jacques (1988), *The Ear of the Other*, Lincoln, NE: University of Nebraska Press.

Descartes, René (1897–1914), *Meditationes de prima philosophia*, in *Œuvres*, ed. Charles Adam and Paul Tannery, 11 vols, Paris: Cerf.

Dietzsch, Steffen (1994), 'Nietzsches französischer Traum', unpublished paper presented at the conference 'Nietzsche in Cosmopolis. Der französische Nietzsche', Stiftung Weimarer Klassik, Weimar, December 1994.

D'Iorio, Paolo (1993), 'La superstition des philosophes critiques: Nietzsche et Afrikan Spir', *Nietzsche-Studien*, **22**, 257–94.

D'Iorio Paolo (1994), 'L'image des philosophes préplatoniciens chez le jeune Nietzsche', in Tilman Borsche, Federico Gerratana and Aldo Venturelli (eds), '*Centauren-Geburten*', Berlin: de Gruyter, pp. 383–417.

D'Iorio, Paolo (1995), 'Cosmologie de l'Éternel Retour', *Nietzsche-Studien*, **24**, 62–123.

Djuric, Mihailo and Josef Simon (eds) (1986), *Kunst und Wissenschaft bei Nietzsche*, Würzburg: Königshausen und Neumann.

Du Bois-Reymond, Emil (1848–60), *Untersuchungen über thierische Electricität*, Berlin: Reimer.

Duhem, Pierre (1996), *Essays in the History and Philosophy of Science*, Indianapolis: Hackett.

Dühring, Eugen (1865), *Der Werth des Lebens*, Breslau: Trewendt.

Dühring, Eugen (1880), *Robert Mayer, der Galilei des neunzehnten Jahrhunderts*, Chemnitz: Schmeitzner.

Eisler, Rudolf (1902), *Nietzsches Erkenntnistheorie und Metaphysik: Darstellung und Kritik*, Leipzig: Hermann Haacke.

Emerson, R.W. (1983), *Essays and Lectures*, New York: Library of America.

Ewald, Oskar (1909), 'Darwin und Nietzsche', *Zeitschrift für Philosophie und philosophische Kritik*, Ergänzungsband **1**, 159–79.

Faraday, Michael (1839-55), *Experimental Researches into Electricity*, London: Taylor.

Farber, Marvin (1967), *Phenomenology and Existence: Toward a Philosophy within Nature*, New York: Harper.

Fechner, Gustav Theodor (1860), *Elemente der Psychophysik*, Leipzig: Breitkopf & Härtel.

Ferrier, David (1878), *The Localisation of Cerebral Disease, Being the Gulstonian Lectures of the Royal College of Physicians*, London: Smith, Elder & Co.

Feuerbach, Ludwig (1903–11), *Sämtliche Werke*, ed. Wilhelm Bolin and Friedrich Jodl, 10 vols, Stuttgart: Fromann.

Feuerbach, Ludwig (1967–2001), *Gesammelte Werke*, ed. Werner Schuffenhauer, 14 vols, Berlin: Akademie-Verlag.

Feuerbach, Ludwig (1986), *Principles of the Philosophy of the Future*, Indianapolis: Hackett.

Feyerabend, Paul (1987), *Farewell to Reason*, London: Verso.

Finlay, Mark R. (1992), 'Quackery and Cookery: Justus von Liebig's Extract of Meat and the Theory of Nutrition in the Victorian Age', *Bulletin of the History of Medicine*, **66**, 404–18.

Fleck, Ludwik (1979), *The Genesis and Development of a Scientific Fact*, trans. F. Bradley and T. Trenn, Chicago: University of Chicago Press.

Florey, Ernst and Olaf Breidbach (eds) (1993), *Das Gehirn – Organ der Seele? Zur Ideengeschichte der Neurobiologie*, Berlin: Akademie Verlag.

Flourens, Pierre (1842), *Recherches expérimentales sur les propiétés et les fonctions du système nerveux dans les animaux vertébrés*, 2nd edn, Paris: Balliere.

Förster-Nietzsche, Elisabeth (1895–1904), *Das Leben Friedrich Nietzsches*, Leipzig: Naumann.

Frank, Phillipp (1941), *Between Physics and Philosophy*, Cambridge, MA: Harvard University Press.

Funke, Otto (1878–79), *Lehrbuch der Physiologie für akademische Vorlesungen und zum Selbststudium*, 6th edn, Leipzig: Voss.

Gasser, Reinhard (1997), *Nietzsche und Freud*, Berlin: de Gruyter.

Gayon, Jean (1999), 'Nietzsche and Darwin', in Jane Maienschein and Michael Ruse (eds), *Biology and the Foundation of Ethics*, Cambridge: Cambridge University Press, pp. 154–97.

Gebhard, Walter (1984), *'Der Zusammenhang der Dinge': Weltgleichnis und Naturverklärung im Totalitätsbewußtsein des 19. Jahrhunderts*, Tübingen: Niemeyer.

Gemes, Ken (1992), 'Nietzsche's Critique of Truth', *Philosophy and Phenomenological Research*, **52**, 47–65.

Gerber, Gustav (1885), *Die Sprache als Kunst*, 2nd edn, Berlin: Gaertner, 1885.

Gerhardt, Volker (1984), 'Von der ästhetischen Metaphysik zur Physiologie der Kunst', *Nietzsche-Studien*, **13**, 374–93.

Golomb, Jacob, Ronald Lehrer and Weaver Santaniello (eds) (1999), *Nietzsche and Depth Psychology*, Albany, NY: SUNY Press.

Gotzes, Heinrich (1904), 'Nietzsche und der Darwinismus', *Natur und Glaube*, **7**, 337–49.

Green, Michael S. (2002), *Nietzsche and the Transcendental Tradition*, Urbana: University of Illinois Press.

Greene, Mott, *Natural Knowledge in Preclassical Antiquity*, Baltimore: Johns Hopkins Press, 1992.

Gregory, Frederick (1977), *Scientific Materialism in Nineteenth-Century Germany*, Dordrecht: Reidel.

Grottewitz, Curt (1897), 'Nietzsches Herrenmoral und die Naturwissenschaft', *Das Magazin für Litteratur*, **66**, 1519–24.

Haas, Ludwig (1932), 'Der Darwinismus bei Nietzsche', PhD thesis, University of Gießen.

Hagner, Michael (1992), 'The Soul and the Brain Between Anatomy and "Naturphilosophie" in the Early Nineteenth Century', *Medical History*, **36**, 1–33.

Hagner, Michael (1993), 'Die elektrische Erregbarkeit des Gehirns: Zur Konjunktur eines Experiments', in Hans-Jörg Rheinberger and Michael Hagner (eds), *Die Experimentalisierung des Lebens: Experimentalsysteme in den biologischen Wissenschaften 1850/1950*, Berlin: Akademie Verlag, pp. 97–115.

Hagner, Michael (1994), 'Hirnbilder: Cerebrale Repräsentationen im 19. und 20. Jahrhundert', in Michael Wetzel and Herta Wolf (eds), *Der Entzug der Bilder: Visuelle Realitäten*, Munich: Fink, pp. 145–60.

Hagner, Michael (1994), 'Lokalisation, Funktion, Cytoarchitektonik: Wege zur Modellierung des Gehirns', in Michael Hagner, Hans-Jörg Rheinberger and Bettina Wahrig-Schmidt (eds), *Objekte, Differenzen, Konjunkturen: Experimentalsysteme im historischen Kontext*, Berlin: Akademie Verlag, pp. 121–50.

Haley, Bruce (1978), *The Healthy Body and Victorian Culture*, Cambridge, MA: Harvard University Press.

Hamann, Johann Georg (1949–57), *Sämtliche Werke: Historisch-kritische Ausgabe*, ed. Josef Nadler, 6 vols, Vienna: Herder.

Hamann, Johann Georg (1872), *Schriften und Briefe*, ed. Moritz Petri (Hanover: Meyer).

Hamburger, Michael (1952), *Hölderlin*, London: Harvill.
Harman, P.M. (1982), *Energy, Force, and Matter: The Conceptual Development of Nineteenth-Century Physics*, Cambridge: Cambridge University Press.
Hartmann, Eduard von (1869), *Philosophie des Unbewussten*, Berlin: Duncker.
Hayman, Ronald (1980), *Nietzsche: A Critical Life*, London: Weidenfeld.
Heidegger, Martin (1961), *Nietzsche I*, Neske: Pfullingen.
Heidelberger, Michael (1993), *Die innere Seite der Natur: Gustav Theodor Fechners wissenschaftlich-philosophische Weltauffassung*, Frankfurt am Main: Klostermann.
Heilbron, John (1979), *Electricity in the Seventeenth and Eighteenth Centuries: A Study of Early Modern Physics*, Berkeley, CA: University of California Press.
Heller, K.D. (1964), *Ernst Mach: Wegbereiter der modernen Physik*, Vienna: Springer.
Heller, Peter (1972), '"Chemie der Begriffe und Empfindungen." Studie zum 1. Aphorismus von "Menschliches, Allzumenschliches I"', *Nietzsche-Studien*, **1**, 210–33.
Hellpach, Willy (1939), *Geopsyche*, 5th edn, Leipzig: Engelmann.
Helmholtz, Hermann von (1855), *Ueber das Sehen des Menschen*, Leipzig: Vose.
Helmholtz, Hermann von (1895), *Popular Lectures on Scientific Subjects*, London: Longmans.
Helmholtz, Hermann von (1962), *Popular Scientific Lectures*, New York: Dover Publications.
Helsloot, Niels (1999), *Vrolijke wetenschap: Nietzsche als vriend*, Baarn: Agora.
Herbart, Johann Friedrich (1824–25), *Psychologie als Wissenschaft, neu gegründet auf Erfahrung, Metaphysik und Mathematik*, Königsberg: Unzer.
Hesiod (1966), *Theogony*, ed. M.L. West, Oxford: Clarendon Press.
Hinman, Lawrence M. (1982), 'Nietzsche, Metaphor, and Truth', *Philosophy and Phenomenological Research*, **43**, 179–99.
Hitzig, Eduard (1874), *Untersuchungen über das Gehirn: Abhandlungen physiologischen und pathologischen Inhalts*, Berlin: Hirschwald.
Horn, Anette (2000), *Nietzsches Begriff der décadence: Kritik and Analyse der Moderne*, Frankfurt am Main: Peter Lang.
Houlgate, Stephen (1986), *Hegel, Nietzsche and the Criticism of Metaphysics*, Cambridge: Cambridge University Press.
Huxley, T.H. (1894–1908), *Collected Essays*, 9 vols, London: Macmillan.
Janaway, Christopher (1989), *Self and World in Schopenhauer's Philosophy*, Oxford: Clarendon Press.
Janaway, Christopher (1991), 'Nietzsche, the Self, and Schopenhauer', in Keith Ansell Pearson (ed.), *Nietzsche and Modern German Thought*, London: Routledge, pp. 119–42.
Janz, Curt Paul (1978), *Friedrich Nietzsche. Biographie*, Munich: Hanser.
Juranville, Alain (1973), *Physique de Nietzsche*, Paris: Denoël/Gonthier.
Kant, Immanuel (1974), *Logic*, trans. Robert Hartman and Wolfgang Schwarz, New York: Bobbs-Merrill.
Kaufmann, Walter (1974), *Nietzsche: Philosopher, Psychologist, Antichrist*, Princeton, NJ: Princeton University Press.
Kee, Alistair (1999), *Nietzsche Against the Crucified*, London: SCM Press.

Kelly, Alfred (1981), *The Descent of Darwin: The Popularization of Darwinism in Germany, 1860–1914*, Chapel Hill, NC: University of North Carolina Press.
Kerenyi, Carl (1951), *The Gods of the Ancient Greeks*, London: Thames and Hudson.
Kittay, Eva Feder (1987), *Metaphor: Its Cognitive Force and Linguistic Structure*, Oxford: Clarendon Press.
Kleinpeter, Hans (1913), 'Ernst Mach und Friedrich Nietzsche', *Neue Freie Presse*, 23 February.
Kleinpeter, Hans (1913), *Der Phänomenalismus, eine naturwissenschaftliche Weltanschauung*, Leipzig: Barth.
Knight, David (1986), *The Age of Science*, Oxford: Blackwell.
Kofman, Sarah (1998), *Socrates: Fictions of a Philosopher*, trans. Catherine Porter, London: Athlone Press.
Kremer-Marietti, Angèle (1992), *Nietzsche et la rhétorique*, Paris: PUF.
Krummel, Richard Frank (1974), *Nietzsche und der deutsche Geist*, vol. 2, Berlin: de Gruyter.
Lachterman, David (1993), *The Ethics of Geometry*, London: Routledge.
Lacoue-Labarthe, Philippe (1971), 'Le détour (Nietzsche et la rhétorique)', *Poétique*, **5**, 53–76.
Lampert, Laurence (1993), *Nietzsche and Modern Times: A Study of Bacon, Descartes and Nietzsche*, New Haven, CT: Yale University Press.
Lampl, Hans Erich (1986), 'Ex Oblivione: Das Féré-Palimpsest', *Nietzsche-Studien*, **15**, 225–49.
Lange, Friedrich Albert (1879), *The History of Materialism and Criticism of its Present Importance*, trans. Ernest Chester Thomas, London: Trübner. Republished as a 3rd edn in 1925 by Routledge and Kegan Paul.
Lange, Friedrich Albert (1866), *Geschichte des Materialismus und Kritik seiner Bedeutung in der Gegenwart*, Iserlohn: Baedeker.
Leary, David E. (1978), 'The Philosophical Development of the Conception of Psychology in Germany, 1780–1850', *Journal of the History of the Behavioral Sciences*, **14**, 113–21.
Lehrer, Ronald (1994), *Nietzsche's Presence in Freud's Life and Thought: On the Origins of a Psychology of Dynamic Unconscious Mental Functioning*, Albany, NY: SUNY Press.
Leiter, Brian (1994), 'Perspectivism in Nietzsche's *Genealogy of Morals*', in Richard Schacht (ed.), *Nietzsche, Genealogy, Morality: Essays on Nietzsche's 'Genealogy of Morals'*, Berkeley, CA: University of California Press, pp. 334–57.
Leiter, Brian (2002), *Routledge Philosophy Guidebook to Nietzsche on Morality*, London: Routledge.
Lewes, G.H. (1874–79), *Problems of Life and Mind*, 5 vols, London: Trübner.
Lichtenberg, Georg Christoph (1967–72), *Schriften und Briefe*, 4 vols, Munich: Hanser.
Liebmann, Otto (1865), *Kant und die Epigonen. Eine kritische Abhandlung*, Stuttgart: Schober.
Liebmann, Otto (1876), *Zur Analysis der Wirklichkeit. Philosophische Untersuchungen*, Strassburg: Trübner.
Liebmann, Otto (1882), *Gedanken und Thatsachen*, Strassburg: Trübner.

Lima, Joseph and Tracy B. Strong (forthcoming), 'Telling the Dancer from the Dance: On the Relevance of Ordinary Language Philosophy for Political Theory', in Andrew Norris (ed.), *Stanley Cavell and Political Theory*.

Lindsay, James (1915–16), 'Eugenics and the Doctrine of the Superman', *Eugenics Review*, **7**, 247–62.

Lindsay, Robert Bruce (1973), *Julius Robert Mayer: Prophet of Energy*, Oxford: Pergamon Press.

Löb, Walter (1908), 'Naturwissenschaftliche Elemente in Nietzsches Gedanken', *Deutsche Rundschau*, **137**, 264–69.

Löwenfeld, Leopold (1887), *Moderne Behandlung der Nervenschwäche*, Wiesbaden: Bergmann.

Lohff, Brigitte (1990), *Die Suche nach der Wissenschaftlichkeit der Physiologie in der Zeit der Romantik*, Stuttgart: Fischer.

Lotze, Rudolph Hermann (1851), *Allgemeine Physiologie des körperlichen Lebens*, Leipzig: Weidmann.

Lotze, Rudolph Hermann (1852), *Medicinische Psychologie, oder Physiologie der Seele*, Leipzig: Weidmann.

Mach, Ernst (1886), *Beiträge zur Analyse der Empfindungen*, Jena: Fischer.

Mach, Ernst (1897), *Contributions to the Analysis of the Sensations*, trans. C.M. Williams, Chicago: Open Court.

Mach, Ernst (1898), *Popular Scientific Lectures*, trans. Thomas J. McCormack, 3 edn, Chicago: Open Court.

Mach, Ernst (1959), *The Analysis of Sensations*, New York: Dover.

Magnus, Bernd (1978), *Nietzsche's Existential Imperative*, Bloomington: Indiana University Press.

Marx, Karl and Frederick Engels (1975-), *Collected Works*, 49 vols, London: Lawrence and Wishart.

Maxwell, James Clerk (1873), *A Treatise on Electricity and Magnetism*, Oxford: Clarendon Press.

Meijers, Anthonie (1988), 'Gustav Gerber und Friedrich Nietzsche: Zum historischen Hintergrund der sprachphilosophischen Auffassungen des frühen Nietzsche', *Nietzsche-Studien*, **17**, 369–90.

Meinel, Christoph (1991), *Karl Friedrich Zöllner und die Wissenschaftskultur der Gründerzeit: Eine Fallstudie zur Genese konservativer Zivilisationskritik*, Berlin: SIGMA.

Meinert, C.A. (n.d.), *Wie nährt man sich gut und billig? Ein Beitrag zur Ernährungsfruge*, Mainz: Mittler.

Mencken, H.L. (1923), *Prejudices. Third Series*, London: Jonathan Cape.

Meyering, Theo C. (1989), *Historical Roots of Cognitive Science: The Rise of a Cognitive Theory of Perception from Antiquity to the Nineteenth Century*, Dordrecht: Kluwer.

Mill, John Stuart (1947), *A System of Logic, Ratiocinative and Inductive*, London: Longmans.

Mill, John Stuart (1969), *Collected Works of John Stuart Mill*, Toronto: University of Toronto Press.

Mittasch, Alwin (1944), *Friedrich Nietzsches Stellung zur Chemie*, Berlin: Verlag Chemie.

Mittasch, Alwin (1950), *Friedrich Nietzsches Naturbeflissenheit*, Heidelberg: Springer.

Mittasch, Alwin (1952), *Nietzsche als Naturphilosoph*, Stuttgart: Kröner.

Moles, Alistair (1989), 'Nietzsche's Eternal Recurrence as Riemannian Cosmology', *International Studies in Philosophy*, **21**, 21–35.

Moles, Alistair (1990), *Nietzsche's Philosophy of Nature and Cosmology*, New York: Peter Lang.

Montinari, Mazzino (1984), 'Nietzsche lesen: Die Götzen-Dämmerung', *Nietzsche-Studien*, **13**, 69–79.

Moore, Greg (1998), 'Beiträge zur Quellenforschung', *Nietzsche-Studien*, **27**, 535–51.

Moore, Gregory (2002), *Nietzsche, Biology, and Metaphor*, Cambridge: Cambridge University Press.

Morrison, Robert G. (1997), *Nietzsche and Buddhism: A Study in Nihilism and Ironic Affinities*, Oxford: Oxford University Press.

Mügge, Maximilian A. (1909–10), 'Eugenics and the Superman: A Racial Science and a Racial Religion', *Eugenics Review*, **1**, 184–93.

Müller-Lauter, Wolfgang (1978), 'Der Organismus als innerer Kampf: Der Einfluß von Wilhelm Roux auf Friedrich Nietzsche', *Nietzsche-Studien*, **7**, 189–223.

Müller-Lauter, Wolfgang (1999), *Nietzsche: His Philosophy of Contradictions and the Contradictions of His Philosophy*, Urbana and Chicago: University of Illinois Press.

Nagel, Thomas (1981), 'What is it Like to be a Bat?', in Fred D. Miller and Nicholas D. Smith (eds), *Thought Probes: Philosophy through Science Fiction*, Eaglewood Cliffs, NJ: Prentice-Hall.

Nehamas, Alexander (1985), *Nietzsche: Life as Literature*, Cambridge, MA: Harvard University Press.

Nola, Robert (1987), 'Nietzsche's Theory of Truth and Belief', *Philosophy & Phenomenological Research*, **47**, 525–62.

Oehler, Max (ed.) (1942), *Nietzsches Bibliothek*, Weimar: Wagner.

Oersted, Hans Christian (1806), 'Über die Art, wie sich Electricität fortpflanzt', *Neues allgemeines Journal der Chemie*, **6**, 292–302.

Oersted, Hans Christian (1820), 'Experiments on the Effect of a Current of Electricity on the Magnetic Needle', *Annals of Philosophy*, **16**, 273–76.

Okochi, Ryogi (1988), 'Nietzsches Naturbegriff aus östlicher Sicht', *Nietzsche-Studien*, **17**, 108–24.

Oppenheim, Janet (1991), *'Shattered Nerves': Doctors, Patients, and Depression in Victorian England*, New York: Oxford University Press.

Orsucci, Andrea (1992), 'Beiträge zur Quellenforschung', *Nietzsche-Studien*, **22**, 371–88.

Orsucci, Andrea (1994), 'Unbewußte Schlüsse, Anticipationen, Übertragungen: Über Nietzsches Verhältnis zu Karl Friedrich Zöllner und Gustav Gerber', in Tilman Borsche, Federico Gerratana, and Aldo Venturelli (eds), *'Centauren-Geburten': Wissenschaft, Kunst und Philosophie beim jungen Nietzsche*, Berlin: de Gruyter, pp. 193–207.

Parkes, Graham (1994), *Composing the Soul: Reaches of Nietzsche's Psychology*, Chicago: University of Chicago Press.

Pasley, Malcolm (1978), 'Nietzsche's Use of Medical Terms', in Malcolm Pasley (ed.), *Nietzsche: Imagery and Thought*, London: Methuen, pp. 123–58.
Pavy, F.W. (1867), *A Treatise on the Function of Digestion, its Disorders, and their Treatment*, London: Churchill.
Peacock, Andrew (1982), 'The Relationship between the Soul and the Brain', in F. Clifford Rose and William F. Bynum (eds), *Historical Aspects of the Neurosciences*, New York: Raven Press, pp. 83–98.
Peirce, Charles Sanders (1955), *Philosophical Writings of Charles Sanders Peirce*, ed. Justus Buchler, New York: Dover.
Pera, Marcello (1992), *The Ambiguous Frog: The Galvani-Volta Controversy on Animal Electricity*, Princeton, NJ: Princeton University Press.
Perkins, Richard (1987), 'Nietzsche's *opus alchymicum*', *Seminar*, **23** (3), 216–26.
Pfotenhauer, Helmut (1985), *Die Kunst als Physiologie: Nietzsches ästhetische Theorie und literarische Produktion*, Stuttgart: Metzler.
Plato (1961), *The Collected Dialogues of Plato*, ed. Edith Hamilton and Huntington Cairns, Princeton, NJ: Princeton University Press.
Porter, James I. (2000), *The Invention of Dionysus: An Essay on the Birth of Tragedy*, Stanford, CT: Stanford University Press.
Porter, James I. (2000), *Nietzsche and the Philology of the Future*, Stanford, CT: Stanford University Press.
Posth, Sebastian (2003), 'Der meteorologische Komplex bei Nietzsche', unpublished MA thesis, University of Bochum.
Principe, Lawrence (1998), *The Aspiring Adept: Robert Boyle and his Alchemical Quest*, Princeton, NJ: Princeton University Press.
Przybyslawski, Artur (2002), 'Nietzsche Contra Heraclitus', *Journal of Nietzsche Studies*, **23**, 88–95.
Quintilian (1966–69), *Institutio Oratoria*, 4 vols, London: Heinemann.
Rabinach, Anson (1990), *The Human Motor. Energy, Fatigue and the Origins of Modernity*, Berkeley, CA: University of California Press.
Radkau, Joachim (1998), *Das Zeitalter der Nervosität: Deutschland zwischen Bismarck und Hitler*, Munich: Hanser.
Rée, Paul (1870), *Der Ursprung der moralischen Empfindungen*, Chemnitz: Schmeitzner.
Rée, Paul (1903), *Philosophie: Nachgelassenes Werk*, Berlin: Duncker.
Rée, Paul (2003), *Basic Writings*, ed. and trans. Robin Small, Urbana and Chicago: University of Illinois Press.
Rey, Abel (1927), *Le retour éternel de la philosophie de la physique*, Paris: Flammarion.
Richardson, William J. (1968), 'Heidegger's Critique of Science', *The New Scholasticism*, **42**, 511–36.
Richter, Claire (1911), *Nietzsche et les théories biologiques contemporains*, Paris: Mercure de France.
Riehl, Alois, *Philosophische Studien*, Leipzig: Quelle & Meyer.
Ritter, Johann Wilhelm (1798), *Beweis, dass ein beständiger Galvanismus den Lebensprozess in dem Thierreiche begleite*, Weimar: Verlag d. Industrie-Comptoirs.
Ritter Johann Wilhelm (1805), *Das elektrische System der Körper: Ein Versuch*, Leipzig: Reclam.

Rolph, W.H. (1882), *Biologische Probleme, zugleich als Versuch einer rationellen Ethik*, Leipzig: Engelmann.
Ross, Andrew (ed.) (1996), *Science Wars*, Durham, NC: Duke University Press.
Ross, Sydney (1962), '*Scientist*: The Story of a Word', *Annals of Science*, **18** (2), 65–88.
Rousseau, Jean-Jacques (1974), *Essai sur l'origine des langues*, ed. Angèle Kremer-Marietti, Paris: Aubier Montaigne.
Salaquarda, Jörg (1999), '"Art is More Powerful than Knowledge": Nietzsche on the Relationship Between Art and Science', *New Nietzsche Studies*, **3** (3/4), 1–11.
Sarasin, Philipp (2001), *Reizbare Maschinen. Eine Geschichte des Körpers, 1765–1914*, Frankfurt am Main: Suhrkamp.
Sarolea, Charles (1917), *German Problems and Personalities*, London: Chatto and Windus.
Schacht, Richard (1983), *Nietzsche*, London: Routledge.
Schacht, Richard (1995), *Making Sense of Nietzsche: Reflections Timely and Untimely*, Urbana and Chicago: University of Illinois Press.
Schallmayer, Wilhelm (1903), *Verberbung und Auslese im Lebenslauf der Völker*, Jena: Gustav Fischer.
Scherer, Wilhelm (1874), *Vorträge und Aufsätze zur Geschichte des geistigen Lebens in Deutschland und Oesterreich*, Berlin: Weidmannsche.
Schlechta, Karl and Anni Anders (1962), *Friedrich Nietzsche: Von den verborgenen Anfängen seines Philosophierens*, Stuttgart: Fromann.
Schnädelbach, Herbert (1984), *Philosophy in Germany, 1831–1933*, Cambridge: Cambridge University Press.
Schopenhauer, Arthur (1960–65), *Sämtliche Werke*, ed. Wolfgang Freiherr von Löhneysen, 5 vols, Frankfurt am Main: Insel.
Schreber, Daniel (1899), *Medical In-Door Gymnastics*, London: Williams and Norgate.
Schrift, Alan D. (1985), 'Language, Metaphor, Rhetoric: Nietzsche's Deconstruction of Epistemology', *Journal of the History of Philosophy*, **23**, 371–95.
Schwartz, Stephen P. (1993), 'The Status of Nietzsche's Theory of Will to Power in Light of Contemporary Philosophy of Science' in *International Studies in Philosophy*, **25**, 85–92.
Serres, Michel (1988), 'Corruption – *The Antichrist*: A Chemistry of Sensations and Ideas', trans. Chris Bongie, in Thomas Harrison (ed.), *Nietzsche in Italy*, Saratoga, CA: ANMA Libri, pp. 31–52.
Shapin, Steven and Simon Schaffer (1985), *Leviathan and the Air Pump: Hobbes, Boyle and the Experimental Life*, Princeton, NJ: Princeton University Press.
Siemens, Werner (1883), 'Ueber die Zulässigkeit der Annahme eines elektrischen Sonnenpotentials und dessen Bedeutung zur Erklärung terrestischer Phänomene', *Annalen der Physik und Chemie*, **20** (9), 108–36.
Sigmund, Carl (1875), *Südliche klimatische Curorte*, 3rd edn, Vienna: Braumüller.
Small, Robin (2001), 'Disturbing Thoughts and Eternal Perspectives: Some Uses of Symbolism in Nietzsche', *New Nietzsche Studies*, **3** (3/4), 29–44.
Small, Robin (2001), *Nietzsche in Context*, Aldershot: Ashgate.
Sommer, Manfred (1985), *Husserl und der frühe Positivismus*, Frankfurt am Main: Klostermann.

Spencer, Herbert (1851), *Social Statics*, London: John Chapman.
Spencer, Herbert (1861), *Education: Intellectual, Moral, and Physical*, London: Williams and Norgate.
Spencer, Herbert (1879), *The Data of Ethics*, London: Williams and Norgate.
Stack, G.J. (1981), 'Nietzsche and Boscovich's Natural Philosophy', *Pacific Philosophical Quarterly*, **62**, 69–87.
Stack, George J. (1983), *Lange and Nietzsche*, Berlin: de Gruyter.
Stack, G.J. (1993), 'Nietzsche's Earliest Essays', *Philosophy Today*, **37**, 154–56.
Stack, George (1994), *Nietzsche: Man, Knowledge, and Will to Power*, Hollowbrook: Durango.
Strong, Tracy B. (1989), 'Aesthetic Authority and Tradition: Nietzsche and the Greeks,' *History of European Ideas*, **11**, 989–1007.
Strong, Tracy B. (1990), 'Nietzsche's Political Aesthetics' in M. Gillespie and T. Strong (eds), *Nietzsche's New Seas*, Chicago: University of Chicago Press, pp. 153–74.
Strong, Tracy B. (1996), 'Nietzsche and the Song in the Self', *New Nietzsche Studies*, **1** (1/2), 1–15.
Strong, Tracy B. (1999), *Friedrich Nietzsche and the Politics of Transfiguration*, Champaign: University of Illinois Press.
Sünner, Rüdiger (1986), *Ästhetische Szientismuskritik: zum Verhältnis von Kunst und Wissenschaft bei Nietzsche und Adorno*, Frankfurt am Main: Peter Lang.
Swift, Paul (2000), 'On Teleology since Kant', *Nietzscheana*, **8**, pp. 1–20.
Tauber, Alfred I (1994), 'A Typology of Nietzsche's Biology', *Biology And Philosophy*, **9**, 25–44.
Teichmüller, Gustav (1882), *Die wirkliche und die scheinbare Welt: Neue Grundlegung der Metaphysik*, Breslau: Koebner.
Thomasius, Christian (1691), *Einleitung zu der Vernufft-Lehre*, Halle: Salfeld.
Thomson, William (1882), *Mathematical and Physical Papers*, Cambridge: Cambridge University Press.
Thomson, William and P.G. Tait (1871), *Handbuch der theoretischen Physik*, trans. Hermann von Helmholtz and Georg Wertheim, Braunschweig: Vieweg.
Tille, Alexander (1895), *Von Darwin bis Nietzsche*, Leipzig: Naumann.
Tönnies, Ferdinand (1990), *Der Nietzsche-Kultus: Eine Kritik*, Berlin: Akademie-Verlag.
Toulmin, Stephen (1979), 'The Inwardness of Mental Life', *Critical Inquiry*, **6**, 1–16.
Turner, R.S. (1971), 'The Growth of Professional Research in Prussia, 1818 to 1848: Causes and Context', *Historical Studies in the Physical Sciences*, **3**, 137–82.
Vaihinger, Hans (1920), *Die Philosophie des 'Als ob': System der theoretischen, praktischen und religiösen Fiktionen der Menschheit auf Grund eines idealistischen Positivismus*, 5th and 6th edns, Leipzig: Meiner.
Vaihinger, Hans (1935), *The Philosophy of 'Asif'*, tr. C.K. Ogden, London: Routledge & Kegan Paul.
van Tongeren, Paul J.M. (2000), *Reinterpreting Modern Culture: An Introduction to Friedrich Nietzsche's Philosophy*, West Lafayette, IN: Purdue University Press.
Volkmann, Richard (1872), *Die Rhetorik der Griechen und Römer in systematischer Übersicht*, Berlin: Ebeling & Plahn.

Volta, Alessandro (1918–29), *Le Opere di Alessandro Volta: Edizione Nazionale*, 7 vols, Milan: Hoepli.

Volz, Pia (1990), *Nietzsche im Labyrinth seiner Krankheit: Eine medizinisch-biographische Untersuchung*, Würzburg: Königshausen und Neumann.

Wackernagel, Wilhelm (1873), *Poetik, Rhetorik und Stilistik: Academische Vorlesungen*, ed. Ludwig Sieber, Halle: Verl. d. Buchh. d. Waisenhauses.

Wahrig-Schmidt, Bettina (1988), '"Irgendwie–jedenfalls physiologisch". Friedrich Nietzsche, Alexandre Herzen (fils) und Charles Féré 1888', *Nietzsche-Studien*, **17**, 434–64.

Wartofsky, Marx W. (1977), *Feuerbach*, Cambridge: Cambridge University Press.

Weber, F. Parkes (1901), *Climatology, Health Resorts, Mineral Springs*, London: Rebman.

Weiner, Marc A. (1997), *Richard Wagner and the Anti-Semitic Imagination*, Lincoln, NE: University of Nebraska Press.

Westphal, Kenneth (1984), 'Nietzsche's Sting and the Possibility of Good Philology', *International Studies in Philosophy*, **16**, 71–90.

Westphal, Kenneth (1984), 'Was Nietzsche a Cognitivist?', *Journal of the History of Philosophy*, **22**, 343–63.

Whewell, William (1840), *The Philosophy of the Inductive Sciences, founded upon their History*, London: Parker.

Whitlock, Greg (1996), 'Roger Boscovich, Benedict de Spinoza and Friedrich Nietzsche: The Untold Story', *Nietzsche-Studien*, **25**, 200–20.

Whorton, James C. (2000), *Inner Hygiene. Constipation and the Pursuit of Health in Modern Society*, Oxford: Oxford University Press.

Wilamowitz-Moellendorf, Ulrich von (2000), 'Future Philology! A Reply to *The Birth of Tragedy* by Friedrich Nietzsche', *New Nietzsche Studies*, **4** (1/2), 1–32.

Wilcox, John T. (1974), *Truth and Value in Nietzsche: A Study of His Metaethics and Epistemology*, Ann Arbor, MI: University of Michigan Press.

Willey, Thomas E. (1978), *Back to Kant: The Revival of Kantianism in German Social and Historical Thought, 1860–1914*, Detroit: Wayne State University Press.

Wise, M.N. (1981), 'German Concepts of Force, Energy and the Electromagnetic Ether: 1845–1880', in G.N. Cantor and M.J.S. Hodge (eds), *Conceptions of Ether*, Cambridge: Cambridge University Press, pp. 269–307.

Wundt, Wilhelm (1877), 'Philosophy in Germany', *Mind*, **2**, 509.

Young, Robert M. (1990), *Mind, Brain and Adaptation in the Nineteenth Century: Cerebral Localisation and its Biological Context from Gall to Ferrier*, 2nd edn, Oxford: Oxford University Press.

Zeitlin, I.M. (1994), *Nietzsche: A Re-Examination*, Cambridge: Cambridge University Press.

Zöllner, Johann Carl Friedrich (1872), *Über die Natur der Cometen: Beiträge zur Geschichte und Theorie der Erkenntnis*, Leipzig: Engelmann.

Index

For Product Safety Concerns and Information please contact our EU
representative GPSR@taylorandfrancis.com
Taylor & Francis Verlag GmbH, Kaufingerstraße 24, 80331 München, Germany

www.ingramcontent.com/pod-product-compliance
Ingram Content Group UK Ltd.
Pitfield, Milton Keynes, MK11 3LW, UK
UKHW022132190926
475702UK00013B/1769

* 9 7 8 1 1 3 8 2 7 7 5 8 8 *